AN INTRODUCTION TO THE GEOGRAPHY OF HEALTH

In the second edition of *An Introduction to the Geography of Health*, Helen Hazen and Peter Anthamatten explore the ways in which geographic ideas and approaches can inform our understanding of health. The book's focus on a broad range of physical and social factors that drive health in places and spaces offers students and scholars an important holistic perspective on the study of health in the modern era.

In this edition, the authors have restructured the book to emphasize the theoretical significance of ecological and social approaches to health. Spatial methods are now reinforced throughout the book, and other qualitative and quantitative methods are discussed in greater depth. Data and examples are used extensively to illustrate key points and have been updated throughout, including several new extended case studies such as water contamination in Flint, Michigan; microplastics pollution; West Africa's Ebola crisis; and the Zika epidemic. The book contains more than one hundred figures, including new and updated maps, data graphics, and photos.

The book is designed to be used as the core text for a health geography course for undergraduate and lower-level graduate students and is relevant to students of biology, medicine, entomology, social science, urban planning, and public health.

Helen Hazen is a broadly trained geographer, who focuses on issues of health and environment. She is currently a teaching associate professor at the University of Denver, USA, where she teaches classes on health and sustainability, as well as other aspects of geography.

Peter Anthamatten studied geography and public health as a graduate student and is currently an associate professor at the University of Colorado, Denver, USA. His research interests are primarily around the geography of health, with a focus on children's behavioral and nutritional environments, but he also enjoys working on geographic education. Peter teaches the geography of health, cartography, and spatial statistics.

AN INTRODUCTION TO THE GEOGRAPHY OF HEALTH

Second edition

Helen Hazen
and
Peter Anthamatten

Routledge
Taylor & Francis Group

LONDON AND NEW YORK

Second edition published 2020
by Routledge
2 Park Square, Milton Park, Abingdon, Oxon, OX14 4RN

and by Routledge
52 Vanderbilt Avenue, New York, NY 10017

Routledge is an imprint of the Taylor & Francis Group, an informa business

First edition published by Routledge 2011

British Library Cataloguing-in-Publication Data
A catalogue record for this book is available from the British Library

Library of Congress Cataloging-in-Publication Data
Names: Hazen, Helen, 1975- author. | Anthamatten, Peter, author. |
    Routledge (Firm)
Title: An introduction to the geography of health / Helen Hazen and
    Peter Anthamatten.
Description: Second Edition. | New York : Routledge, 2020. | First edition published
    in 2011 with Peter Anthamatten as principal author. | Includes bibliographical
    references and index.
Identifiers: LCCN 2019041345 (print) | LCCN 2019041346 (ebook) |
Subjects: LCSH: Medical geography—Textbooks.
Classification: LCC RA792 .A58 2020 (print) | LCC RA792 (ebook) |
    DDC 614.4/2—dc23
LC record available at https://lccn.loc.gov/2019041345
LC ebook record available at https://lccn.loc.gov/2019041346

ISBN: 978-0-367-10964-6 (hbk)
ISBN: 978-0-367-10965-3 (pbk)
ISBN: 978-0-429-02411-5 (ebk)

Typeset in Adobe Garamond Pro and Futura
by Apex CoVantage, LLC

For our nieces and nephews

CONTENTS

TABLES

BOXES

PREFACE

As we strive to cope with unprecedented and rapid changes in the planet's geography, an understanding of the significance of connections between health and the physical and social environment is more critical than ever before. Climate change, landscape conversion, population growth, and the rapid flow of people, goods, and information that have accompanied globalization have all had significant impacts on human health in complex and nuanced ways. Additionally, aging populations and other demographic changes are placing new stresses on healthcare and social support systems. In this context, there is growing recognition of the value of approaching public health issues from an integrated social science perspective, and great interest among undergraduate students in holistic approaches to health topics. We believe that a geographic approach to health offers a useful perspective on these issues, considering how changing relationships between people and their environments—both physical and social—influence human health.

The intent of this book is to provide undergraduate and early graduate students with an accessible introduction to the geography of health. The book is designed to be used as the core text for a health geography course, covering the theoretical and methodological backbone of the sub-discipline, as well as introducing students to relevant case studies. The book gives equal weight to ecological and social approaches to the topic, as well as providing discussion of spatial techniques. Individual chapters begin by introducing the reader to theoretical background before turning to case studies that draw out key themes and encourage students to apply what they have learned. Throughout, considerable reference is made to original scholarly research in order to direct the reader to additional resources to pursue specific topics further.

While we focus on topics that may be found in many public health textbooks, we strive to emphasize how *geographic* methods specifically can provide critical insight to health questions. Purposefully, we cite scholars from a wide range of disciplinary backgrounds (not just geographers), illustrating how approaches are used both within and beyond the disciplinary confines of geography. Recognizing that health geography has experienced a recent proliferation of work at its theoretical boundaries, particularly developments in the realms of social and cultural theory, we have made an effort to introduce students to these research frontiers. Finally, we have provided updated examples of the ways in which geospatial technologies can contribute to the study of health topics, as we guide students to be critical users of data—a fundamental twenty-first-century skill.

We have extensively updated the text in this second edition, including the addition of a variety of new case studies. Major new sections include the recent Zika and Ebola epidemics, microplastic pollution, the US opioid epidemic, Flint, Michigan's water crisis, the "Latino Paradox," and the global obesity pandemic. We have also given extra attention to the health-related impacts of major processes of global changes such as climate change and biodiversity loss.

ACKNOWLEDGMENTS

As with any extended work, this book could not have been completed without the support of family, friends, and colleagues. We are deeply indebted to our families and long-suffering spouses without whose support, hours of donated childcare, and patience this project would not have been possible. We also thank Connie Weil, who introduced us both to the subject we have grown to love. This work was substantially improved by many people who provided critical feedback, including several of our students and colleagues who used the first edition of the book, our friend and colleague Heike Alberts, and our anonymous peer reviewers.

ABBREVIATIONS

CDC   Centers for Disease Control and Prevention (US)
EARSS   European Antimicrobial Resistance Surveillance System
EPA   Environmental Protection Agency (US)
FDA   Food and Drug Administration (US)
FEMA   Federal Emergency Management Agency (US)
GDP   Gross Domestic Product
GPEI   Global Polio Eradication Initiative
IMF   International Monetary Fund
NICE   National Institute for Health and Care Excellence (UK)
OECD   Organization for Economic Co-operation and Development
PAHO   Pan American Health Organization
PPGIS   Public Participation GIS
UN   United Nations
UNICEF United Nations Children's Fund
USAID   United States Agency for International Development
WHO   World Health Organization
WRI   World Resources Institute

1

INTRODUCTION

In many ways, human health has never been better. Great strides have been made in reducing the **prevalence** of malnutrition and the **incidence** of infectious diseases such as polio, measles, and cholera. Rising incomes and targeted development efforts have led to improved access to healthcare, education, and other services, enabling many people to enjoy higher living standards than ever before. At the same time, efforts to improve **public health** through basic interventions such as sanitation and vaccination have greatly decreased **infant mortality rates**. As a result, **life expectancies** have been generally rising worldwide, even in low-income regions.

These improvements are inconsistent across global regions, however, with economic progress and attendant health improvements in some countries diverting attention away from continued economic and health challenges in many parts of the **Global South** (Box 1.1). In addition, there are stark divides *within* many countries between the newfound health and wealth of a rapidly growing middle class and the stagnation of health outcomes among poorer, often rural households. For example, while undernutrition is declining at the global scale, it remains a problem for many of the world's poorest people. In 2018, an estimated 224 million children

---

Box 1.1

THE GLOBAL NORTH AND GLOBAL SOUTH

Geographers, economists, and political scientists have long recognized the significance of economic development to many aspects of human society and have categorized the countries of the world to reflect these differences, using labels such as "First World," "Third World," "developed countries," and "less-developed countries." Increasingly, however, scholars and policymakers have become wary of the value judgments inherent in these comparative terms—for instance, the notion that *developed* countries have already succeeded is implicit in the term "developed," with *less-developed* countries

continued

Box 1.1 *continued*

appearing wanting, by contrast. To address these concerns, many scholars today prefer the terms **Global North** and **Global South**. The Global North—including Western Europe, Canada, the US, Australia, New Zealand, and parts of East Asia—is recognized by its economic affluence, while the world's remaining countries comprise the Global South and are characterized by their marginalized position in the global economy.

These categories reflect the significant impact of economic development on health, with the countries of the Global North traditionally having higher **life expectancy**, lower **infant mortality**, and lower **fertility** than the Global South. Health issues do not always neatly divide the Global North and South, however. In particular, parts of the Global South have made great strides in lowering infant mortality and fertility, while concurrently extending lifespans. Thailand, Lebanon, and Chile, for instance, have fertility rates similar to Western Europe, while lifespans in countries as diverse as Panama, Bahrain, and New Caledonia are comparable to the US (PRB 2019).

Another traditional distinction lies in the significance of infectious versus non-infectious disease, with the countries of the Global South tending to have higher rates of infectious disease than the North. By contrast, mortality in the Global North is largely associated with non-infectious diseases such as cancer and heart disease. Even here, the distinction is increasingly blurred, however, with the Westernization of lifestyles in the Global South leading to the rising incidence of non-infectious diseases, while certain infectious diseases appear to be making a comeback in the affluent North.

One further complication is that income inequalities *within* countries are often greater than those *among* countries, with affluent populations in the Global South often having very similar health outcomes to those living in the Global North. Furthermore, certain sub-populations in wealthy countries may actually have poorer health outcomes than Global South averages. For instance, Sri Lanka and the state of Kerala in India both report lower infant mortality rates than some minority populations in the United States. We therefore acknowledge the importance of terms like Global North and Global South in highlighting important trends, while recognizing the need to apply these terms judiciously.

under age 5 were underweight or experienced growth deficiencies as a result of malnutrition and around 45 percent of deaths in this young age group were tied to nutritional deficiencies (WHO 2018). Pockets of undernutrition persist even in the world's most affluent countries, as evidenced by the need for free food banks in wealthy countries such as the US, Australia, and the UK.

At the other end of the spectrum, there has been a dramatic increase in so-called **diseases of affluence**, associated with the adoption of westernized lifestyles across the world, including in the Global South. As more people live sedentary lives with easy access to abundant processed food, rates of obesity and other **chronic** diseases such as cancer and heart disease are rapidly increasing across the globe. The Worldwatch

Institute announced that there were more over-nourished than under-nourished people in the world for the first time in the year 2000 (Gardner and Halweil 2000), and this trend has continued. Many low- and middle-income countries are now experiencing a **double burden of disease** as their healthcare systems struggle to cope with both the traditional threats of malnutrition and infectious disease, as well as chronic conditions associated with Western lifestyles among increasingly affluent, urbanized, and sedentary populations.

Many countries of the **Global North** are also seeing a growing socioeconomic divide. As affluent populations see more people than ever before reach the age of 100, some poor and marginalized populations are actually experiencing declining life expectancies. Wealthy populations have traditionally experienced better health than the poor because of better living conditions and nutrition, as well as more time, better education, and more **social capital**. Today, the rich have maintained many of these benefits, while middle- and working-class communities often struggle with stresses associated with an increasingly competitive work environment related to processes of **globalization**. Poorer communities often have the resources to obtain abundant food and a sedentary lifestyle but may not have the time or means to invest in longer-term health goals, resulting in higher rates of lifestyle diseases such as obesity and diabetes. The stresses of daily life can become overwhelming for some, leading to destructive behaviors such as drug abuse, suicide, and alcoholism. The so-called **deaths of despair** associated with these behaviors have hit the US particularly hard, with drug overdose and suicide held responsible for much of a recent and unprecedented decline in life expectancy in the US (Case and Deaton 2015; AAFP 2019).

Such inequalities highlight the hierarchies of power that affect health at a variety of scales, from the household to the global. Why does a businessman in Brazil have a life expectancy of 80, while his maid's is only 60? Why is a baby born to an African American woman in Chicago more likely to be born prematurely than a white baby born in the same community? Why is a head of household in northern Ghana well nourished while some of his children remain undernourished, or a girl in an urban slum in India less likely to be taken to a doctor than her brother? In all these cases, we must consider how individuals are embedded in a web of cultural, economic, and political structures that enable or constrain their ability to meet their health needs. For example, a poor farmer in Mali may be constrained in her ability to grow sufficient food to feed her family owing to competition for land from commercial peanut growers, her inability to borrow money as a woman, and increasingly sporadic rains. The **proximate** (immediate) cause of her children's poor health is malnutrition—but social structures of society must be explored to fully understand the challenges her family faces. A study of these broader structural factors could consider **patriarchal** structures of the local community, government agricultural policies, Mali's marginal position in the global economy, and

even the global political and economic forces that have allowed climate change to continue unchecked to the point where it is now affecting rainfall patterns in Saharan Africa. Influences on health can thus be recognized at a variety of **geographic scale**, with potential points of intervention at all of them.

This scalar framework broadens our approach to understanding what drives health and disease, encouraging us to think beyond clinical solutions that target individuals. In particular, **structural factors** such as **race**, socioeconomic status, and political power are widely understood to be important influences on health and healthcare. These factors have long been recognized to act materially (e.g., poverty can lead to poor access to food or healthcare and thus poor health) but are increasingly understood to have measurable physiological effects associated with psycho-social factors such as the stress of experiencing discrimination. As our understanding of these connections grows, communities are exploring social and political approaches to tackling health problems that address structural factors. The United Nations' identification of eight Millennium Development Goals in 2000 (to be achieved by 2015) and the Sustainable Development Goals of 2015 are illustrative of these efforts. While several of the goals address health issues directly—for example, reducing child mortality and improving maternal health—others aim to improve human well-being by addressing broad structural issues through strengthening institutions and reducing inequalities.

In an era of climate change and rapid landscape conversion, our world's changing ecology also creates new challenges (Figure 1.1). Changes in climate, vegetation patterns, species distributions, and how humans interact with their environments—particularly as we move from rural to urban lifestyles—have profound implications for health. Ecological changes have been held at least partly responsible for a resurgence in some infectious diseases such as malaria and Lyme disease, as well as the emergence of some newly identified pathogens, including Ebola, HIV, and avian flu. In several countries in sub-Saharan Africa, life expectancies actually declined in the first part of the twentieth century (in some countries to under 40 years) as a result of the HIV/AIDS epidemic. Although the use of antiretroviral therapies

Figure 1.1 Clear-cutting, South Australia

Landscape conversion has profound impacts related to changes in vegetation patterns, hydrology, soil composition, and micro-climates. All of these factors can affect human health.

for HIV infection has reversed this trend, it provides a timely reminder of the ever-present potential of infectious diseases to generate devastating pandemics.

WHAT IS THE GEOGRAPHY OF HEALTH?

Just as the study of life is the focus of biology, geographers focus on the importance of *space* and *place*. With an emphasis on *how and why things vary across space* and *how people interact with their physical and social environments*, health geographers consider not only spatial patterns such as disease distributions but also how people's health is influenced by their relationship with their physical and social environments. For example, health

geographers might ask why malaria occurs only in certain parts of the world. What is it about the natural and built environments of these places that makes people vulnerable to malaria? How do human activities in these places influence vulnerability to malaria?

Many of these topics fall in the domain of public health. In contrast to clinical medicine, which focuses on the health of individuals, public health examines population-wide processes and emphasizes health-promoting behaviors that will benefit entire communities. Public health study often involves exploring how individuals interact with one another and their environment, and frequently places an emphasis on **preventative health** rather than curative approaches. Although health geography overlaps with public health research,

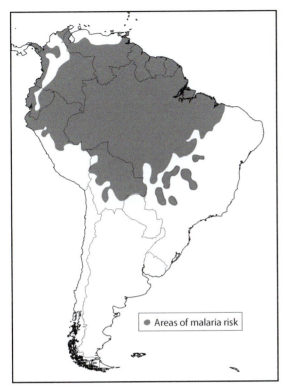

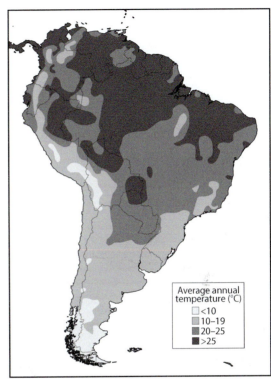

Figure 1.2 Malaria risk in South America
Source: Adapted from Guerra et al. (2008)

Figure 1.3 Average annual temperature in South America
Source: Adapted from Legates and Willmott (1990)

the geography of health is distinct in its emphasis on **space** and **place**, grounding geographers in spatial questions. In addition, while public health research is often **quantitative**, many health geographers now consider experiential questions that demand **qualitative** research, such as why particular places make us feel as we do, or how a person's position in society affects their health.

Geographers acknowledge a clear distinction between space and place in their analyses. **Space** refers to where things are located, and so spatial questions consider factors such as distribution, direction, and connectivity. **Place** refers to a unique locale and its social significance from the values imbued by individuals and communities, often requiring more qualitative analyses. For instance, your house may be little more than a location to most people, but it has significance to you because of your memories and emotional attachments. In this way, a "house" becomes your "home"—a space becomes a place.

Geographers frequently turn to mapping as a tool for identifying and exploring *spatial* patterns. The maps in Figures 1.2 through 1.7, for instance, enable us to consider how temperature, precipitation, elevation, population density, and national affluence (**GDP**) are related to the distribution of malaria across South America. A good first step for a map analysis is to identify the main patterns. The map of malaria risk, drawn from nationally reported incidence data (Guerra et al. 2008), shows that malaria is concentrated in the northern part of South America, focused in the Amazon basin (Figure 1.2). We can then examine the other maps to think about what characterizes the parts of the continent at risk of malaria.

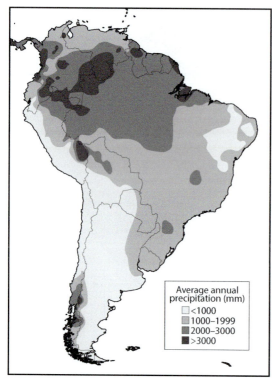

Figure 1.4 Average annual precipitation in South America
Source: Adapted from Legates and Willmott (1990)

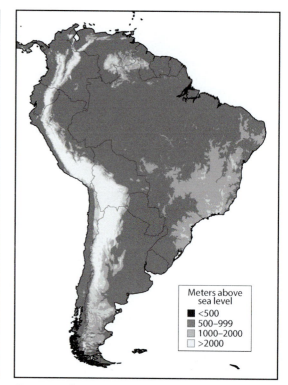

Figure 1.5 Elevation in South America
Source: Data from United States Geological Survey (2019)

The factors that most closely **correlate** with malaria risk appear to be temperature and precipitation—malaria is associated with hot, wet conditions (Figures 1.3 and 1.4). Temperature and rainfall affect both vegetation cover and the likelihood of standing water, influencing the range of the mosquito that transmits malaria. Elevation also appears to be significant, with the Andes Mountains standing out as largely free of malaria in the tropical zone that otherwise seems to support malaria transmission (Figure 1.5). Juxtaposing elevation with precipitation and temperature, it appears that higher elevations are associated with lower temperature and, in this example, also lower precipitation, reducing the likelihood of malaria transmission.

Population density and malaria incidence show a **negative correlation** (*high* population density is associated with *low* malaria risk) (Figure 1.6). We could hypothesize

that something about sparsely populated rural areas may be significant—perhaps through providing good habitat for the mosquitos that spread malaria. We can also glean additional important information from the population density map, such as where high-density populations at the edge of the current range of the vector might be at risk in the future if climate change extends the range of mosquitos.

Income does not show a clear association with malaria risk at the national scale, although the high-income Southern Cone countries (Chile, Argentina, and Uruguay) appear to have little malaria (Figure 1.7). In combination with the other maps that show a strong association between malaria and climate, we could hypothesize that these regions have little malaria for climatic rather than economic reasons. It is possible that malaria and national affluence may nonetheless be

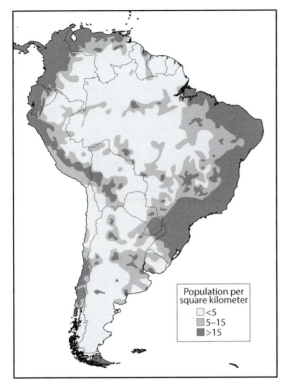

Figure 1.6 Population density in South America

Source: Data from Socioeconomic Data and Applications Center (SEDAC) (2012)

Figure 1.7 Income (GDP per capita) in South America (US$)

Source: Data from World Bank (2018)

related, although we cannot definitively identify a particular cause-and-effect mechanism from these maps. For instance, affluent countries may be able to afford better malaria control programs, or countries that have traditionally had less malaria may have been able to grow more affluent. The visual display of the map therefore enables us to identify spatial patterns that we might not otherwise notice in order to develop hypotheses for further exploration, but we cannot state definitively that one factor *caused* another without further analysis.

Using *place-based* approaches to the study of health prompts us to consider the social significance of particular locations. Figure 1.8, for instance, shows women collecting water from a public water source in Morocco, generating a variety of cultural questions that could

deepen our understanding of water and health issues in this community. For instance, what structural factors related to global economic institutions or local politics could explain why these women rely on a public standpipe rather than having water piped straight into their own homes? How are gendered structures of society relevant to the practice of collecting water in this community? What is the cultural significance of the decorations on this public fountain? Could a deeper understanding of the local culture around water help us to understand how people use water in this community and why they collect it from the places that they do? Although these questions may not appear to apply as directly to an outbreak of waterborne disease as an investigation of the pathogen load of the water, they may nonetheless prove critical in getting to the root causes of disease.

Figure 1.8 Women collecting water, Morocco
Credit: Heike Alberts

THE EVOLUTION OF HEALTH GEOGRAPHY: CHANGING APPROACHES TO HEALTH AND DISEASE

Viewing health in cultural context prompts us also to consider how ideas about health and disease have changed over time. Many societies traditionally attributed poor health to supernatural explanations and so curing illness relied on pleas to gods. In South India, for example, Mariyamman has long been recognized as the goddess of smallpox, who both caused and cured the affliction (van Voorthuizen 2004). In some cases, health has also been attributed a moral explanation, with poor health seen as a punishment for misdeeds. Such beliefs remain important for many people, even as scientific explanations for disease become widely accepted.

Greek philosophers were among the first to propose that the world could be explained using empirical observation. The capacity to study and hypothesize about the environment revolutionized the study of medicine. Some early philosophers argued that geographic differences were responsible for patterns in disease and that people's relationships with their environment could be manipulated to influence health. A key origin of this perspective was the *Hippocratic Corpus*. Although this body of work is associated with the philosopher Hippocrates (who lived circa 460–360BC), it was likely written over a period of several hundred years by multiple authors. The *Hippocratic Corpus* discusses medical ethics, holistic medicine, and environmental influences on health. The section entitled *On Airs, Waters, and Places* is among the earliest known documents to make an explicit connection between disease and place. Although there are errors in the text, it represents a philosophical milestone in connecting disease with place. Studies of ancient Chinese and Indian medical literature suggest that these cultures similarly made connections between disease and place (Magner 2005).

Hippocrates is often described as an early proponent of **environmental determinism**, which posits that health conditions, body types, and personalities are directly attributed to environmental factors, especially climate. Although the environment clearly influences our health in many ways, the suggestion that human health and behaviors can be fully explained by environmental factors has a tarnished history. Specifically, the idea that environmental constraints could limit the potential of some populations was used to justify racist policies, particularly colonization. As a result, many people now squarely reject ideas that are considered environmentally deterministic. Most modern scholars argue instead that the environment offers opportunities and imposes constraints, but that human free-will and ability, or human **agency**, allows people to make choices within this framework. While it may be tiring to work on a hot, humid day, it is not impossible, and there is certainly no reason to believe that tropical populations are destined to be poorer than European ones just because of differences in climate! From the mid-twentieth century onwards,

therefore, much work in health geography adopted a theoretical framework that considers how humans interact with their physical and social environments as cultural beings, rather than simply viewing them as hapless pawns of their environment.

Health researchers made an important theoretical advance when they began to use studies of ecology to inform their understandings of health and disease. In this perspective, sometimes called **disease ecology**, humans are viewed as one part of an integrated disease cycle, rejecting the idea that humans and the environment are separate entities. Humans are considered alongside infectious **agents**, disease **vectors**, and other animal **hosts**, so that the environmental conditions that are required for effective transmission of a disease can be understood. For instance, a disease ecologist might study how the malaria parasite, its mosquito vector, and human host interact in ways that enable the disease to spread.

The study of disease ecology was aided by technological advances in medical science. In particular, improvements in optical lenses paved the way for the discovery of microscopic agents of disease such as bacteria. Once infectious agents had been identified, people could more fully comprehend how humans fit into natural cycles and begin to map out ways to stop disease transmission. The disease cycle of schistosomiasis (also called "bilharzia"), for example, involves the pathogen spending time in a human host, a snail, and fresh water to complete its life cycle. Careful study of the activities of each organism in the disease cycle suggests that we could stop the transmission of schistosomiasis by improving sanitation to prevent the pathogen being excreted into open water, removing snails from the ecosystem with a pesticide, or discouraging people from wading in snail-infested waters.

The nineteenth-century discovery that microbes invade human bodies, which can result in disease (sometimes called **germ theory**), prompted important changes in medicine and public health practice. These changes included treatment of sewage, vaccination, antibiotics, chlorination of water, and sterilization of medical equipment, which dramatically increased human lifespans, particularly through lowering infant mortality. The separation of sewage from drinking water—sanitation— probably had a greater effect on reducing the spread of infectious disease and increasing human lifespans than anything that doctors have done before or since (Figure 1.9).

Over time, the idea that poor health could be attributed to particular disease **agents** led to a simple cause-and-effect perspective. This **doctrine of specific etiology** suggests that disease originates from a specific cause, such as a pathogen or toxin. The study of medicine began to focus heavily on seeking connections between specific causative agents and particular symptoms, rather than considering broader influences on health such as mental fitness or environmental conditions. Although there is clear scientific merit to this **biomedical perspective**, which ultimately serves as the foundation of Western medicine, it has been criticized for divorcing humans from their physical and social environments and setting the stage for a **reductionist** view of health, which ignores broader context.

In response, many definitions of health now include both physical and mental wellness and, in some cases, also social well-being. The World Health Organization (WHO), for example, defines health as "a state of complete physical, mental and social well-being and not merely the absence of disease or infirmity" (World Health Organization 1946, 1). This definition is not without its problems—scholars have noted that the definition is so utopian as to preclude any individual from ever actually achieving "health" (Mayer and Meade 1994). Nonetheless, reframing health in broader terms encourages us to be open to alternative perspectives that consider health in more **holistic** ways.

Other scholars look beyond biomedical perspectives to explore structural influences on disease, including social, economic, and political contexts. For example, in a biomedical perspective, tuberculosis (TB) is considered to be the result of infection by the TB bacterium. However, a structural approach to health might also incorporate the influence of poverty, marginalization, and government policies as critical to explaining TB incidence.

Social and cultural theory has also gained currency in studies of health. Concepts such as race, gender, and marginalized populations have become important in these "post-medical" approaches to health and healthcare. These approaches reject the reductionist view that

health and disease can be explained by the presence or absence of a pathogen and instead consider the role of a person's broader social and psychological life. The individuality of human experience is recognized as profoundly influencing health and well-being, and the meanings of places to different people are recognized as significant. Place in this context includes not only attributes of the landscape or built environment, but also how the people who live there interact with and respond to their surroundings, as well as the feelings that place engenders, and even the significance of a person's social situation or "place-in-the-world" (Eyles 1985). These intangible aspects related to the social significance of place are sometimes referred to as "**sense of place**" (Curtis 2004).

In this broader framework of place-based approaches to health, we can think about why some landscapes encourage healing, how certain places become symbolic of disease, or how mental health can be influenced by attachment to place. For instance, in a study of four smallpox epidemics that affected San Francisco in the mid-1800s, Craddock (1995) argues that Chinatown became associated with the smallpox epidemic as xenophobic attitudes among dominant white groups led to the portrayal of Chinese communities as the source of smallpox, reinforcing negative stereotypes that were already established between immigrants and disease. This kind of work explores **social constructions** of place and exemplifies **postmodern** approaches to the geography of health, in which scholars encourage us to consider the ways in which definitions of health and disease are created by people, with different cultural frameworks leading to starkly different ideas about how to maintain health and treat disease (see Kearns 1993).

This shifting emphasis of health geography from a focus on disease towards a broader conceptualization of health is now well recognized among health geographers. In this context, the term "medical geography"—the traditional name for the sub-discipline—has been contested in response to what some see as an overemphasis on disease and biomedical approaches at the expense of considering influences such as social context and individual experience on health. Robin Kearns initiated this discussion in the early 1990s, arguing that a "geography of health" should be recognized alongside medical

Figure 1.9 Traditional Inca sanitation channel, Ollantaytambo, Peru

Efforts to improve sanitation became particularly important when people began living at high densities in towns and cities. Early urban cultures such as the Inca were in the vanguard of developing innovative sanitation solutions for their towns, some of which still function today.

geography to reflect new "post-medical" concerns and the importance of engaging with social theory and cultural geography (Kearns 1993, 140). Kearns and Moon identify three distinct themes that have emerged within this new geography of health: "the emergence of 'place' as a framework for understanding health, the adoption of self-consciously sociocultural theoretical positions, and the quest to develop critical geographies of health" (Kearns and Moon 2002, 606). More recently, others have called for further extending health research into the humanities, emphasizing the lived experiences of health and healthcare and more nuanced spatial analyses (de Leeuw et al. 2018). As several scholars have pointed out, however, accepting the value of a post-medical geography of health should not undermine the value of traditional approaches of medical geography such as **disease ecology** (Kearns and Moon 2002; Kearns 1993). The term "health geography" has now become an inclusive title that refers both to the traditional themes of study associated with medical geography and the expansion of the sub-discipline into "post-medical" approaches. We use the term "health geography" in the title of this text to reflect this emphasis on both traditions.

This evolution of geographic approaches within health studies is exemplified by changing approaches to the study of **healthcare**. Traditionally, geographic approaches to healthcare focused on *spatial* questions, such as where best to locate a new hospital. Over time, interest developed in *structural* aspects of healthcare—such as how best to provide care given government spending priorities, economic constraints, and cultural preferences. Most recently, geographic work on healthcare has expanded to include more *critical* questions related to equality of access and ethics of care, often incorporating qualitative methods to address questions related to individual experiences of healthcare and healing.

The importance of *space* and *place* in health studies has, therefore, changed over time. Early interest in place, following the Hippocratic tradition, focused on the influence of the natural environment on health, with places viewed simply as containers for attributes such as a particular climate and vegetation. The quantitative revolution of the 1960s led to a renewed emphasis on space, considering how factors such as location, proximity, and connectivity are significant to health. Approaches to health geography have since expanded to emphasize also social processes and people's subjective experience of different places. These perspectives on the study of health and place are reflected, albeit imperfectly, in the major sections of this book, with the first section on ecological approaches emphasizing the long-standing traditions of disease ecology, while the social approaches of the second part of the book include many of the themes that have emerged more recently related to social constructions of health. Geospatial methods are emphasized throughout the book as a tool for geographic study, including a chapter devoted explicitly to discussion of spatial approaches (Chapter 6).

MODELING A GEOGRAPHIC APPROACH: UNDERSTANDING THE 2013–2016 EBOLA OUTBREAK USING GEOGRAPHIC TECHNIQUES

To provide an example of how we can apply geographic methods to a health issue, we now consider the devastating 2013–2016 Ebola outbreak in West Africa. This was the biggest outbreak of Ebola virus disease (formerly Ebola hemorrhagic fever) on record, leading to more cases than in all previous outbreaks combined. This raises several important questions: Why did this outbreak occur? Why was it so much more devastating than previous outbreaks of the disease? What can this tell us about the likelihood of a future **pandemic** of Ebola?

Ebola is caused by a virus that normally circulates in animals. Symptoms of the disease include fever, fatigue, muscle aches, vomiting, rash, impaired liver and kidney function, and internal and external bleeding (WHO 2019). Although **case fatality rates** have varied among outbreaks, approximately half the people infected with the virus die from the disease. This high mortality rate, in combination with its disturbing **hemorrhagic** symptoms, helps explain why Ebola is so widely feared.

Identified in 1976 near the Ebola River in a region that is now part of the Democratic Republic of Congo, the disease has emerged periodically in short-lived outbreaks, mostly within the tropical rainforest regions of central

Africa until the 2013–16 outbreak in West Africa (Kaner and Schaack 2016; WHO 2019) (Figure 1.10). At least five different strains of the virus are known to have caused outbreaks of disease (WHO 2019). The virus is believed to circulate in wild animals but appears to cause disease primarily in humans, other primates, and antelopes (Kaner and Schaack 2016). Fruit bats of the Pteropodidae family are thought to be the main natural **reservoir** of the Ebola virus (WHO 2019). A **spillover event** occurs when a virus spreads from its usual animal host into the human population through direct contact with an animal's body fluids, often when an animal is butchered to be eaten. Once the virus has entered the human population, Ebola then spreads through direct contact with body fluids while tending patients or during preparation of bodies for burial, although sexual transmission is also known to occur (WHO 2019; CDC 2017).

All Ebola cases in the 2013–2016 West African event can be traced to a spillover event involving contact between a bat and an 18-month-old boy in Guinea in December 2013 (Kaner and Schaack 2016). Over the next two years, 28,616 cases of Ebola virus disease were reported, leading to 11,310 deaths in the West African countries of Guinea, Liberia, and Sierra Leone, dwarfing all previous outbreaks (CDC 2017). The disease spread more widely than any previous event, crossing three international borders in West Africa and causing a handful of additional cases and deaths as far away as Europe and the US. Previously recorded outbreaks had been small in number and spatially limited, burning themselves out in just a few weeks or months. By contrast, the 2013 outbreak persisted for two years and spread outside the remote rural communities that had historically been affected by Ebola into densely populated urban settings.

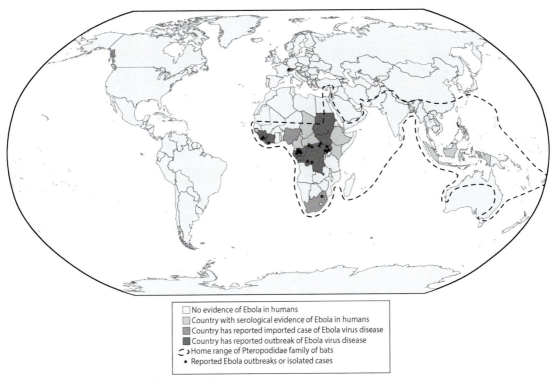

☐ No evidence of Ebola in humans
☐ Country with serological evidence of Ebola in humans
☐ Country has reported imported case of Ebola virus disease
☐ Country has reported outbreak of Ebola virus disease
⊂ ⊃ Home range of Pteropodidae family of bats
★ Reported Ebola outbreaks or isolated cases

Figure 1.10 Distribution of Ebola virus, 2014
Source: Adapted with permission from the WHO (2014)

How can a geographic approach to health contribute to analysis of this event? First, it is clear that Ebola risk is closely tied to environmental context. The disease occurs only after people have direct contact with forest animals, especially bats, and so the initial spillover event occurs in a tropical rainforest region. An ecological approach enables us to assess the ways in which people are put at risk through interactions with forest animals and one another. In particular, the escalating trade in bush meat (wild animals hunted for meat) provides opportunities for people to be exposed to the virus. Growing populations and forest loss exacerbate the problem as wild animals are brought into closer contact with human communities owing to habitat fragmentation, increasing the likelihood of spillover events (WHO 2015).

We can then take a step back to explore different scales of analysis by asking what economic and political structures encourage habitat fragmentation and local dependencies on bush meat. At the national scale, for instance, government policies may provide few opportunities for economic advancement for remote rural communities, leaving them dependent on natural resources, including wild animals and timber from forests. At the international scale, we can consider how global trade networks leave many African countries on the margins of the global economy, again impoverishing local communities and perpetuating their dependence on local resources.

Additionally, we can isolate cultural aspects of the disease through analysis of the social environment. Medical anthropologists have noted that some traditional West African burial practices are especially problematic with respect to Ebola, particularly the custom of washing bodies before burial and showing respect to the dead through physical contact (WHO 2015). The fact that West Africa had not experienced an outbreak before this event was also critical because the region lacked public health institutions that were proficient at identifying and responding to Ebola outbreaks (WHO 2015). This enabled the disease to spread widely before containment measures were put into place, rapid and heavy-handed containment strategies having been instrumental in limiting previous Ebola outbreaks.

Finally, we can see how increasingly mobile human populations may have caused the recent outbreak to spread more widely than previous outbreaks. The rainforest belt of Africa was a relatively remote and isolated place in the 1970s and '80s, holding epidemics in check, but increasing human mobility meant that, in the 2013–2016 outbreak, infected individuals traveled as far as the United States and Europe before they were identified. Air transportation also brought the disease to the huge population of Lagos, Nigeria, although luckily the disease was contained there too. The disease also rapidly made its way along rural-urban trade routes via ground transportation, making this outbreak among the first to reach dense urban settlements. Indeed, the capital cities of the three countries at the epicenter of the outbreak were all profoundly affected (WHO 2015). A tradition of returning to native villages to die for burial near ancestors further contributed to the movement of infectious individuals (WHO 2015). Understanding the mobility patterns of affected West African populations was critical for ensuring that this localized **epidemic** did not generate a global pandemic.

So, how likely is a future pandemic, given what we learned from this outbreak? Previous outbreaks have shown that containment is possible when it comes to Ebola. The disease is only infectious when a person has obvious symptoms (WHO 2019) and so basic public health interventions that prevent people coming into contact with contaminated body fluids are effective. The most recent epidemic suggests that increased human movement is elevating risk, however. With rapid airline transportation, previous containment strategies may prove insufficient. Nonetheless, effective public health systems that could quickly identify and isolate patients were able to control the disease, even in the 2013–2016 outbreak, as exemplified by containment efforts in the US and Europe. Once again, we should turn to political and economic structures as pivotal factors in explaining how Ebola managed to take hold in West Africa, where poor communities lacked the public health infrastructure to monitor and control the disease effectively. As the WHO (2019) states: "Good outbreak control relies on applying a package of interventions, namely case management, infection prevention and control practices, surveillance and contact tracing, a good laboratory service, safe and dignified burials and social mobilisation." None of these requirements is beyond the scope of public health managers, given sufficient funding, but that funding has not been forthcoming. Another outbreak in the Democratic Republic of the Congo and Uganda was already developing in 2019, suggesting that more regular, larger outbreaks might be the new pattern for Ebola. An effective treatment for Ebola may also be on the horizon, however, which would once again change the balance of power between pathogen and host.

STRUCTURE OF THIS BOOK

This book is divided into two parts to reflect the theoretical distinction between ecological and social approaches to health, while recognizing that that there is considerable overlap between these approaches. As we explore in Chapter 2, ecological approaches to health focus on how the environment influences health, including aspects of the natural, built, and social environment. Although the social environment is included as one aspect of an ecological approach, the approaches taken by social theorists are theoretically distinct and worth exploring in a separate section. We therefore save consideration of social approaches for the second part of the book.

In each section, we combine theoretical material that informs that approach with case studies. We also outline common techniques for data collection and analysis for each approach. We emphasize the significance of spatial methods as a unique contribution offered by geographers to the study of health in Chapter 6.

Ecological approaches

Ecological approaches focus on humans as biological entities, recognizing that people are part of interdependent ecological systems. One of the commonest areas of study is the relationship between humans and their natural and built environments. The following are examples of studies that would fall clearly within this approach:

- the impact of the development of irrigation on the spread of schistosomiasis,
- the role of occupational exposure to particulate matter in the development of lung disease in miners,
- the significance of climate change to the incidence of respiratory conditions.

Social approaches

Social approaches to health consider how human health and well-being are influenced by factors such as economic and political systems, cultural norms, identity, and social status. The consideration of social aspects of space enables the investigation of factors such as social exclusion and individuals' positions within social hierarchies. The following studies would fall within this framework:

- language as a barrier to healthcare access for Turkish immigrants in Germany,
- the influence of disparities in wealth on the distribution of healthcare providers,
- the role of community gardens as places of healing.

Combining ecological and social approaches

The conclusion of the book (Chapter 11) demonstrates the connections and interdependencies among these theoretical and methodological approaches with two case studies that exemplify the power of synthesizing information from multiple sources and incorporating a diversity of approaches. First, we consider efforts towards disease control and eradication. Using the example of the successful campaign to eradicate smallpox and analysis of subsequent efforts to eradicate polio, we argue that an effective combination of ecological, social, and spatial approaches is essential to tackling modern health challenges successfully. In a second case study, we turn to the obesity epidemic that is occurring in many high-income, and increasingly also middle- and low-income, countries. We argue that obesity cannot be effectively tackled unless both the physical and social components of the disease are recognized and addressed.

DISCUSSION QUESTIONS

1   Discuss the geographic aspects of a current health-related news story, as demonstrated in the case of Ebola. Do you believe that ecological, social, or spatial approaches, or a combination

of approaches, would be useful in exploring this topic further?

2   Reread the discussion concerning the maps of malaria. Find a map on the Internet of the distribution of a different disease that you are familiar with and see if you can develop some hypotheses about its distribution. The WHO provides some useful maps at www.who.int/gho/map_gallery/en/ as a starting point.

3   Do you think that the **doctrine of specific etiology** is sufficient for explaining complex phenomena such as malaria, cancer, depression, or alcoholism? Why or why not? What other factors might be important?

4   We note that sanitation has probably had a more significant impact on human health and lifespan than any clinical intervention. Think of some other public health interventions (e.g., vaccination, quarantine, screening campaigns) and consider them in terms of:

   a   the number of people who benefit,

   b   the cost-benefit ratio of the intervention,

   c   potential infringement of individual rights.

5   What is health? What causes disease? How are health and disease place-specific?

6   How would you describe the sub-discipline of health geography to a layperson?

FURTHER READING

Emch, M., E. Root, and M. Carrel. 2017. *Health and Medical Geography*. 4th ed. Guilford Publications.

Kearns, R. A. 1993. "Place and Health—Towards a Reformed Medical Geography." *Professional Geographer* 45: 139–47.

Kearns, R., and G. Moon 2002. "From Medical to Health Geography: Novelty, Place and Theory After a Decade of Change." *Progress in Human Geography* 26: 605–25.

Mayer, J., and M. Meade. 1994. "A Reformed Medical Geography Reconsidered." *Professional Geographer* 46: 103–6.

[WHO] World Health Organization. 2015. "One Year into the Ebola Epidemic: A Deadly, Tenacious and Unforgiving Virus." www.who.int/csr/disease/ebola/one-year-report/introduction/en/.

VIDEO RESOURCES

"Outbreak." Frontline, May 15, 2015. PBS. More information at: www.pbs.org/wgbh/frontline/film/outbreak/.

"Rx for Survival." 2005. PBS. More information at: www.pbs.org/wgbh/rxforsurvival/.

REFERENCES

[AAFP] American Academy of Family Physicians. 2019. "CDC Data Show U.S. Life Expectancy Continues to Decline." Accessed June 9, 2019. www.aafp.org/news/health-of-the-public/20181210lifeexpectdrop.html.

Case, Anne, and Angus Deaton. 2015. "Rising Morbidity and Mortality in Midlife Among White Non-Hispanic Americans in the 21st Century." *Proceedings of the National Academy of Sciences* 112 (49): 15078–83. doi:10.1073/pnas.1518393112.

[CDC] Centers for Disease Control. 2017. "Ebola Outbreak in West Africa Epidemic Curves." Accessed August 29, 2019. www.cdc.gov/vhf/ebola/history/2014-2016-outbreak/cumulative-cases-graphs.html.

Craddock, S. 1995. "Sewers and Scapegoats—Spatial Metaphors of Smallpox in 19th-Century San-Francisco." *Social Science & Medicine* 41 (7): 957–68.

Curtis, Sarah. 2004. *Health and Inequality: Geographical Perspectives*. London and Thousand Oaks, CA: Sage Publications.

de Leeuw, Sarah, Courtney Donovan, Nicole Schafenacker, Robin Kearns, et al. 2018. "Geographies of Medical and Health Humanities: A Cross-Disciplinary Conversation." *GeoHumanities* 4 (2): 285–334. doi:10.1080/2373566X.2018.1518081.

Eyles, John. 1985. *Senses of Place*. Warrington, Cheshire, England: Silverbrook Press.

Gardner, Gary T., and Brian Halweil. 2000. *Underfed and Overfed: The Global Epidemic of Malnutrition, Worldwatch Paper, 150*. Washington, DC: Worldwatch Institute.

Guerra, C. A., P. W. Gikandi, A. J. Tatem, A. M. Noor, et al. 2008. "The Limits and Intensity of Plasmodium falciparum Transmission: Implications for Malaria Control and Elimination Worldwide." *PLoS Medicine* 5 (2): 300–11. doi:10.1371/journal.pmed.0050038.

Kaner, Jolie, and Sarah Schaack. 2016. "Understanding Ebola: The 2014 Epidemic." *Globalization and Health* 12 (1): 53. doi:10.1186/s12992-016-0194-4.

Kearns, R. A. 1993. "Place and Health—Towards a Reformed Medical Geography." *Professional Geographer* 45 (2): 139–47.

Kearns, R., and G. Moon. 2002. "From Medical to Health Geography: Novelty, Place and Theory After a Decade of Change." *Progress in Human Geography* 26 (5): 605–25. doi:10.1191/0309132502ph389oa.

Legates, David R., and Cort J. Willmott. 1990. "Mean Seasonal and Spatial Variability in Gauge-Corrected, Global Precipitation." *International Journal of Climatology* 10 (2): 111–27. doi:10.1002/joc.3370100202.

Magner, Lois N. 2005. "A History of Medicine." Taylor & Francis. www.netlibrary.com/urlapi.asp?action=summary&v=1&bookid=158950.

Mayer, J. D., and M. S. Meade. 1994. "A Reformed Medical Geography Reconsidered." *Professional Geographer* 46 (1): 103–06.

[PRB] Population Reference Bureau. 2019. "2018 World Population Data Sheet." Accessed June 20, 2019, from www.prb.org/Publications/Datasheets/2009/2009wpds.aspx.

[SEDAC] Socioeconomic Data and Applications Center. 2012. Gridded Population of the World (GPW), v4.

United States Geological Survey. 2019. EarthExplorer.

van Voorthuizen, A. 2004. "Mariyamann's Sakti: The Miraculous Power of a Smallpos Goddess." In *Women and Miracle Stories: A Multidisciplinary Exploration*, edited by A. Korte, 248–70. Boston, MA: Brill.

[WHO] World Health Organization. 1946. "Constitution of the World Health Organization." New York: WHO Interim Commission.

[WHO] World Health Organization. 2014. "Geographic Distribution of Ebola Virus Disease Outbreak in Humans and Animals." Accessed August 29, 2019. www.who.int/csr/disease/ebola/global_ebolaoutbreak risk_20140818-1.png?ua=1.

[WHO] World Health Organization. 2015. Accessed June 9, 2019. www.who.int/csr/disease/ebola/one-year-report/introduction/en/.

[WHO] World Health Organization. 2018. "Malnutrition." Accessed June 9, 2019. www.who.int/news-room/fact-sheets/detail/malnutrition.

[WHO] World Health Organization. 2019. "Ebola Virus Disease." Accessed August 28, 2019. www.who.int/news-room/fact-sheets/detail/ebola-virus-disease.

World Bank. 2018. "GNI per Capita, PPP (Current International $)." Last modified July 19, 2018. Accessed August 28, 2019. https://data.worldbank.org/indicator/NY.GNP.PCAP.PP.CD?locations=EU.

SECTION I

ECOLOGICAL APPROACHES
TO HUMAN HEALTH

Ecological approaches to human health consider human beings as part of a broader ecosystem. Disease ecologists focus on interactions between humans and the environments in which they live and work, helping to describe and explain patterns of health and disease across space.

Humans interact with their environment in many ways that make them more or less susceptible to ill health. For instance, staying out too long in cold weather can lead to hypothermia; exposure to the sun may promote the development of skin cancer. Not all connections are this obvious, however. One of the main ways in which disease ecology has been useful in explaining disease patterns is through consideration of how characteristics of the environment influence where disease-causing organisms, or the vectors that carry them, can live. For instance, many diseases are restricted to tropical climates where year-round warm temperatures allow vectors such as mosquitoes to thrive. Warm temperatures can also speed up the reproduction rates of microorganisms such as viruses and bacteria, as well as the invertebrates that transmit them, leading to more rapid transmission of disease among humans. Analyzing relationships between people and infectious agents of disease was one of the first focuses of disease ecologists and remains a fundamental part of disease ecology today.

Many early ecological studies in health geography emerged from research in the tropics during the colonial period when European doctors and geographers noticed how cultural and environmental conditions in the tropics influenced disease outcomes. The susceptibility of Europeans to tropical diseases became a major spur to medical research and innovation in disease ecology. For example, Jacques May, sometimes referred to as the "father of medical geography," noted that the outcomes of his patients in Asia often differed from what he was led to believe should occur from his European textbooks, feeding his interest in understanding how environment influences human health (see May 1958). May subsequently helped found the American Geographical Society's Department of Medical Geography in 1948 with the goal of compiling an atlas of disease.

Life cycle diagrams, which illustrate how disease-causing organisms interact with the animals that host and transmit them, can help to clarify the environmental requirements for transmission of a disease. Such requirements could include an appropriate **ambient** temperature for the **pathogen** to reproduce effectively, specific habitat characteristics for a vector to thrive, and a source of water in which the vector can breed. Once a disease cycle has been effectively mapped out in this way, it is then possible to consider ways in which the cycle can be broken to try to stop transmission of the disease.

Human ecology was developed in the twentieth century to study these relationships between humans and their environment. In this context, "environment" includes not only the physical aspects of one's surroundings but also the **social environment**. Because the study of social environments is often facilitated by different approaches and tools than consideration of the built and natural landscape, we save most of our discussion of these themes until the social approaches section of the book. Bear in mind, however, that consideration of the social environment is also pertinent to a human ecology approach.

REFERENCE

May, J. 1958. *The Ecology of Human Disease.* New York: MD Publications.

2

ECOLOGICAL APPROACHES
TO HUMAN HEALTH

One of the clearest ways that geographical approaches can be applied to health topics is by thinking about interactions between people and their environments. A recent WHO report suggested that just under one quarter of all deaths around the world are attributable to environmental causes (Prüss-Üstün et al. 2016). In this chapter, we focus on how people's relationships with the physical aspects of their environment can encourage or discourage disease. We consider several ways to conceptualize the relationship between disease and environment, including ecological approaches to human health, landscape epidemiology, and disease cycles. These themes are extended in Chapter 4, where we consider recent changes in the disease landscape associated with environmental change.

ECOLOGY OF HUMAN HEALTH

An ecological approach to health considers humans as biological entities within broader disease cycles. This approach requires that we consider whole ecosystems, particularly the roles played by organisms that harbor and transmit **pathogens** and how their

environmental requirements influence the range and transmission of disease. Jacques May (1958) coined the term "ecology of human disease" to describe how people interact with their environment in ways that support or damage health. For example, people living in environments that are favorable to mosquitos may experience high rates of mosquito-borne diseases such as malaria. An individual's risk of contracting malaria is influenced by more than just the natural environment, however. Factors such as whether houses have screened windows, whether people use insecticides, and how much time people spend outdoors may also be significant. We must therefore consider both the physical and social environment to fully understand disease patterns.

Physical aspects of the environment can be divided into natural and built components. The natural environment includes landscapes such as forests and mountains, water bodies such as lakes and oceans, and the earth's atmosphere and weather, which can influence the human body in a variety of ways (Figure 2.1). For instance, microorganisms that live in our natural environment can invade our bodies and cause disease, airborne particles such as pollen and dust can lead to

Figure 2.1 The natural environment

The natural environment can influence health in a variety of ways. In this environment in the Rocky Mountains of the US, for instance, human health may be affected by pathogens such as Giardia *in the water, vectors such as ticks in the vegetation, and extreme weather events such as snowstorms.*

respiratory illness and allergic reactions, and ultraviolet (UV) radiation from the sun can damage our skin and eyes. The field of **biometeorology** considers the many ways in which the weather and atmosphere influence our health (Box 2.1).

The **built environment** refers to human-constructed parts of the landscape such as buildings, dams, and roads (Figure 2.2). Although the term is frequently used in contrast to the natural environment, the distinction is often blurred. For example, agricultural fields and urban parks could be considered built spaces because they are human constructed, but they are often viewed as part of the natural environment. The natural and built environments can perhaps best be viewed as a continuum from mostly natural spaces such as the Amazon rainforest to highly humanized spaces such as Hong Kong, with most places falling somewhere in between. We nonetheless recognize a distinction between many of the potential human health impacts of the built versus the natural environment. Much of our concern with the natural environment focuses on the impact of *living* agents of disease, or **pathogens** (e.g., bacteria, viruses, fungi), and their ability to cause infectious diseases. By contrast, the built environment is more often seen as a source of potentially harmful *non-living* agents of disease, or **geogens** (e.g., asbestos, traffic exhaust fumes, lead). Other facets of the built environment can affect health, however, such as the danger of falling masonry during an earthquake or the relative "walkability" of a neighborhood.

Box 2.1

BIOMETEOROLOGY AND THUNDERSTORM ASTHMA

The field of biometeorology explores the many ways in which the earth's atmosphere and weather interact with its biosphere. One important aspect of this research is the influence of atmospheric forces and weather on human health. Although philosophers and scholars have contemplated the impact of weather and climate on human health for generations, there has recently been increased interest in the topic as climate change makes it especially relevant.

A variety of connections have been suggested between weather and health, with factors such as levels of ambient ozone, ultraviolet radiation, and the humidity of air known to affect the human body. Changes in rates of asthma, heart disease, respiratory diseases, and migraines have all been associated with climatic conditions; mood disorders and mental health problems have been connected to seasons of low light; studies have even suggested possible associations between air pressure and arthritis. Here we present the case study of **thunderstorm asthma** as an example of research in this field.

Thunderstorm asthma is not yet well understood but has been reported in a variety of places over the past twenty years, including England, Canada, Italy, Iran, the US, and Australia (Elliot et al. 2014; Schumacher 2017; Grundstein et al. 2008). Although emergency room visits for asthma appear to rise slightly after many thunderstorms, the phenomenon appears to be worst when associated with storms with high rainfall and strong winds (Grundstein et al. 2008), which can create mega-events that threaten to overwhelm emergency services. One such event occurred on November 21, 2016, when a large spring thunderstorm over Melbourne, Australia, led to an estimated 42 percent increase in the caseload for emergency medical services and a 432 percent increase in emergency services provided for acute respiratory symptoms (Andrew et al. 2017). Within a few days, 8,500 people had received hospital treatment associated with the event and ten had died (Thien 2018; ABC 2016a, 2016b). Many of the individuals seeking treatment had no history of asthma, although a large proportion reported a past history of hay fever (Schumacher 2017).

Research suggests that an acute allergic reaction to pollen may be to blame for these outbreaks of thunderstorm asthma. It is believed that the rain associated with strong thunderstorms can cause pollen grains to expand and rupture, releasing allergenic particles small enough to get much deeper into the lungs than pollen grains can under normal conditions (Taylor et al. 2002; Suphioglu et al. 1992). This inflames the lining of the airways of people with a sensitivity to pollen, leading to symptoms of asthma such as wheezing (Wark et al. 2002).

The particular geography of Melbourne may make it especially prone to thunderstorm asthma. Located south of a large area of grasslands, strong northerly winds during storms tend to carry grass pollen across Melbourne (Schumacher 2017). The spring of 2016 was Victoria's tenth wettest on record, leading to unusually rapid plant growth in the months preceding the event, and much higher than normal airborne grass pollen concentrations on the day of the event (Thien 2018). This, combined with a line of thunderstorms that created a powerful gust front that pushed the pollen over Melbourne during the evening commute (Thien 2018), effectively created the perfect storm. As climate change increases the risk of strong thunderstorms developing, future outbreaks of thunderstorm asthma are expected (Grundstein et al. 2008).

The **social environment** refers to aspects of human behavior and organization that comprise the social structures in which we live. These include political and economic frameworks, as well as cultural norms

Figure 2.2 The built environment

> *Urban areas represent one of the most extreme forms of the built environment. Health impacts may relate to toxic building materials, high levels of air pollution, or even the mental health impacts of a lack of green space.*

such as dietary practices and gender roles. Power plays an important role in the social environment at a variety of scales. At global and regional scales, power relations drive who controls land-use decisions and who has access to capital, influencing human interactions with environments and, therefore, disease patterns. At national scales, government and legal frameworks mediate access to healthcare and education, with important health implications. At the household scale, power may be exercised via ideas around gender roles, which can influence the type of food or medical care offered to family members. In this chapter, we consider the social environment in terms of the cultural practices that influence people's interactions with their natural and built environments, leaving discussion of other social aspects to the second part of the book.

Disease transmission

People are continually exposed to organisms and substances that can cause disease (Table 2.1). These **agents of disease** can be divided into living agents of disease—**pathogens** and non-living environmental factors—**geogens** (Table 2.2).

Table 2.1 Basic categorization of hazards to physical health, with examples

Pathogens	Geogens	Physical hazards
Bacteria	Coal dust	Trauma
Viruses	Asbestos	Extreme heat/cold
Fungi	Lead	Radiation

Table 2.2 Basic categorization of pathogens

Infectious agent	Description	Examples	Notes
Bacteria	Diverse group of unicellular organisms	*Salmonella typhi* (typhoid), *Clostridium tetani* (tetanus)	Susceptible to antibiotics (although many bacteria are now resistant to some antibiotics)
Rickettsiae	Rickettsiae are similar to bacteria but cannot survive outside host cells	Typhus is caused by a rickettsia	Most infectious disease events are caused by bacteria or Rickettsiae
Viruses	Genetic material surrounded by a protein coat; as they cannot independently metabolize or reproduce, viruses are not technically living organisms	Influenza viruses; *Herpes* viruses; HIV; rabies	Not susceptible to antibiotics, although antiviral medications can treat some infections; can recombine rapidly to form new genetic strains leaving therapies ineffective
Fungi	Members of the kingdom Fungi; distinct from animals and plants; reproduce primarily using spores	Ringworm; dermatitis; yeasts of the *Candida* genus cause infections of the throat and vagina	Fungi are found throughout our environment and bodies; usually kept in check by the immune system and competition with other microorganisms; fungal diseases are often problematic in immuno-suppressed individuals or when competing microorganisms have been destroyed by antibiotics
Protozoa	Unicellular organisms	Malaria; trypanosomiasis (sleeping sickness); *Giardia lamblia* causes diarrheal disease	Often acquired through contaminated food/water or via a vector
Helminths	Simple invertebrate animals	Tapeworms; schistosomiasis; *Trichinella spiralis*, which causes trichinosis	Many helminths have complex life cycles requiring one or more hosts; most helminths are not fatal but take a toll on a person's health
Prions	Infectious particles that consist only of protein	Bovine spongiform encephalopathy, BSE ("mad cow disease"); Creutzfeldt-Jakob disease	Poorly understood; all known prion diseases seem to affect the brain; some prion diseases appear to be inherited, others are apparently caused by eating infected tissue

Note: Although there is debate over the extent to which viruses and prions can be said to be "alive," we list them as pathogens as their mode of operation most closely resembles other agents in this category.

Source: Data from NIH (1999), Walters (2014)

Physical hazards such as temperature extremes, trauma, and radiation can also directly damage the human body.

Communicable diseases such as tuberculosis and malaria can be passed from one person to another (Figure 2.3). The ease with which a disease can be transferred from one person to another is referred to as its

transmissibility. Diseases that can be passed easily from person to person through air, water, or personal contact are described as **contagious**. Diseases resulting from physical hazards, exposures, or deficiencies such as malnutrition that cannot be transferred from one individual to another are **non-communicable**.

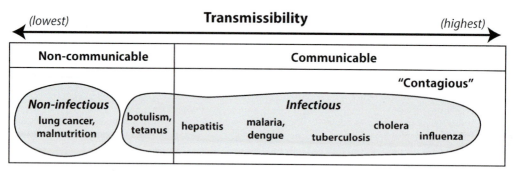

Figure 2.3 Categorization of diseases by transmissibility

Figure 2.4 Public health poster warning of the danger of amoebic meningitis in geothermal water, Rotorua, New Zealand

Diseases caused by living organisms are described as **infectious** and are usually communicable. Indeed, in common usage, the term "infectious" often implies that a disease can be transmitted from person to person. A few infectious agents are not normally transmitted from one person to another, however. Tetanus, for instance, is contracted from infection with tetanus bacteria from the local environment through a wound. The disease causes painful muscle spasms and can be fatal. Another pathogen, *Naegleria fowleri*, can be found worldwide in warm water. If the pathogen enters the brain via the nose when the head is submerged, it can cause amoebic meningitis (Figure 2.4). Although human infections are very rare, they are usually fatal. In both these cases, the agent is infectious (it is a living agent) but is considered non-communicable because the normal route of infection is not from person to person.

The introduction of an infectious agent into the body may not initially cause symptoms. At this early stage, the disease is **sub-clinical**, meaning that it is not possible to determine whether a person is infected from observable symptoms or clinical tests. If the infectious agent can replicate successfully in the body, it may eventually become detectable—for instance, by high levels of virus in the blood (**viremia**). When an infection is detectable with appropriate tests, the disease is considered **clinical**.

The severity and combination of symptoms exhibited after infection by a pathogen can vary among individuals. The symptoms associated with infection may range from no symptoms at all (an **asymptomatic** case), to mild symptoms like cough or fever, to severe symptoms such as paralysis, brain damage, or even death. Polio, for instance, can lead to a wide range of symptoms. Many people infected with the poliovirus are asymptomatic or experience only a mild, flu-like illness. For others, the disease can be devastating, with a small proportion of cases leading to infection of the central nervous system and paralysis—this can prove fatal if the paralysis affects a vital system such as breathing (WHO 2019b). How a person responds to an infectious agent will depend on the **virulence** (severity) of the strain of the disease they contract, the person's genetic resistance to the pathogen, and the strength of the physiological response that the body mounts in response to the infection.

Many symptoms of disease, such as sneezing or coughing, facilitate the transmission of pathogens, although even asymptomatic individuals may be able to transmit a disease to others. Individuals who are asymptomatic but can transmit a disease are known as **carriers**. The potential for carriers to spread a disease can be high because people may not take precautions against spreading disease if they do not realize they are infectious. In one famous case in the early 1900s, a carrier known as Typhoid Mary passed typhoid on to at least 51 people while working as a cook (Box 2.2).

Box 2.2

TYPHOID MARY

Typhoid is an infectious disease spread in conditions of poor sanitation. A person infected with the disease sheds the pathogen in urine and feces, which can be passed to others through contaminated food or drink. Individuals with typhoid can become **carriers**, particularly since a small proportion of the people who recover from the disease continue to excrete the pathogen for months or even years afterwards.

Mary Mallon may be the most famous disease carrier in history. Mary worked as a cook in New York in the early 1900s—an ideal situation for spreading food- and waterborne disease. She is believed to have infected at least 51 people with typhoid, three of whom died (Brooks 1996). Public health authorities identified her as the source of several typhoid outbreaks and informed her of her potential to spread the disease. Mary became notorious for refusing to stop working as a cook and vehemently denying her role in spreading the disease. Subsequently, she was forcibly quarantined twice and ultimately died in **quarantine** after 26 years in isolation (Brooks 1996).

It is important to remember that disease transmission patterns were poorly understood at this time, and few believed that a healthy person could transmit disease. Mary probably had few employment options and may have experienced prejudice on a daily basis, both as a member of the working class and as an Irish immigrant, which probably contributed to her defiance at being told that she was spreading disease.

In an ecological framework, a person or animal infected with a pathogen is referred to as the pathogen's **host**. The host provides resources that help the pathogen thrive and reproduce. Pathogens have evolved various ways to facilitate movement among hosts. Some pathogens require direct contact between an infected and a **susceptible** person, minimizing the time spent in the hostile conditions of the external environment. Sexual transmission provides a clear case of such a transmission route, but other forms of physical contact such as kissing or touching can also efficiently spread pathogens. Other pathogens, including the measles and influenza viruses, can survive for a limited time in the air and infect others if they are inhaled. These are all considered to be forms of **direct transmission** because the pathogen is transferred rapidly from one host to another without passing through other organisms or spending time on objects.

Indirect transmission occurs when a pathogen travels between hosts via other organisms or objects in the external environment. Inanimate objects that commonly harbor pathogens are known as **fomites**. Objects touched by many people within a short period of time such as door handles or computer keyboards are particularly effective at transmitting disease in this way. Antibacterial coatings have been developed for some common fomites, such as toilet flush handles, to prevent disease transmission.

Food or water can also introduce pathogens into the human body, another form of indirect transmission. The **fecal-oral route** of transmission—the spread of feces-borne bacteria from one individual to the mouth of another—typically involves tainted food or water. This route of transmission is normally associated with pathogens that infect the digestive system like cholera and dysentery.

In another form of indirect transmission, pathogens use a separate species—a **vector**—to transmit the pathogen between hosts. Common vectors include mosquitos, ticks, and flies. The range of vector-borne diseases is often defined by very specific environmental requirements, such as appropriate ambient temperatures, as the vector must be able to thrive for disease transmission to be sustained.

The types of transmission just described are all instances of **horizontal transmission**—the transfer of a pathogen from one independent host to another. Some pathogens can also be transmitted by **vertical transmission**—from mother to child during pregnancy, birth, or breastfeeding. A mother can transmit HIV to her child during birth, for instance, although drugs have been developed to help prevent this from occurring.

Adaptation

The concept of adaptation can also help explain spatial patterns of disease in human ecology. Every individual is subject to a unique set of place-specific exposures from his or her environment. For example, exposure to extreme cold is more likely in Canada than in Kenya; exposure to vehicle emissions is more likely near a major intersection than in a rural area. Over time, individuals and populations may adapt to conditions to which they are **chronically** exposed, increasing their tolerance of local insults to health. At the population scale, we can recognize groups of people with similar adaptations. For example, Zambians are more likely to have some resistance to malaria than Germans; similarly, a Zambian farmer is more likely to have some resistance to malaria than a Zambian office worker.

Adaptation to local conditions can be genetic, physiological, or behavioral. **Genetic adaptation** refers to processes of natural selection resulting in a concentration of genes favorable to local conditions in a population. Individuals who exhibit genetic resistance to a disease are likely to possess a survival advantage and are therefore more likely to survive long enough to pass their genes to future generations. Over several generations, the prevalence of that favorable gene in the population can thereby increase. For instance, genes that provide resistance to plague were favored in many European populations over the course of centuries, owing to repeated outbreaks of the disease after its introduction from the steppes of Asia. As a result, the prevalence of genes for resistance to plague increased in Europe, reducing the impact of later outbreaks.

Genetic adaptation is traditionally held to occur at the scale of whole populations, with no change to the **genomes** of individuals. However, recent evidence suggests that individual genomes may change in response to environmental conditions under certain circumstances.

The emerging field of **epigenetics** explores the way in which genes may be "switched on" or "switched off" in response to environmental stimuli. One of the best studied examples is the Dutch Hunger Winter of 1944–1945, when epigenetic changes appear to have occurred in babies that were in utero during the famine, causing health changes all the way into adulthood (Box 2.3).

Physiological adaptation refers to changes in an individual's body in response to an external stressor. For example, ultraviolet light triggers the skin to produce melanin, which darkens the skin; low oxygen concentrations stimulate our bodies to produce more red blood cells; and physical exertion strengthens muscles. A particularly significant example of physiological adaptation

---

Box 2.3

EPIGENETICS AND THE DUTCH HUNGER WINTER

with Hannah Langford

Intensive research on DNA and the human **genome** since the 1950s has revealed the ways in which our genes make us all individuals, driving our appearance, our physical capabilities, even our personalities—the observable characteristics that we call our **phenotype**. Genes do not fully define us, however. For example, we know that tall parents are likely to produce children with a genome that codes for tallness, and yet their children will only be tall given sufficient nutrition to reach that maximum height. Even identical twins, who share the same genome, have slightly different phenotypes—different fingerprints, for example—related to differences in the environments in which they grew up. In this framework, human bodies can be viewed as a result of an interaction between genetics and the environment, with the environment enabling or constraining the expression of particular genes.

Mounting evidence is showing that environmental exposures can actually change the way in which our genes are expressed in a way that is heritable. Environmental conditions during early development may be especially significant (Lillycrop and Burdge 2014). The environment does not directly change our genome (i.e., our genes are not mutating) but instead influences how our genes are voiced via the **epigenome**. The epigenome is a host of chemical compounds and proteins that attach to the genome and influence how genes are expressed—turning genes on or off, controlling the production

of particular proteins, affecting how genetic information is interpreted by our cells (National Human Genome Research Institute 2016). Significantly, subsequent generations are able to inherit these epigenetic changes via so-called epigenetic transgenerational inheritance.

Some evidence for this idea comes from the case of the Dutch Hunger Winter of September 1944 to May 1945. Near the end of World War II, German blockades led to famine in parts of the Netherlands, with some Dutch communities consuming only 30 percent of normal calorie intake during the famine; 20,000 people died (Zimmer 2018; Carey 2011). Given the excellent records of the Dutch health system and the clear dates of the start and end of the famine, researchers were able to follow the outcomes of babies who were in utero during the famine and compare them with babies whose mothers received adequate nutrition during pregnancy. They found that individuals exposed to famine during critical early in-utero development became heavier than average as adults and often developed health problems in middle age, including higher rates of obesity, diabetes, and schizophrenia (Zimmer 2018; Lumey, Poppel, and Vaiserman 2013). Individuals in utero during the famine also proceeded to have, on average, lower-birthweight babies themselves, suggesting that impacts of the famine were being passed into the next generation (Lumey, Poppel, and Vaiserman 2013).

continued

Box 2.3 *continued*

Exactly how this epigenetic impact occurs is still not yet fully understood, but the study of epigenetics is burgeoning in medical research. Studies of identical twins suggest that epigenetic differences build up over the life course (Fraga et al. 2005), helping to explain how a disease with a strong genetic component such as asthma can occur in one identical twin but not the other (Meymandi 2010). Other studies have explored the potential role of epigenetics in the development of schizophrenia and reproductive diseases (Nilsson and Skinner 2015; Roth et al. 2009).

---

is the production of antibodies. After exposure to a pathogen, the body's immune system learns to recognize that pathogen, allowing the body to mount a rapid immunological response if subsequently exposed to it. This makes people less susceptible to diseases that they have already encountered and explains why young children are especially prone to infectious diseases. Where malaria is endemic in Africa, for instance, most children suffer periodic bouts of malaria in childhood until they develop some physiological resistance to the disease. People who lack previous exposure to a pathogen and are therefore unable to mount a rapid immune response are sometimes referred to as **immunologically naïve**.

Stressors to the body, such as age, malnutrition, psychological strain, or co-infection with another disease, can weaken the immune system. HIV/AIDS became a devastating pandemic in the 1980s precisely because of its ability to damage individuals' immune systems, making people more susceptible to other infections such as tuberculosis and pneumonia. Some previously rare medical conditions were able to take hold in populations with high rates of HIV infection. For example, Kaposi's sarcoma is a type of cancer associated with a strain of herpes virus, which became increasingly common among individuals with immune systems weakened by HIV infection (American Cancer Society 2018).

Behavioral adaptation to disease refers to cultural practices that hinder the spread of disease. Covering one's mouth when sneezing is a straightforward example. Over time, behavioral adaptations to disease can become cultural norms and may no longer be recognized as health related. For instance, traditions that discourage eating pork may originally have developed to avoid contracting infections from pigs, but over time pork avoidance has become accepted as a cultural practice in some communities, with little awareness of its health implications.

Genetic, physiological, and behavioral adaptations may all occur simultaneously. For instance, being at high elevation requires the body to adapt to conditions of low oxygen availability. The body begins to make *physiological* adjustments within minutes of arrival at high elevation, stimulating deeper breathing and a faster heart rate. Within a few hours, urination increases in response to changes in the blood's acid/base balance, which is why dehydration can be a problem at altitude. These physiological changes are often supplemented with *behavioral* adaptations. For instance, upon arriving at high elevation in the Andes, a traveler would be wise to choose to walk rather than run for the first few days to avoid breathlessness. Over the next several days, additional physiological changes occur as the body acclimatizes, including an increase in the production of red blood cells so that the blood can carry more oxygen. For populations that have been living at high altitude for generations, *genetic* adaptations have also occurred, giving these populations innate abilities for high-altitude living. Research suggests that populations in different regions exhibit different routes of genetic adaptation. In the Andes, people have adapted to carry more oxygen in each blood cell by increasing the amount of hemoglobin in the blood. Tibetans, by contrast, compensate by taking more breaths per minute and have blood vessels of greater than average diameter to allow more blood flow (Mayell 2004).

A basic understanding of adaptation can shed light on how diseases circulate within populations. When a disease affects a population at higher than expected rates, it is considered an **epidemic** (or a **pandemic** if it spreads across a wide area of the globe). This occurs when a pathogen is introduced into a population where few people have immunity to that disease. Where an entire population is immunologically naïve, a dramatic **virgin soil epidemic** can occur because many susceptible hosts are present (Crosby 1976). For example, the rapid spread of European diseases among Native American and Pacific populations following voyages of conquest in the 1400s occurred because indigenous populations had little immunity to the introduced diseases. Mortality rates probably exceeded 90 percent in some places, devastating communities and destroying entire cultures.

The passage of an epidemic can be represented as an S-shaped curve (Figure 2.5). At first, the disease spreads slowly as only a few people are infected to act as sources of contagion. As more people become infected, the disease spreads more rapidly. Eventually, the curve levels off because most of the population has by now developed physiological immunity, is genetically resistant to the pathogen, or has died, and so the epidemic finally burns itself out.

With long-term exposure to a pathogen, adaptations can build up in the host population until a large proportion of the host population is resistant. The disease is then restricted to circulating at a low level, with small outbreaks occurring only when a large enough pool of new **susceptible** individuals has accrued. At this point the disease is considered **endemic**. Children often serve as this new susceptible population as adults are immune to the disease from previous exposure. Over time, infants are born who have never been exposed to the disease and, once this susceptible group is large enough for the disease to circulate, an outbreak can occur. Diseases that circulate in this way are often referred to as **diseases of childhood**. Childhood diseases can be modeled with respect to how quickly this new susceptible population takes to build up in a population. For instance, in the absence of effective vaccination programs but given a large enough population, measles circulates in cycles, with outbreaks approximately every four years (Baker 2007).

The notion of **herd immunity** is significant here. Herd immunity is the idea that once a large enough proportion of a population is immune to a disease, the likelihood of an infected individual encountering a susceptible person is so small that transmission of the disease will stop. The proportion of the population required for this to occur varies by disease. In the case of measles, approximately 95 percent of a population must be vaccinated to achieve herd immunity (Baker 2007, 66). Many vaccination programs aim to achieve herd immunity so that individuals too young, too weak, or otherwise unable to receive the vaccination will be protected by high rates of vaccination in the community.

Unfortunately, vaccination rates in many populations that were well protected in the past are falling below this critical threshold, with predictable consequences. In Dublin, Ireland, for example, a measles outbreak flared up in the year 2000 after vaccination rates had fallen to 79 percent. This outbreak led to at least 355 cases of the disease, 111 hospital admissions, and three deaths (Baker 2007, 64–66). Similarly, a pertussis (whooping cough) outbreak in California led to a 418 percent increase in pertussis cases between January and June 2010, compared with the same period the previous year. Infants too young to be fully vaccinated were at particularly high risk (Winter et al. 2010), illustrating the importance of maintaining herd immunity to protect vulnerable populations. The resurgence of measles associated with falling vaccination rates has been identified as a public health crisis by the World Health Organization.

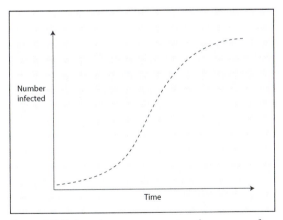

Figure 2.5 S-shaped curve representing the passage of an epidemic

CULTURAL ECOLOGY OF DISEASE

Although ecological approaches may prompt us to think of humans as victims of their environment, it is important to bear in mind that people actually have considerable influence over *how* they interact with their environment. Disease ecologists employ the study of **cultural ecology** to examine how cultural practices influence the likelihood of a disease circulating in a population.

Some practices, such as immunization programs, are deliberately designed to inhibit the spread of disease. These practices are embedded in complex cultural frameworks that are influenced by a population's attitudes, experiences, and cultural norms. The history of handwashing illustrates the cultural nature of these sorts of behavioral adaptations (Box 2.4). Much of what is considered within the cultural ecology of disease is not the result of conscious efforts to improve health, however, but instead supports or impairs health *coincidentally*. For instance, nomadic lifestyles may limit gastrointestinal diseases as people move away from areas soiled with human waste during their migrations (Briceño-Leon 2007).

The cultural ecology of Chagas disease (American trypanosomiasis)

The influence of everyday practices on human health is well illustrated by how changing living patterns influenced the range and incidence of Chagas disease in the Amazon. Chagas disease is caused by a protozoan, *Trypanosoma cruzi*, most commonly transmitted to people by insect vectors of the subfamily *Triatominae* ("kissing bugs"), although the disease can also be transmitted through infected food, organ transplants, or from

---

Box 2.4

IGNAZ SEMMELWEIS AND THE IMPORTANCE OF HANDWASHING

It was not until the nineteenth century that handwashing became standard practice in medical settings. While working on the maternity ward of Vienna Hospital, Hungarian physician Ignaz Semmelweis realized that women attended by doctors during childbirth had a significantly higher likelihood of dying of "childbed fever" (puerperal fever) than women attended by midwives. Semmelweis instituted an aggressive policy of handwashing on his maternity ward, following the hypothesis that these poor outcomes were related to the autopsies that doctors performed alongside their obstetric work. This dramatically reduced rates of maternal death from "childbed fever." Despite this, Semmelweis's theories were not widely accepted until after his death, when Louis Pasteur's work with microorganisms explained the underlying mechanisms behind Semmelweis's success. "Childbed fever" was subsequently found to be a bacterial infection.

The story was not a happy one for Semmelweis. His forceful campaign to convert others to his techniques widely alienated the medical establishment and bore little fruit. Pained by what he saw as the needless deaths of thousands of women, he became increasingly disturbed and was eventually hospitalized in an asylum where he died.

Today, handwashing has become standard medical practice as well as an important broader hygiene practice. Research has reinforced the importance of handwashing. For instance, in one study in urban squatter settlements in Karachi, Pakistan, a team of researchers educated more than 900 families about effectively washing their hands with soap after defecation and prior to preparing food. Children living in the households that received this education and free soap were subsequently found to have a 53 percent lower incidence of diarrhea compared with children living in control communities (Luby et al. 2004). Despite evidence such as this, some public health experts argue that health practitioners have become slapdash about simple hygiene measures and need to return to Semmelweis's seminal rules (Gawande 2004).

mother to fetus. Infection initially causes mild symptoms, including fever and body aches, but can ultimately lead to heart and digestive problems.

Although the disease's historic range extends from Mexico to Argentina (WHO 2003), Chagas disease was not recorded in humans in the Amazon until 1969 (Shaw, Lainson, and Fraiha 1969). Briceño-León (2007, S35) argues that the absence of Chagas disease from the Amazon until recently was largely due to "the way indigenous peoples had related to their environment and the land occupation patterns these communities had maintained for centuries." These cultural patterns made many traditional villages unfavorable for the Chagas vector, greatly limiting the potential for transmission of the disease. An abundance of vector species and mammal species infected with the pathogen are indigenous to the Amazon, further supporting the idea that it is not the natural environment of the Amazon that puts human populations at risk, but rather the way humans interact with their environment. This theory is also supported by the fact that cases of the disease began to appear in the Amazon only with the arrival of a large influx of non-indigenous settlers.

Anthropological studies of several indigenous communities suggest that four traditional cultural practices were significant in limiting disease transmission before the 1960s: human dwelling type, continuous mobility, lack of domestic animals, and the existence of an abundance of food sources for the Chagas vector outside human settlements (Briceño-Leon 2007). The large, open communal huts common to many indigenous Amazonian communities prevent vector colonization because they lack cracks and crevices where the vector can shelter (Schofield et al. 1990). The high mobility of many indigenous Amazonian communities further reduces opportunities for the vector to become established in houses because dwellings are abandoned before infestations have time to build up (Briceño-Leon 2007). Few indigenous Amazonian communities traditionally kept domestic animals, which can attract disease vectors. Finally, until recently, most Amazonian communities lived in relatively intact ecosystems where pathogen and vector species were held in check by a diversity of other species (Abad-Franch 2005).

Changes precipitated by more recent developments in the Amazon have altered these patterns. Deforestation is considered one of the most important factors (Coura et al. 2002). Upon losing their natural habitat and food sources, Chagas vectors are driven to human settlements to search for sustenance and shelter. The more stable, denser human populations in areas recently populated by non-native agricultural colonists have proved ideal for transmission of the disease. In particular, the style of housing common in these communities—with unfinished mud walls, dirt floors, and thatched roofs providing abundant crevices—has created a favorable environment for the vector (Coura et al. 2002). The domestic animals that accompanied these new human populations also help support populations of vectors (Briceño-Leon 2007). Efforts to control the disease must now be instituted in parts of the Amazon, owing to the breakdown of these traditional, coincidental forms of vector control. Chagas disease has also increased its range into North America, Europe, and parts of the Western Pacific as human mobility has introduced the disease into new regions (WHO 2019a).

DISEASE CYCLES

One important factor in the human ecology of disease is the ways in which humans interact with other species. Many human illnesses evolved from animal diseases, including probably tuberculosis, measles, and diphtheria from cattle; influenza from pigs and ducks; leprosy from water buffalo; and plague, typhus, and perhaps smallpox from rodents, although disagreement persists over some of these links (Lieberman 2013). As such, Diamond (1997) hypothesizes that communities with a tradition of keeping domestic animals have historically been afflicted by a greater variety of pathogens than communities with few domestic animals. Since most animals were domesticated in Eurasia, Old World populations would have had higher levels of resistance to their pathogens, perhaps explaining why Native American populations were so susceptible to Old World pathogens, yet few diseases spread in the opposite direction (Diamond 1997). The role of animals in transmitting disease to humans continues to be important, as evidenced by concerns regarding the likelihood of **spillover events** causing avian flu and Ebola epidemics. One widely cited

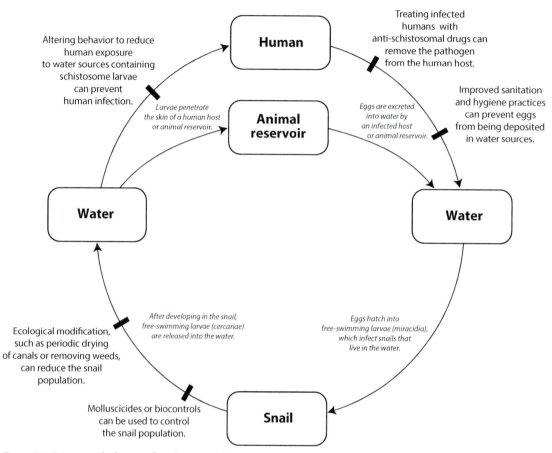

Figure 2.7 Disease cycle diagram for schistosomiasis

LANDSCAPE EPIDEMIOLOGY

The term "**landscape epidemiology**" was coined by Russian geographer E.N. Pavlovsky (1966) to refer to the ways in which regions impart patterns to disease distributions via factors such as vegetation, geology, and climate. For instance, radon gas emitted naturally from some types of rock can lead to lung cancer—a problem in parts of Southwest England and the Rocky Mountain region of the US. Exposure to high levels of ultraviolet radiation from the sun can lead to skin cancer, a significant cause for concern in Australia where the predominantly light-skinned population is not genetically adapted to high levels of solar radiation. Landscape

epidemiology also considers the role of environment in vector-borne diseases. For transmission of a vector-borne disease to continue, environmental conditions must be amenable to host, vector, and pathogen. Vector-borne diseases, therefore, provide a good example of the intersection of geography and health, and are a common topic of landscape epidemiology studies.

Temperature influences the distribution of many diseases. There is a latitudinal gradient in the richness of human pathogen species, with more pathogens near the Equator, probably related to the higher temperatures and precipitation at these low latitudes (Guernier, Hochberg, and Guegan 2004). Additionally, the temperature requirements of vectors restrict the range of many pathogens. For instance, dengue fever is limited to places with a minimum average monthly temperature of approximately 10°C (50°F) (WHO 1997), the lowest temperature at which populations of the disease's main mosquito vector, *Aedes aegypti*, can persist (Figure 2.8). As the map illustrates, temperature is not the only factor influencing the range of dengue fever, however. Other environmental factors, particularly availability of water, also influence whether an environment can support the dengue vector. Most of central Australia is not at risk for dengue, for instance, because there is insufficient precipitation for the vector to survive. Human activities influence the distribution of the disease as well. In some places, there may be no human hosts present. In other places, the disease may be controlled by human interventions such as mosquito control, rapid treatment of infected individuals, or public education campaigns to alter dengue-related behaviors. Humans are not hapless victims of their environments but have considerable **agency** to reduce disease transmission.

Many vector species cannot survive the winter in the cooler high and mid-latitudes, where temperatures drop seasonally, limiting the range of many "tropical diseases."

Ticks often serve as vectors of disease in cooler latitudes because some species can overwinter as an egg or larva. Additionally, some bacteria and *Rickettsia* species can be passed from adult ticks to tick eggs via **vertical transmission**, enabling the persistence of these pathogens through cold winters as well.

Water and disease

Water is critically important to many disease cycles, with a variety of pathogens infecting water supplies where poor sanitation allows excreted waste to contaminate drinking water. The acronym WASH (water, sanitation, and hygiene) is used as a reminder of the critical connections between water and health. According to the Global Burden of Disease Study, diarrhea was still a leading cause of death worldwide for all age groups in 2015 and was the fourth-leading cause of death among children under the age of 5 (Troeger et al. 2017). Although improvements in sanitation have led to a significant decrease in diarrheal disease over the past 25 years, diarrheal diseases remain an important source of preventable disease in South Asia and sub-Saharan

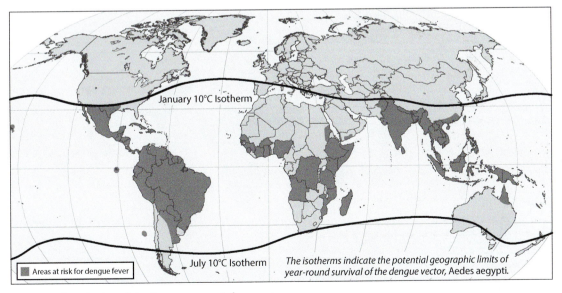

January 10°C Isotherm

July 10°C Isotherm

The isotherms indicate the potential geographic limits of year-round survival of the dengue vector, Aedes aegypti.

Areas at risk for dengue fever

Figure 2.8 Distribution of the dengue vector, *Aedes aegypti*

Source: Adapted from WHO Regional Office for South-East Asia (2011, 4)

Africa, in particular (Troeger et al. 2017). Deaths from diarrheal disease are heavily concentrated in the Global South, where mortality rates may be hundreds of times higher than in affluent countries (Figure 2.9), but diarrhea remains a source of **morbidity** (illness) in the Global North, creating a significant burden on healthcare systems. In the US, for instance, the Centers for Disease Control (CDC) reports that *Giardia*, *Legionella*, Norovirus, *Shigella*, and *Campylobacter* are the top five causes of diarrheal outbreaks from public water supplies, despite having public water systems that are generally very safe by global standards (CDC 2014).

Many pathogens and vectors require specific water conditions to be able to thrive, and so even slight changes to water characteristics can influence disease rates. The bacteria that causes cholera (*Vibrio cholerae*), for example, can survive in water, but only where microenvironmental factors such as salinity and aquatic flora and fauna are appropriate (Drasar and Forrest 1996). Disease cycles are, therefore, often place specific, and knowledge of a disease cycle in one region may be insufficient for understanding the same disease's life cycle

elsewhere. To provide one example, a malaria control program implemented in 1907 in northwest Argentina failed because it was based on experiences with malaria eradication in Italy, where a different mosquito species with different breeding preferences served as the vector. Draining swamps, which had successfully controlled malaria in Italy, was of little use in controlling malaria in northwest Argentina where the vector bred instead in clean, shallow water (Carter 2007).

The availability of water also influences vegetation patterns and, by extension, vector habitats and breeding sites. Lack of water has traditionally served as an important barrier to vector-borne diseases—the Sahara Desert stands out in Africa as a region without endemic malaria, for example. The expansion of irrigation has therefore had a significant impact on many vector-borne diseases through providing new or improved vector habitats (Figure 2.10). Increased incidence of schistosomiasis, malaria, Rift Valley Fever, mosquito-borne encephalitis, and onchocerciasis have all been linked to the development of irrigation systems. Rates of diarrheal diseases and infestations of

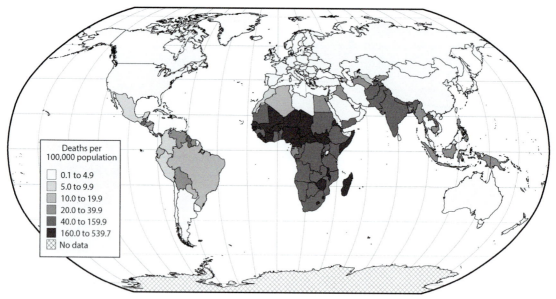

Figure 2.9 Child mortality from diarrhea (deaths per 100,000 population), 2015

Source: Data from Troeger et al. (2017)

Figure 2.10 Irrigated land, Bali, Indonesia

Standing water in rice fields can provide habitat for a variety of potential pathogen and vector species, including mosquitos and parasitic worms.

parasites such as flukes and worms may also increase when irrigation systems are introduced (Jobin 1999).

Landscape epidemiology of schistosomiasis

A variety of studies have linked increased incidence of schistosomiasis to irrigation projects. Studies like these underscore the importance of carefully considering the implications of large-scale development projects such as dams, which have been a traditional focus of many economic development strategies but may have unintended consequences for human health and well-being.

Schistosomiasis affects about 200 million people worldwide, with transmission of the disease reported in 78 countries in 2016 (WHO 2018b). Although prevalence of the disease has declined substantially over the past several decades, it remains a significant and probably underestimated global health burden (King and Galvani 2018). The disease cycle involves several species of parasitic worms (schistosomes). One species causes mild irritation in the form of "swimmer's itch"; infection with other species endemic to parts of Africa, Asia, and Latin America leads to the far more serious disease schistosomiasis (NIH 1999), which we consider here. As described earlier, the schistosome completes its life cycle in snails and vertebrate hosts such as people or water buffalo (see Figure 2.7). Free-swimming larvae infect humans—the **primary host**—while they swim or wade in water. The larvae penetrate a person's skin and make their way into blood vessels and then reproduce and release eggs in the gut or bladder. Some eggs are excreted in human urine and feces and, where sanitation is poor, can find their way back into open water. Once in the water, the eggs hatch and the larvae infect water snails—the **intermediate host**. The schistosomes replicate asexually in the snails before returning to the water, where they can infect people once again.

Many schistosome eggs are not released from their human host but remain in the body where they can cause an immunological reaction (WHO 2018b). As many as half of all eggs released into the body may become lodged in tissues, causing physical damage (Skelly 2008). Schistosome infection results in fatigue and fever and, if left untreated, can lead to organ damage and even death. In children, prolonged infection can cause cognitive and growth deficits. The disease takes a considerable toll on the health and productivity of the already poor populations most often affected.

It has been widely reported that the construction of the Aswan High Dam in Egypt in the 1960s led to increases in the prevalence of certain forms of schistosomiasis in the region, providing an early example of the application of **landscape epidemiology**. The trapping of silt behind the dam increased the clarity of water downstream, which, combined with slower water velocity, increased the likelihood of schistosomes infecting snails (Malek 1975). The reduced water current resulting from construction of the dam was also associated with flourishing snail populations (Abu-Zeid 1990). The number of snails is generally linked to the abundance of weeds, which provide food for the snail and places to lay eggs (Jobin 1999) and so vegetation changes may also have had an impact. The year-round flow of water from the dam eliminated the need for winter closure of irrigation canals, enabling transmission of the disease over a greater part of the year (Malek 1975). Changing environmental conditions also led to a change in the ratio of two different types of schistosomiasis infection, with an increase in *Schistosoma mansoni* infections, which primarily cause intestinal problems, relative to *Schistosoma haematobium*, which cause urinary infections. This change had significant public health implications because treatment for the urinary form of the disease was cheaper and simpler

than the intestinal form of the disease, which developed early drug resistance (Abdel-Wahab et al. 1979).

More recently, the construction of the Three Gorges Dam in China raised concerns that ecological changes may occur there too, potentially reversing the success of schistosomiasis control measures that had enabled the Chinese government to reduce prevalence of the disease from twelve million human infections in 1949 to a few hundred thousand cases at the turn of the millennium (Minter 2005). Before the construction of the dam, two distinct disease cycles were found on either side of the two sides of the natural barrier formed by the gorges. Upstream, in the mountains of Sichuan Province, the infection cycle primarily involved humans and was facilitated by the tradition of using human waste as soil fertilizer, resulting in infection rates as high as 60 percent in some areas. Below the dam around Poyang Lake, where agriculture was traditionally heavily irrigated and water buffalo were the major host of schistosomiasis, the prevalence of human infections was only 10 to 12 percent (Minter 2005). When floodgates were closed in June 2003, Poyang Lake initially developed new snail habitat and snail infections spiked, leading to concerns over future human infections.

Early fears of dramatic increases in infection rates appear not to have been realized in this case, however. Systematic reviews of the literature suggest that snail densities and schistosome infection rates actually declined in the period 2003 to 2015 (Li et al. 2017). Although there is no doubt that the dam has significantly altered the ecology of the Three Gorges region, this may have actually made conditions less favorable for the *Oncomelania* snail associated with the disease's transmission, illustrating the sensitivity of ecological systems and how difficult they can be to model. These ecological changes, combined with concerted efforts to control the disease by the Chinese government, probably explain the decline in incidence of the disease (Zhou et al. 2016; Li et al. 2017).

While schistosomiasis transmission appears to be under control in China, evidence from sub-Saharan Africa suggests that the disease's significance there may be increasing for different reasons. Recent evidence suggests that some forms of schistosome infection may increase an individual's susceptibility to HIV infection, as well as potentially increasing the speed of progression and transmissibility of HIV (Secor 2012). This increased susceptibility to HIV appears to be especially significant in women, possibly contributing to the high female-to-male ratio of HIV infection in many communities in sub-Saharan Africa (Brodish and Singh 2016).

Public health workers have attempted a variety of techniques to interrupt the schistosomiasis disease cycle. A safe anti-schistosomal drug has been available since the 1970s. The drug is cheap and has few side effects, but there is evidence that some schistosome populations have developed resistance to it. Efforts to develop a vaccine against the parasite have not yet been successful, but there is hope that a vaccine could eventually lead to eradication of the disease (Tebeje et al. 2016). For now, however, schistosomiasis control depends largely upon manipulating the environment and human behavior, with the application of molluscicides to remove snails still the key control mechanism (Tebeje et al. 2016). Improving sanitation to prevent untreated human waste from mixing with the water supply is also effective. Other approaches include limiting snail habitat through lining river channels with concrete and removing weeds, periodically drying out water channels to kill off snails, and encouraging people to stay out of snail-infested water.

The urban environment

Urban environments are among the most dramatically human-altered landscapes. Over half the global population already lives in cities, and this is set to rise to 70 percent by 2050 (WRI 2018), fueled largely by rural-to-urban migrations in the Global South, particularly in Asia and sub-Saharan Africa where urban populations are low by global standards (Stimson 2013). This mass migration is stimulated by perceptions of affluence and opportunity in the city, with many poverty-stricken rural dwellers looking to the city for jobs and services such as education and healthcare.

While it is true that cities are often the location of much of a country's wealth and prosperity, they are also places of substantial inequality, with much of the wealth in the hands of an affluent elite in many countries. Removed from rural areas, where householders had some influence over their food supply and living conditions, new rural-urban migrants in the Global South

may find themselves **food insecure** and resorting to self-built housing when faced with high rents (Figure 2.11). Although cities offer abundant job opportunities, other job-seekers are likely to outcompete recent rural-urban migrants with limited education and little work experience. Furthermore, the rapid rate of urban growth in many poorer countries has outpaced the development of infrastructure, leading to problems of poor sanitation, traffic congestion, and lack of services. Poverty and poor living conditions are therefore a significant problem in the **megacities** of the Global South (Figure 2.12). About 30 percent of the global population lived in slums in 2014 (Table 2.3). Although this figure decreased from 46 percent in 1990, it nonetheless represents a large absolute number of people living in poverty because of the rapid growth in urban population that took place over this period (Habitat III 2018).

Urban challenges in the Global South are not without precedent, of course. Rapid urbanization driven by the Industrial Revolution historically generated overcrowding and overextended infrastructure in Europe and North America too. Sometimes referred to as the "**urban penalty**," the poor conditions associated with rapid urbanization and industrialization have significant impacts on health (Curtis 2004), including high exposure to airborne pollutants, high rates of traffic accidents, and increases in harmful behaviors such as drug and alcohol use (Stimson 2013). High population densities also facilitate the spread of pathogens, particularly those associated with airborne and waterborne disease. Avoiding outbreaks of infectious disease in highly concentrated urban populations requires sophisticated public health and sanitation systems that necessitate considerable investment. This level of investment has been achieved in many cities in the Global North, and urban populations in affluent countries now enjoy some of the best health ever seen in human populations.

Recent conflict in Yemen illustrates the precariousness of urban public health systems, however. Since 2016, several major cholera outbreaks in Yemen have collectively infected more than one million people, with over ten thousand new cases occurring per week in peak periods—the worst outbreak of cholera in modern history (Reuters 2017; Lyons 2017). Although less than 30 percent of Yemenis live in cities, the cholera

Figure 2.11 Shanty town, Cape Town, South Africa
Credit: Heike Alberts

Shanty towns are common in cities of the Global South, where a crisis of affordable housing means that many new rural-urban migrants have little choice but to build their own houses. Shanty towns are often found on the outskirts of the city or on undesirable land such as in areas prone to flooding or near polluting industrial facilities.

outbreak largely occurred in cities where populations had the pre-existing vulnerabilities of overcrowding and centralized water supplies that left them susceptible to outbreaks of waterborne disease (Erickson 2017). In this case, conflict led to the breakdown of sanitation and public health systems (including direct airstrikes on water facilities), leading to epidemic disease in a population already weakened by large-scale population movements and malnutrition (Reuters 2017; Lyons 2017).

Rapid urbanization has also had significant effects on the transmission of some vector-borne diseases, including dengue fever, malaria, and lymphatic filariasis. In these cases, the vectors have proved capable of adapting to, and even thriving in, urban environments. Dengue fever provides perhaps the clearest case of a disease whose vector has adapted to the urban environment.

Dengue fever: adapting to the urban environment

Dengue fever is caused by a virus, transmitted by several species of the mosquito genus *Aedes*. An infected mosquito can pass the virus to a susceptible human host

Figure 2.12 Cairo, Egypt
Credit: Heike Alberts

Rapid urbanization in the Global South over the past fifty years has created many megacities, with problems such as over-crowding, pollution, and inadequate sanitation having significant health implications.

when feeding or probing for a blood meal, passing the virus into the human bloodstream from its saliva. The human host is capable of infecting mosquitos for as long as the virus remains in the bloodstream (Figure 2.13).

Dengue fever was reported on all inhabited continents throughout the nineteenth and early twentieth centuries (WHO 1997). Both vector and virus must, therefore, have been widespread for the past two hundred years. During most of this time, dengue was seen as a "benign, non-fatal disease of visitors to the tropics" (Gubler and Clark 1995, 55). More recently, however, a global pandemic has emerged, with a thirty-fold increase in the disease over the past fifty years (WHO 2019c),

including not only classic forms of dengue but also two more serious manifestations of infection: dengue hemorrhagic fever and dengue shock syndrome. The WHO estimates that half of the global population is now at risk of contracting the disease (WHO 2018a), with perhaps 400 million dengue infections occurring each year, of which approximately fifty to one hundred million show clinical symptoms (Bhatt et al. 2013; Shepard et al. 2016). The total annual cost of dengue infections to the global economy has been estimated at US$8.9 billion (Shepard et al. 2016). The actual number of infections is difficult to estimate, however, given the large number of cases that probably go unreported and the likelihood

Table 2.3 Proportion of urban populations living in slums by region, 2014

Region	Proportion of urban population living in slums (%)
Sub-Saharan Africa	55
Southern Asia	31
Southeastern Asia	27
Eastern Asia	25
Western Asia	25
Oceania	24
Latin American and the Caribbean	20
Northern Africa	11
Worldwide	*30*

Note: A group of individuals is defined by UN-HABITAT as living in a slum if their dwelling lacks one or more of the following: security of tenure, adequate access to improved water, adequate access to improved sanitation and other infrastructure, structural quality of housing, and sufficient living area.

Source: Data from United Nations (2015)

that many cases are confused with other tropical diseases (WHO 2018a). One reason for this dramatic increase in incidence is the vector's success in adapting to urban areas; indeed, dengue fever is now considered to be a primarily urban disease (WHO 2018a).

Dengue is most successfully transmitted in high-density human populations owing to the preferences of its vector, *Aedes aegypti*. Although several species of mosquito can transmit the virus, one of the commonest is *Aedes aegypti*, which is increasingly being recognized as an urban species. *Aedes aegypti's* limited flight range also means that many potential hosts must be available in a small area for effective transmission. That is not to say that outbreaks cannot occur in rural areas; they can and frequently do. However, a relatively large and densely settled population is required for dengue to become endemic.

One of the main reasons for *aegypti's* success among human populations is its preference for breeding in small pools of water. Originally adapted to breed in small pockets of water trapped in vegetation, the species now frequently breeds in containers used or discarded by people. The abundance of non-biodegradable packaging in urban areas may be a primary reason for the huge increase in potential mosquito breeding sites. In one study in Yangon (Rangoon), Myanmar, 98 percent of all *aegypti* larvae were found in domestic water storage containers (Sebastian et al. 1990). Larvae are also commonly found in buckets, vases, water tanks, discarded cans, and tires. Tires are an especially significant breeding site because the shipping of tires containing *aegypti* larvae has spread the vector widely (Pliego Pliego et al. 2018).

Efforts to control dengue fever remain focused on manipulating the environment in ways that make it less amenable to the dengue vector (Bhatt et al. 2013). The first dengue vaccine was approved for use in the US in 2019, but its use is not yet widespread and it remains to be seen how successful it will be. Treatment of the disease relies on alleviating symptoms while the body fights the virus, although efforts are under way to develop more effective anti-viral treatments. Bed nets are of little use for dengue as the mosquito feeds during the day (O'Neil 2015). Traditionally, dengue control has therefore relied heavily on insecticides to control the mosquito population. Unfortunately, increasing levels of pesticide resistance in *aegypti* populations is rendering this method less effective.

One other avenue that is being actively explored is the use of a species of bacteria known as *Wolbachia* as a biocontrol for dengue fever. *Wolbachia* infections occur naturally in a variety of insect species and were initially investigated for their potential to shorten the lifespan of disease-causing mosquitos like *Aedes aegypti*, which are not natural hosts of *Wolbachia* (O'Neil 2015). Once *aegypti* mosquitos were infected with *Wolbachia* in the lab, however, *Wolbachia* appeared to compromise the ability of the dengue virus to replicate within the mosquito and so the mosquitos proved less able to transmit the disease to people (Silva et al. 2017; O'Neil 2015). Today, *Wolbachia* infection of mosquitos is being investigated as a way to control a variety of mosquito-borne diseases, including dengue fever, chikungunya, and Zika virus (all vectored by *Aedes aegypti*), as well as malaria with its *Anopheles* vectors. It has proved challenging to infect *aegypti* with *Wolbachia* and then maintain the infection in the wild *aegypti* population over generations,

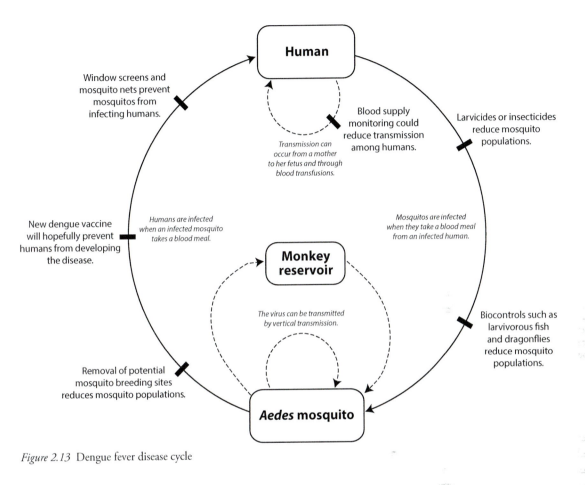

Figure 2.13 Dengue fever disease cycle

but researchers have now managed to successfully infect mosquitos in the lab with a strain of *Wolbachia* that can subsequently be passed between *aegypti* under natural conditions (O'Neil 2015). Experiments have now established wild populations of *Wolbachia*-infected *Aedes aegypti* mosquitos in several countries, and *Wolbachia* has been shown to circulate in the wild population for several years after the initial release of lab-infected mosquitos (World Mosquito Program 2018). To date, there is insufficient evidence to suggest how this approach might affect dengue incidence in human populations, however, owing to the challenges of implementing robust, safe experiments (O'Neil 2015).

There are several potential advantages to this method. After the initial up-front costs of releasing the lab-infected mosquitos, the *Wolbachia* infection should be passed from generation to generation of mosquito at no additional cost, potentially providing a long-term solution to dengue fever. In addition, this method avoids the problems of pesticide resistance in mosquitos as well as the potential toxic effects of insecticides being introduced into the environment. Nonetheless, there are always significant concerns with releasing a biological agent into the environment. Before release, careful experiments were done to ensure that *Wolbachia* would not be harmful to non-target species or affect people if they were bitten by *Wolbachia*-infected mosquitos (O'Neil 2015). Nonetheless, ecological systems are complex and so the possibility of unforeseen consequences persists.

CONCLUSION

As human beings are critically dependent on their environment, so too are pathogens and vectors. An ecological approach to disease offers several key contributions to understanding these complex relationships. First, treating humans as part of an interconnected disease cycle encourages greater consideration of all aspects of the cycle, rather than focusing on humans in isolation. Ecological approaches also prompt us to consider individuals as part of larger populations and to seek solutions at these broader scales. Finally, examining different aspects of a disease cycle brings the role of environment to the fore. These three factors force us to think beyond clinical approaches to the study of disease, emphasizing solutions that employ broader mechanisms such as vector control and landscape change.

Ecological approaches also offer a cautionary tale of the impacts of environmental disruption. While it is tempting to ignore the impacts of environmental change on remote places and obscure species, it is hard to deny the potential human costs associated with changes in the disease landscape. We turn to this topic in Chapter 4. First, though, we continue our exploration of pathogens and people as biological entities by focusing on population-scale processes that can further inform ecological approaches to health and disease.

DISCUSSION QUESTIONS

1 Using specific examples, consider how the built environment influences human health.

2 Considering what you now know about transmission of infectious diseases and the case of Typhoid Mary, is forcible quarantine ever justified?

3 Discuss how a landscape epidemiology approach might inform a campaign to control schistosomiasis.

4 Explain the notion of the "urban penalty" with respect to health.

5 Consider how cultural ecology was significant to the 2013–16 outbreak of Ebola, given your understanding of the disease from the previous chapter.

FURTHER READING

Diamond, J. M. 1997. "The Lethal Gift of Livestock." In *Guns, Germs and Steel*. W.W. Norton & Co.

Galea, S., N. Freudenberg, and D. Vlahov. 2005. "Cities and Population Health." *Social Science & Medicine* 60: 1017–33.

O'Neill, S. 2015. "How a Tiny Bacterium Called Wolbachia Could Defeat Dengue." *Scientific American*, June 1. www.scientificamerican.com/article/how-a-tiny-bacterium-called-wolbachia-could-defeat-dengue/.

Mayer, J. D. 2000. "Geography, Ecology and Emerging Infectious Diseases." *Social Science & Medicine* 50 (7–8): 937–52.

Ritchie, H., and M. Roser. 2019. "Urbanization." *Our World InData.org*. https://ourworldindata.org/urbanization.

VIDEO RESOURCES

"Guns, Germs and Steel" by Jared Diamond. 2005. PBS. More information at: www.pbs.org/gunsgermssteel/.

"Rx for Survival: Deadly Messengers." 2005. PBS. More information at: www.pbs.org/wgbh/rxforsurvival/series/about/episodes.html.

REFERENCES

Abad-Franch, F. 2005. "Complejidad ecológica y enfermedad de chagas en la Amazonia." Segunda (II) reunión de la Iniciativa Intergubernamental de Vigiliancia y Prevención de la enfermedad de Chagas en la Amazonia (AMCHA II), Cayenne.

[ABC] Australian Broadcasting Corporation. 2016a. "Thunderstorm Asthma Claims Fifth Life in Victoria." Accessed June 20, 2019. www.abc.net.au/news/2016-11-26/thunderstorm-asthma-claims-fifth-life-in-melbourne/8060674.

[ABC] Australian Broadcasting Corporation. 2016b. "Thunderstorm Asthma: Eighth Person Dies After Melbourne Freak Weather Event." Accessed June 20, 2019.

www.abc.net.au/news/2016-11-29/thunderstorm-asthma-eighth-person-dies-from-melbourne/8074776.

Abdel-Wahab, M. F., G. T. Strickland, A. Elsahly, N. Elkady, et al. 1979. "Changing Pattern of Schistosomiasis in Egypt 1935–79." *Lancet* 2 (8136): 242–44.

Abu-Zeid, M. 1990. "Environmental Upgrading of Irrigation Systems to Control Schistosomiasis." *Waterlines* 9 (2): 31–35.

American Cancer Society. 2018. "What Is Kaposi Sarcoma?" Accessed January 8, 2019. www.cancer.org/cancer/kaposi-sarcoma/about/what-is-kaposi-sarcoma.html.

Andrew, E., Z. Nehme, S. Bernard, M. J. Abramson, et al. 2017. "Stormy Weather: A Retrospective Analysis of Demand for Emergency Medical Services During Epidemic Thunderstorm Asthma." *British Medical Journal* 359: 8. doi:10.1136/bmj.j5636.

Baker, R. 2007. *Quiet Killers: The Fall and Rise of Deadly Diseases*. Sutton Publishing.

Bhatt, Samir, P. W. Gething, O. J. Brady, J. P. Messina, et al. 2013. "The Global Distribution and Burden of Dengue." *Nature* 496: 504. doi:10.1038/nature12060.

Briceño-Leon, R. 2007. "Chagas Disease and Globalization of the Amazon." *Cadernos De Saude Publica* 23: S33–40. doi:10.1590/s0102-311x2007001300005.

Brodish, P. H., and K. Singh. 2016. "Association Between Schistosoma haematobium Exposure and Human Immunodeficiency Virus Infection Among Females in Mozambique." *American Society of Tropical Medicine and Hygiene* 94 (5): 1040–44. doi:10.4269/ajtmh.15-0652.

Brooks, J. 1996. "The Sad and Tragic Life of Typhoid Mary." *Canadian Medical Association Journal = journal de l'Association medicale canadienne* 154 (6): 915–16.

Carey, N. 2011. *The Epigenetics Revolution: How Modern Biology Is Rewriting Our Understanding of Genetics, Disease and Inheritance*. Icon Books Limited.

Carter, E. D. 2007. "Development Narratives and the Uses of Ecology: Malaria Control in Northwest Argentina, 1890–1940." *Journal of Historical Geography* 33 (3): 619–50. doi:10.1016/j.jhg.2006.06.002.

Centers for Disease Control and Prevention. 2014. "Water-Related Diseases and Contaminants in Public Water Systems." Accessed January 9, 2019. www.cdc.gov/healthywater/drinking/public/water_diseases.html.

Chandra, G., S. Chatterjee, and A. Ghosh. 2012. "Role of Dragonfly (Brachytron pratense) Nymph as a Biocontrol Agent of Larval Mosquitoes." *Buletin Penelitian Kesehatan* 34 (4): 147–51.

Coura, J. R., A. C. V. Junqueira, O. Fernandes, S. A. S. Valente, and M. A. Miles. 2002. "Emerging Chagas Disease in Amazonian Brazil." *Trends in Parasitology* 18 (4): 171–76. doi:10.1016/s1471-4922(01)02200-0.

Crosby, Alfred W. 1976. "Virgin Soil Epidemics as a Factor in the Aboriginal Depopulation in America." *The William and Mary Quarterly* 33 (2): 289–99. doi:10.2307/1922166.

Curtis, Sarah. 2004. *Health and Inequality: Geographical Perspectives*. London and Thousand Oaks, CA: Sage Publications.

Diamond, J. 1997. *Guns, Germs and Steel*. Random House.

Drasar, B. S., and B. D. Forrest. 1996. *Cholera and the Ecology of Vibrio cholerae*. London and New York: Chapman & Hall.

Elliot, A. J., H. E. Hughes, T. C. Hughes, T. E. Locker, et al. 2014. "The Impact of Thunderstorm Asthma on Emergency Department Attendances Across London During July 2013." *Emergency Medicine Journal* 31 (8): 675–78. doi:10.1136/emermed-2013-203122.

Erickson, A. 2017. "One Million People Have Contracted Cholera in Yemen. You Should Be Outraged." *The Washington Post*. www.washingtonpost.com/news/worldviews/wp/2017/12/21/one-million-people-have-caught-cholera-in-yemen-you-should-be-outraged/?noredirect=on&utm_term=.82ff4c87f080.

Fraga, Mario F., E. Ballestar, M. F. Paz, S. Ropero, et al. 2005. "Epigenetic Differences Arise During the Lifetime of Monozygotic Twins." *Proceedings of the National Academy of Sciences* 102 (30): 10604–9. doi:10.1073/pnas.0500398102.

Gawande, A. 2004. "Notes of a Surgeon: On Washing Hands." *New England Journal of Medicine* 350 (13): 1283–86. doi:10.1056/NEJMp048025.

Grundstein, A., S. E. Sarnat, M. Klein, M. Shepherd, et al. 2008. "Thunderstorm Associated Asthma in Atlanta, Georgia." *Thorax* 63 (7): 659–60. doi:10.1136/thx.2007.092882.

Gubler, D. J., and G. G. Clark. 1995. "Dengue/Dengue Hemorrhagic Fever: The Emergence of a Global Health Problem." *Emerging Infectious Diseases* 1 (2): 55–57.

Guernier, V., M. E. Hochberg, and J. F. O. Guegan. 2004. "Ecology Drives the Worldwide Distribution of Human Diseases." *PLoS Biology* 2 (6): 740–46. doi:10.1371/journal/pbio.0020141.

Habitat III Policy Unit 10. 2018. Habitat III Policy Paper Framework 10—Housing Policies. Nairobi: UN-Habitat. Nairobi: UN-Habitat. www.nature.com/articles/nature12060#supplementary-information.

Jobin, William R. 1999. *Dams and Disease: Ecological Design and Health Impacts of Large Dams, Canals, and Irrigation Systems.* London and New York: E & FN Spon and Routledge.

Jones, K. E., N. G. Patel, M. A. Levy, A. Storeygard, et al. 2008. "Global Trends in Emerging Infectious Diseases." *Nature* 451 (7181): 990–94. doi:10.1038/nature06536.

King, Charles H., and Alison P. Galvani. 2018. "Underestimation of the Global Burden of Schistosomiasis." *The Lancet* 391 (10118): 307–8. doi:10.1016/S0140-6736(18)30098-9.

Li, F., S. Ma, Y. Li, H. Tan, et al. 2017. "Impact of the Three Gorges Project on Ecological Environment Changes and Snail Distribution in Dongting Lake Area." *PLoS Neglected Tropical Diseases* 11 (7): e0005661.

Lieberman, D. 2013. *The Story of the Human Body: Evolution, Health, and Disease.* Knopf Doubleday Publishing Group.

Lillycrop, K. A., and G. C. Burdge. 2014. "Environmental Challenge, Epigenetic Plasticity and the Induction of Altered Phenotypes in Mammals." *Epigenomics* 6 (6): 623–36. doi:10.2217/epi.14.51.

Luby, S. P., M. Agboatwalla, J. Painter, A. Altaf, et al. 2004. "Effect of Intensive Handwashing Promotion on Childhood Diarrhea in High-Risk Communities in Pakistan—A Randomized Controlled Trial." *Journal of the American Medical Association* 291 (21): 2547–54. doi:10.1001/jama.291.21.2547.

Lumey, L. H., F. W. van Poppel, and A. M. Vaiserman. 2013. "The Dutch Famine of 1944–45 as a Human Laboratory: Changes in the Early Life Environment and Adult Health." In: *Early Life Nutrition, Adult Health and Development*, edited by L. H. Lumey and A. Vaiserman, 59–76. Nova Science Publishers.

Lyons, K. 2017. "Yemen's Cholera Outbreak Now the Worst in History as Millionth Case Looms." *The Guardian.* www.theguardian.com/global-development/2017/oct/12/yemen-cholera-outbreak-worst-in-history-1-million-cases-by-end-of-year.

Malek, E. A. 1975. "Effect of Aswan High Dam on Prevalence of Schistosomiasis in Egypt." *Tropical and Geographical Medicine* 27 (4): 359–64.

May, J. M. 1958. *The Ecology of Human Disease.* MD Publications.

Mayell, H. 2004. "Three High-Altitude Peoples, Three Adaptations to Thin Air." Accessed January 8, 2019. www.nationalgeographic.com/culture/2004/02/high-altitude-adaptations-evolution/.

Meymandi, Assad. 2010. "The Science of Epigenetics." *Psychiatry* 7 (3): 40–41.

Minter, A. 2005. "Breeding Snail Fever." *Scientific American* 293 (1): 21–22.

National Human Genome Research Institute. 2016. "Epigenomics Fact Sheet." Accessed June 12, 2019. www.genome.gov/about-genomics/fact-sheets/Epigenomics-Fact-Sheet.

[NIH] National Institutes for Health. 1999. "Understanding Emerging and Re-emerging Infectious Diseases." Accessed January 8, 2019. www.ncbi.nlm.nih.gov/books/NBK20370/.

Nilsson, E. E., and M. K. Skinner. 2015. "Environmentally Induced Epigenetic Transgenerational Inheritance of Reproductive Disease." *Biology of Reproduction* 93 (6): 145. doi:10.1095/biolreprod.115.134817.

O'Neil, S. 2015. "How a Tiny Bacterium Called Wolbachia Could Defeat Dengue." *Scientific American.* www.scientificamerican.com/article/how-a-tiny-bacterium-called-wolbachia-could-defeat-dengue/.

Pavlovsky, E. N., and N. D. Levine. 1966. *Natural Nidality of Transmissible Diseases: With Special Reference to the Landscape Epidemiology of Zooanthroponoses.* University of Illinois Press.

Pliego Pliego, E., J. Velazquez-Castro, M. P. Eichhorn, and A. Fraguela Collar. 2018. "Increased Efficiency in the Second-Hand Tire Trade Provides Opportunity for Dengue Control." *Journal of Theoretical Biology* 437: 126–36. doi:10.1016/j.jtbi.2017.10.025.

Prüss-Üstün, A., J. Wolf, C. Corvalán, R. Bos, and M. Neira. 2016. "Preventing Disease Through Healthy Environments: A Global Assessment of the Burden of Disease from Environmental Risks." Accessed December 4, 2018. http://apps.who.int/iris/bitstream/handle/10665/204585/9789241565196_eng.pdf;jsessionid=8737A4C469480FC5AFAB51B3A70B59A2?sequence=1.

Reuters. 2017. "Yemen Cholera Outbreak Accelerates to 10,000+ Cases per Week: WHO." Accessed December 14, 2018. www.reuters.com/article/us-yemen-security-cholera/yemen-cholera-outbreak-accelerates-to-10000-cases-per-week-who-idUSKCN1MC23J.

Roth, T. L., F. D. Lubin, M. Sodhi, and J. E. Kleinman. 2009. "Epigenetic Mechanisms in Schizophrenia." *Biochimica et Biophysica Acta* 1790 (9): 869–77. doi:10.1016/j.bbagen.2009.06.009.

Schofield, C., R. Brinceño-Leon, N. Kolstrupp, D. Webb, and G. White. 1990. "The Role of House Design in Limiting Vector-Borne Disease." In *Appropriate Technology in Vector Control*, edited by C. F. Curtis. Boca Raton, FL: CRC Press.

Schumacher, M. J. 2017. "Thunderstorm Asthma." *Internal Medicine Journal* 47 (5): 605–7. doi:10.1111/imj.13411.

Sebastian, A., M. M. Sein, M. M. Thu, and P. S. Corbet. 1990. "Suppression of Aedes aegypti (Diptera: Culicidae) Using Augmentative Release of Dragonfly Larvae (Odonata: Libellulidae) with Community Participation in Yangon, Myanmar." *Bulletin of Entomological Research* 80 (2): 223–32.

Secor, W. E. 2012. "The Effects of Schistosomiasis on HIV/AIDS Infection, Progression and Transmission." *Current Opinion in HIV and AIDS* 7 (3): 254–59. doi:10.1097/COH.0b013e328351b9e3.

Shaw, Jeffrey, Ralph Lainson, and Habid Fraiha. 1969. "Considerações sobre a epidemiologia dos primeiros casos autóctones de doença de Chagas registrados em Belém, Pará, Brasil." *Revista de Saúde Pública* 3: 153–57.

Shepard, D. S., E. A. Undurraga, Y. A. Halasa, and J. D. Stanaway. 2016. "The Global Economic Burden of Dengue: A Systematic Analysis." *The Lancet Infectious Diseases* 16 (8): 935–41. doi:10.1016/s1473-3099(16)00146-8.

Silva, J. B. L., D. Magalhaes Alves, V. Bottino-Rojas, T. N. Pereira, et al. 2017. "Wolbachia and Dengue Virus Infection in the Mosquito *Aedes fluviatilis* (Diptera: Culicidae)." *PLoS One* 12 (7): e0181678. doi:10.1371/journal.pone.0181678.

Skelly, P. 2008. "Fighting Killer Worms." *Scientific American* 298 (5): 94–99.

Sokolow, S. H., E. Huttinger, N. Jouanard, M. H. Hsieh, et al. 2015. "Reduced Transmission of Human Schistosomiasis After Restoration of a Native River Prawn That Preys on the Snail Intermediate Host." *Proceedings of the National Academy of Sciences* 112 (31): 9650–55. doi:10.1073/pnas.1502651112.

Stimson, Gerry V. 2013. "The Future of Global Health Is Urban Health." *Lancet* 382 (9903): 1475. doi:10.1016/S0140-6736(13)62241-2.

Suphioglu, C., M. B. Singh, P. Taylor, R. Bellomo, et al. 1992. "Mechanism of Grass-Pollen-Induced Asthma." *Lancet* 339 (8793): 569–72. doi:10.1016/0140-6736(92)90864-y.

Taylor, P. E., R. C. Flagan, R. Valenta, and M. M. Glovsky. 2002. "Release of Allergens as Respirable

Aerosols: A Link Between Grass Pollen and Asthma." *Journal of Allergy and Clinical Immunology* 109 (1): 51–56. doi:10.1067/mai.2002.120759.

Tebeje, B. M., M. Harvie, H. You, A. Loukas, and D. P. McManus. 2016. "Schistosomiasis Vaccines: Where Do We Stand?" *Parasites & Vectors* 9 (1): 528. doi:10.1186/s13071-016-1799-4.

Thien, F. 2018. "Melbourne Epidemic Thunderstorm Asthma Event 2016: Lessons Learnt from the Perfect Storm." *Respirology* 23 (11): 976–77. doi:10.1111/resp.13410.

Troeger, Christopher, Mohammad Forouzanfar, Puja C. Rao, Ibrahim Khalil, et al. 2017. "Estimates of Global, Regional, and National Morbidity, Mortality, and Aetiologies of Diarrhoeal Diseases: A Systematic Analysis for the Global Burden of Disease Study 2015." *The Lancet Infectious Diseases* 17 (9): 909–48. doi:10.1016/S1473-3099(17)30276-1.

United Nations. 2015. "The Millennium Development Goals Report 2015." New York: United Nations.

Walters, J. 2014. *Six Modern Plagues and How We Are Causing Them*. Island Press.

Wark, P. A. B., J. Simpson, M. J. Hensley, and P. G. Gibson. 2002. "Airway Inflammation in Thunderstorm Asthma." *Clinical and Experimental Allergy* 32 (12): 1750–56. doi:10.1046/j.1365-2222.2002.01556.x.

[WHO] World Health Organization. 1997. *Dengue Haemorrhagic Fever: Diagnosis, Treatment, Prevention, and Control*. Geneva: World Health Organization.

[WHO] World Health Organization. 2003. "Control of Chagas Disease: Second Report of the WHO Expert Committee." In *WHO Technical Report Series*. Geneva: World Health Organization.

[WHO] World Health Organization. 2018a. "Dengue and Severe Dengue." Accessed January 9, 2019. www.who.int/news-room/fact-sheets/detail/dengue-and-severe-dengue.

[WHO] World Health Organization. 2018b. "Schistosomiasis." Accessed January 9, 2019. www.who.int/en/news-room/fact-sheets/detail/schistosomiasis.

[WHO] World Health Organization. 2019a. "Chagas Disease (American Trypanosomiasis)." Accessed August 29, 2019. www.who.int/news-room/fact-sheets/detail/chagas-disease-(American-Trypanosomiasis).

[WHO] World Health Organization. 2019b. "Poliomyelitis (Polio)." Accessed August 11, 2019. www.who.int/topics/poliomyelitis/en/.

[WHO] World Health Organization. 2019c. "What Is Dengue?" Accessed June 20, 2019. www.who.int/denguecontrol/disease/en/.

WHO Regional Office for South-East Asia. 2011 *Comprehensive Guidelines for Prevention and Control of Dengue and Dengue Haemorrhagic Fever*. Geneva: World Health Organization Regional Office for South-East Asia.

Winter, K., K. Harriman, R. Schechter, E. Yamada, J. Talarico, and G. Chavez. 2010. "Notes from the Field: Pertussis-California, January East Asia. June 2010 (Reprinted from MMWR, vol 59, pg 817, 2010)." *Journal of the American Medical Association* 304 (10): 1067.

World Mosquito Program. 2018. "Wolbachia." Accessed January 8, 2019. www.eliminatedengue.com/our-research/wolbachia.

[WRI] World Resources Institute. 2018. "WRI Ross Center for Sustainable Cities." Accessed December 14, 2018. www.wri.org/our-work/topics/sustainable-cities.

Zhou, Yi-Biao, Song Liang, Yue Chen, and Qing-Wu Jiang. 2016. "The Three Gorges Dam: Does It Accelerate or Delay the Progress Towards Eliminating Transmission of Schistosomiasis in China?" *Infectious Diseases of Poverty* 5 (1): 63. doi:10.1186/s40249-016-0156-3.

Zimmer, C. 2018. "The Famine Ended 70 Years Ago, but Dutch Genes Still Bear Scars." *New York Times*, February 5, D5. www.nytimes.com/2018/01/31/science/dutch-famine-genes.html.

3

POPULATION-SCALE PROCESSES

DEMOGRAPHIC CHANGE AND THE EVOLUTION OF PATHOGENS AND VECTORS

A key principle of disease ecology is that there is a close relationship between human populations and their physical and biological environments. In the previous chapter, we considered how the physical environment influences disease patterns; here we turn attention to characteristics of populations that affect health and disease. We begin by considering patterns in the human population, particularly changes in mortality and morbidity over time. We then turn to the evolution of pathogen and vector populations, extending our discussion of antibiotic and pesticide resistance from the previous chapter.

THE DEMOGRAPHIC TRANSITION

Human population size and structure are clearly influenced by patterns of health and disease. Fluctuations in population size were once largely driven by changes in human health that influenced death rates. Periods of improved nutrition and absence of disease led to increases in population size, while food shortages and epidemics resulted in population decline. For instance, the bubonic plague outbreaks that swept across Europe in the Middle Ages may have killed around one third

of the European population (Baker 2007). Despite fluctuations like this, human lifespans have generally increased and the world's population grown rapidly over the past several centuries. It took thousands of years for the human population to reach one billion around the year 1800; since then, the rate of population growth has increased dramatically. Global population reached five billion in 1987 and then added another billion in just twelve years by 1999. By 2012, after just 13 more years, the population reached seven billion. Although population growth is now slowing, global population size is projected to continue to rise, potentially surpassing eleven billion by 2100 (United Nations 2019) (Figure 3.1). As fertility rates have fallen dramatically in many parts of the world, a handful of countries will likely be responsible for most population growth in the future. The UN predicts that between 2017 and 2050, just nine countries will be responsible for half of all population growth, with India, Nigeria, the Democratic Republic of the Congo, and Pakistan at the top of the list (United Nations 2017).

An ecological approach views human population growth as a function of human relationships with their environment. The agricultural revolution yielded some

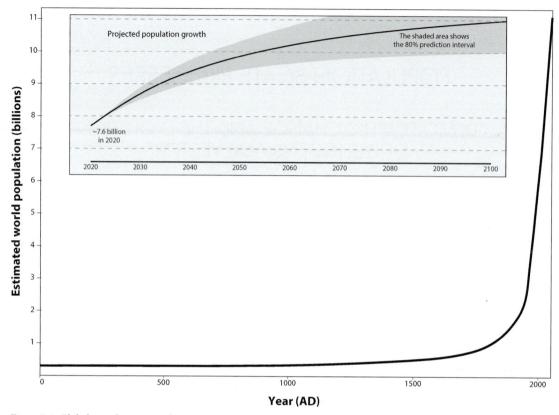

Figure 3.1 Global population growth
Source: Data from United Nations (2019)

of the most dramatic changes in how people interact with their environment. Over time, land use was intensified for food production using irrigation and fertilizers, capturing larger amounts of the earth's energy for human use and increasing global food supplies. In ecological terms, humans used agricultural technologies to increase the land's **carrying capacity**, or the number of individuals that could be supported by a given land area.

Larger food supplies not only provided more fuel for human population growth but also made individuals healthier and more resistant to disease, increasing lifespans. Malnourishment, including micronutrient deficiencies, is a key risk factor for deaths from infectious disease, and so improvements in the food supply allowed more infants to survive childhood and reach

their child-bearing years, further boosting population growth. Human technological innovations also began to address health problems directly. Improvements in sanitation were especially important (Figure 3.2), significantly reducing infant mortality rates. This development was followed by other innovations such as immunization, sterilization, and antibiotics.

Changes in population size and structure are shown in the Demographic Transition Model, which illustrates changes in **mortality** and **fertility rates** over time (Figure 3.3). The model was based on the experience of Western European countries as the Industrial Revolution progressed. In the first stage of the model, reflecting pre-industrialized societies, high birth rates and high death rates maintain a stable population size. In stage

Figure 3.2 Toilet facilities in a shanty town, Cape Town, South Africa
Credit: Heike Alberts

Sanitation is one of the key factors reducing the incidence of infectious disease as communities modernize. Although many communities in the Global South still have inadequate sanitation or lack access to clean water, the situation has improved significantly in many places over the past forty years.

two, death rates begin to drop in response to improved sanitation and better nutrition. As birth rates remain high, the population grows rapidly. In the third stage, birth rates begin to drop as declining infant mortality means that more children reach adulthood, reducing the need for large families. In addition, better education, empowerment of women, and improved social security systems make smaller families desirable. Combined with access to more effective methods of contraception, these factors collectively lead to declining birth rates and

therefore slower population growth. Nonetheless, population continues to grow for some years after birth rates begin to decline as the large number of children who were born when fertility rates were high become parents themselves—a phenomenon known as **demographic momentum**. Subsequently, in stage four of the model, birth rates drop low enough to mirror low death rates, leading population size to stabilize once more.

Although the Demographic Transition Model was designed to show changes over time, it is often used

Table 3.1 Major causes of death in low- and high-income countries, 2016

High-income countries

Ischemic heart disease

Stroke

Alzheimer and other dementias

Trachea, bronchus, lung cancers

Lower respiratory infections

Colon and rectum cancers

Diabetes mellitus

Kidney disease

Breast cancer

Low-income countries

Lower respiratory infections

Diarrheal diseases

Ischemic heart disease

HIV/AIDS

Stroke

Malaria

Tuberculosis

Preterm birth complications

Birth asphyxia and birth trauma

Road injury

Source: WHO (2018c)

number of scientific innovations that could be imported from other countries. Omran noted that the rapid declines in death rates achieved with imported technologies often led to dramatic population growth before a corresponding drop in birth rates occurred.

Critics have noted shortcomings of this model too. As with the Demographic Transition Model, some scholars have argued that the idea of epidemiologic transition implies that each stage is more desirable than the last, with affluent countries appearing more "advanced" than the rest of the world. The scale of analysis of entire countries may also veil significant race, class, and gender differences (see Barrett et al. 1998). Many countries in the Global South appear to have stalled in the middle

of the epidemiologic transition, further challenging the relevance of the model. Still struggling to address infectious diseases among poorer members of their societies, public health systems in many low- and middle-income countries must increasingly contend with diseases of affluence associated with a growing middle class as well. This phenomenon has been referred to as the "**double burden of disease**" (Marshall 2004). For instance, the prevalence of obesity is rising in many countries, even though certain parts of the population remain undernourished. The WHO (2018b) suggests that this nutrition "double burden" is caused by inadequate prenatal, infant, and young child nutrition, followed by exposure to high-fat, energy-dense, micronutrient-poor foods, combined with lack of physical activity. Several scholars have extended the idea to suggest that a "triple burden of disease" exists in some communities, where infectious disease and non-communicable disease are the first two burdens, plus an additional factor. A variety of ideas have been suggested for the additional burden, including undernutrition (Singh et al. 2007; Guerrant et al. 2012), the health consequences of globalization (Frenk and Gomez-Dantes 2011), or reproductive health issues (Ladusingh, Mohanty, and Thangjam 2018).

The distinction between infectious and degenerative diseases may also be more complex than the model implies. Increasingly, infectious agents are being discovered that appear to cause or contribute to the onset of diseases that were once believed to be non-infectious. Although now gaining acceptance, the idea that infectious agents could contribute to "non-communicable" diseases such as cancer was initially controversial and exemplifies how the social environment can influence our understanding of disease and how to treat it. Box 3.1 describes the challenge of changing the widespread belief that ulcers were caused by lifestyle factors after a bacterium associated with the development of stomach ulcers was isolated in the 1980s. More recently, infectious agents have been associated with cervical and liver cancers, certain types of lymphoma, and some types of skin cancer (National Cancer Institute 2015). The finding that human papillomaviruses (HPVs) are responsible for almost all cervical cancers (National Cancer Institute 2015) is now widely accepted and has led to vaccination programs (Box 3.2).

Box 3.1

INFECTIOUS AGENTS OF "NON-INFECTIOUS" DISEASES

For many years, mainstream Western medicine treated stomach ulcers as a chronic "lifestyle" condition. Even though stomach ulcers had been treated successfully with antibiotics in New York City hospitals in the late 1940s, indicating a bacterial cause for the disease, by the 1950s and '60s received wisdom assumed instead that a combination of genetic predisposition and lifestyle factors such as stress, smoking, and alcohol consumption were the main contributory causes.

It was not until the 1980s that the idea of an infectious agent was formally developed. Several researchers had noted that a particular bacterium (*Helicobacter pylori*) was commonly found in ulcers of the stomach, but it was the perseverance of an Australian doctor, Barry Marshall, that gave the infectious hypothesis credibility. In a brave experiment, Marshall infected himself with the bacteria and subsequently developed symptoms associated with stomach ulcers. Further experiments suggested more and more conclusively that antibiotics could be used to treat the infection. Today, infectious causation of peptic and duodenal ulcers is widely accepted by the medical establishment (Ewald 2002).

Box 3.2

THE CULTURAL ECOLOGY OF HPV VACCINATION

Human papillomaviruses (HPVs) are among the commonest sexually transmitted infections worldwide (Sabeena et al. 2018). The discovery that HPV infection is associated with almost all cervical cancers as well as a variety of other cancers associated with the reproductive organs (National Cancer Institute 2019) is a compelling example of how the distinction between infectious and non-infectious disease is less clear than we once believed.

The significance of HPVs and the HPV vaccine reveals yet another example of the impact of **cultural ecology** on health. An effective vaccine against HPVs has been licensed since 2006 (Gallagher, LaMontagne, and Watson-Jones 2018), and the WHO recommends that girls aged 9 to 13 years receive the vaccine before sexual debut (WHO 2017b). By 2008, approximately one quarter of high- and upper-middle income countries had introduced a vaccination program. Even though low-income countries still lag behind more affluent countries, by 2016 almost one quarter of 9- to 14-year-old girls worldwide were being reached by vaccination programs (Gallagher, LaMontagne, and Watson-Jones 2018), with 74 countries including HPV vaccination in their national immunization program by 2018 (Sabeena et al. 2018). Most countries target only girls with the vaccine, but some countries such as the US also encourage boys to be vaccinated to limit circulation of the virus (*ibid.*).

The need for HPV vaccination is arguably greatest in low-income countries because cervical screening programs that can catch subsequent cases of cancer are beyond the reach of underfunded public health systems, and more than 85 percent of cervical cancer deaths occur in low- and middle-income countries (Gallagher, LaMontagne, and Watson-Jones 2018). The high burden of HIV in many low-income countries exacerbates the problem. Vaccine subsidy programs have begun to assist low- and middle-income countries in covering the cost of the vaccine (*ibid.*), but many countries of the Global South have yet to prioritize

continued

Box 3.2   *continued*

HPV vaccination given other pressing public health needs. In low-income contexts, HPV vaccination is often viewed as a long-term investment behind more immediate health crises, especially given that the timeline for seeing declines in cervical cancer rates associated with HPV vaccination is long (Bruni 2017).

Even among high-income countries, coverage with the vaccine has been variable. Some countries, including Scotland and Australia, have achieved coverage of more than 80 percent of girls targeted for vaccination (Gallagher, LaMontagne, and Watson-Jones 2018). Other countries such as Japan, France, Denmark, and the US have struggled to even reach 50 percent coverage in a climate of concerns over vaccine safety, often fueled by negative media coverage and sometimes also a lack of health provider support (*ibid.;* Gilkey et al. 2017). The status of HPV as a sexually transmitted virus has been a barrier to vaccination uptake in many places, with early concerns with the vaccine often focused on the idea that the child was not sexually active and therefore did not need the vaccine, or that vaccination would encourage sexual promiscuity, even though empirical evidence does not support this idea (Larson et al. 2014).

Other commentators have suggested that we should not think in terms of a single epidemiologic transition, but rather of several transitions associated with changes in human relationships with the environment, such as the shift to settled agriculture and the Industrial Revolution. Rethinking the notion of epidemiologic transition in this way enables the consideration of recent and future changes as further transitions. In particular, the recent resurgence of infectious disease may constitute a new "epidemiologic transition" (Armelagos, Brown, and Turner 2005; Barrett et al. 1998). For many, this resurgence of infectious disease reflects ecological imbalance and social upheaval. Walters (2004, 9) suggests that we may be entering a new phase of epidemic disease, "spawned by an unprecedented scale of ecological and social change." Others argue that we are facing an era of "double exposures," in which a negative synergy between environmental change and globalization compounds the stresses that each imposes on human populations (Leichenko and O'Brien 2008). Building on these ideas, we must strive to understand not only how our changing relationship with our environment influences human health, but also the social structures that underlie many of these changes.

HUMAN POPULATION GENETICS AND MIGRATIONS

Analyzing patterns at the population scale encourages us to think also about genetic traits, which become concentrated in some populations owing to their heritability. Histories of isolation or migration—key geographic themes—are important for understanding patterns in genetic conditions and heritable characteristics.

We have already discussed the notion of **genetic adaptation** to disease, noting how infectious diseases such as bubonic plague and measles have placed selection pressures on populations, encouraging the accumulation of genetic adaptations that promote resistance to those diseases. We have also noted that migrations can bring immunologically naïve populations into contact with diseases they have not experienced before. In some cases, isolated communities have been devastated by the introduction of new diseases; in other cases, emigration has brought individuals to places with **endemic** diseases to which they and their ancestors had not previously been exposed. When Europeans moved to the tropics during colonization, for instance, many settlers died from tropical diseases such as malaria and yellow fever.

Non-infectious, genetic conditions are also influenced by human migration patterns. Mutations in the human genome can result in an array of health problems if they damage normal physiological function. The seriousness of gene mutations ranges from a minor condition such as excessive hair growth to life-threatening conditions such as cystic fibrosis (which affects the respiratory system) or Tay-Sachs disease (which affects nerve function). Every human inherits two sets of genetic information, one from each parent. Some genetic disorders, such as Huntington's disease, require only one abnormal gene to be inherited for the disease to manifest. Others, such as sickle cell disease and cystic fibrosis, will occur only if an individual has inherited a copy of the faulty gene from *both* parents—these are known as **autosomal recessive conditions**. People with only one copy of the abnormal gene are referred to as **carriers**, as they can pass the gene to their children but do not show symptoms of the disease.

Geography defines the range of many rare gene mutations as they will only be inherited within a particular population unless migration spreads them more widely. This is particularly important with respect to autosomal recessive conditions because the likelihood that an individual will inherit *two* copies of the same faulty gene is extremely low except in circumstances where population isolation leads to high levels of intermarriage. The Faroe Islands of Northern Europe provide an example of such extreme isolation. Settled by a small number of people of Nordic and British ancestry in the ninth century, the islands have experienced limited immigration and currently have a population of only around fifty thousand people. Autosomal recessive diseases are unusually common in the Faroese population because chance mutations have become concentrated within the population owing to the limited number of potential partners. One example is *carnitine transporter deficiency*, a condition that can lead to sudden death if untreated. Extremely rare in most populations, this disorder occurs in approximately one in three hundred Faroe islanders (Rasmussen et al. 2014). The issue of limited genetic variation in the population is of such importance to health that the Faroese Health Ministry announced in 2012 that it

planned to offer free gene sequencing to Faroese citizens (Borrell 2013).

The ancestral lifeways of groups of people are being explored as another path through which genetic ancestry may influence health. For example, researchers have begun to explore whether genetic patterns underlie why some ethnic groups are at higher risk for lifestyle diseases such as obesity and diabetes. Research on obesity among Pacific islanders (where obesity rates reach over 80 percent in some communities) suggests that gene variants promoting fat storage are found at high levels in some Pacific Island communities. The so-called **thrifty gene hypothesis** suggests that genes promoting fat storage and efficient use of energy, which are beneficial during food scarcity, become a liability in today's world of overabundance (Neel 1962). Pacific Islanders, whose ancestors undertook long migrations, would likely have experienced a strong selection pressure for these "thrifty genes," which would have helped individuals to survive long sea journeys. Evidence supports this idea—one study found that one quarter of all Samoans have a variant of a gene that increased the odds of obesity by 30 to 40 percent, while the gene is almost non-existent in European and African populations and occurs at only low rates among Asians (Minster et al. 2016). Of course, the situation is not entirely genetic, as changing diet and exercise patterns are the **proximate** causes that have led Pacific Islander communities to undergo changes in body mass, and these changes in lifestyle are linked to broader social changes. Genetic factors have nonetheless mediated the way in which Islanders' bodies have responded to changing conditions.

Intriguingly, the higher BMI associated with this genetic variant for obesity among Pacific Islanders is *not* linked to an increased risk of Type II diabetes—which is often so closely connected with obesity that the two conditions are sometimes referred to as "diabesity" (Kalra 2013)—suggesting that our understanding of the genetic components of obesity and diabetes remains far from complete. The link between obesity and diabetes remains strong in other communities, however. Research in collaboration with the Pima Indians of Arizona and Mexico, for instance, provides another compelling example of the "thrifty gene" hypothesis, with Pima communities suffering from high rates of both obesity and diabetes, documented as early as the 1930s. In this case, the rise in obesity was triggered by a sudden change from a traditional agricultural lifestyle to one that relies on processed foods, following the forced curtailment of the Pima's agricultural activities when white settlers began to divert the Pima's water sources in the early 1900s (Schulz and Chaudhari 2015).

DIET AND HEALTH

As illustrated by the "thrifty gene" hypothesis, diet and nutrition provide an additional key point of intersection between genetics and the physical and social environment at the population scale. The rapidity of cultural changes to our diet has outpaced our bodies' ability to adapt genetically to new conditions, leading some to argue that we now see a mismatch between what our bodies evolved to eat and today's Western diets. We use this as a point of introduction to consider diet more generally.

Through most of human history, we lived as hunter-gatherers and so our bodies evolved in ways consistent with the active lifestyles and diets of our Stone Age past. Many human adaptations likely relate to conditions from our early evolutionary past, including color vision, which may have evolved through facilitating the identification of ripe fruits (Jacobs and Nathans 2009); bipedal walking, which may have enabled more efficient foraging; and sweating, which may have helped early humans pursue game over long distances without overheating (Lieberman 2015). Along similar lines, it is reasonable to assume that we should be physiologically better adapted to the diet of our Stone Age ancestors than to today's Western diet (Frassetto et al. 2001). Indeed, today the so-called paleo diet has many advocates who promote the benefits of a Stone Age diet as a healthier alternative to a Western diet, although the desirability of limiting ourselves to a diet of *only* foods available to our Stone Age ancestors is controversial.

As hunter-gatherers, our diet would have been largely based on plants—tubers, fruits, leaves, and seeds—with periodic boosts of protein and micronutrients from meat or seafood when hunting was successful. It has been estimated that 65 percent of calories in a Stone Age diet probably came from fruits, vegetables, legumes, nuts, and honey; with 35 percent contributed by lean game, wild fowl, eggs, and seafood. By contrast, modern diets in the US are estimated to comprise only 17 percent fruits, vegetables, legumes, and nuts; 28 percent meat, fish, and eggs; with the remaining 55 percent of calories coming from foods that would not have been eaten by our early ancestors, including cereals, milk products, sweeteners, separated fats, and alcohol (Mardigan 2014). In combination with an increasingly sedentary lifestyle, today we often find ourselves consuming more calories than our bodies need—our minimally active lifestyles require just 2,000 to 2,500 calories per day; by comparison, Stone Age people were estimated to have needed about 3,000 calories per day to maintain their high levels of physical activity (*ibid.*).

Many of the diet choices we are encouraged to make by public health campaigns are geared towards rebalancing our diet by emphasizing the plant-based, micronutrient-rich, fiber-rich, low-sugar, low-salt diet familiar to our Stone Age predecessors. That is not to argue that we are identical to our Stone Age ancestors, however. Human evolution has continued over the past thousands of years, changing us in subtle and sometimes dramatic ways, including with respect to the food we eat (Jabr 2013). Perhaps most notably, a group of humans with a chance mutation that allowed them to continue to digest milk beyond infancy developed an evolutionary advantage by opening access to an important new food source: dairy. This mutation became so widespread in some populations—most notably Europeans—that today *lactose intolerance*, or the inability to digest milk protein, is considered an abnormality in these populations.

For our hunting and gathering ancestors, the key challenge was periodic shortages of certain nutrients, or food more generally, in seasons or periods of scarcity, which led human bodies to evolve to store scarce nutrients as efficiently as possible and to crave them so that they would be eaten when available. Our enthusiasm for high-fat foods and capacity to store fat provide the best-known examples. Sodium and potassium provide a further instructive example, however. Sodium (found in table salt) was in relatively short supply in the diet of our evolutionary past, and our bodies evolved a tendency to crave it, leading people to develop complex strategies to try to harvest it and ensure a steady supply (Figures 3.4 and 3.5). Potassium, which is found in many fruits and

Figure 3.4 Traditional salt harvesting ponds, Salinas, Peru

Salt harvesting technologies have been developed by cultures around the world and practiced for centuries. Naturally occurring brine is left to evaporate in the sun to yield salt for human consumption.

Figure 3.5 Traditional salt harvesting ponds, Hawaii

vegetables, was rarely in short supply for our ancestors, with their diverse plant-based diets. In contemporary communities, however, limited consumption of fruits and vegetables leaves many people potassium deficient, while manufacturers exploit our taste for sodium by adding salt to processed food, encouraging diets with an overabundance of sodium. Our ancestors probably ate 7,000 mg of potassium but only 600 mg of sodium per day, while a contemporary American diet averages only 2,500 mg of potassium and more than 4,000 mg of sodium (Mardigan 2014). Diets high in sodium or with a high sodium to potassium ratio are associated with high blood pressure, heart disease, and stroke (Drewnowski, Maillot, and Rehm 2012).

Our bodies' tendency to crave and store certain nutritional elements has therefore become a liability in modern contexts. The shift to a Western diet did not happen overnight, however; instead, there has been a progressive alienation of human nutrition patterns from our Stone Age past to the present day. A major change to diet occurred with the development of agriculture, with enormous consequences for what we eat. Some foods proved easier to cultivate than others and so the diversity of foods being eaten decreased, leaving a huge reliance on grains because they were easy to cultivate and produced abundant energy. Diamond (1987) suggests that this development had significant consequences for human health: the starchy, grain-rich diet provided lots of calories but lacked the micronutrient diversity of the traditional hunter-gatherer diet, and the starches in the grains quickly turn to sugars in the mouth, leading to tooth decay. Diamond (1987) argues that while farming could support more people, it resulted in deteriorating nutrition, evidenced by the decline in average height that occurred when communities transitioned from hunting and gathering to agricultural lifestyles. Agricultural

Table 3.2 Micronutrient sources and symptoms from deficiencies

Micronutrient	Food sources	Health impacts from deficiency
Iron	Meat, poultry, fish, lentils	Anemia, impaired cognition, impaired physical performance, maternal and child mortality
Vitamin A	Dairy, fruits, and vegetables	Visual impairment, diarrhea, maternal and child mortality
Iodine	Multiple, deficiency in areas with low levels of iodine in the soil	Goiter, hypothyroidism, cognitive impairment, decreased fertility
Zinc	Cereals and legumes, meat	Non-specific symptoms that can include diarrhea, growth retardation, mental disturbance, recurrent infection
Vitamin B_{12}	Dairy and meat	Anemia, nervous system disorders, damages cognitive function and development
Vitamin C	Fresh fruit and vegetables, offal	Scurvy, follicular hyperkeratosis, hemorrhagic manifestations, swollen joints, swollen bleeding gums and peripheral edema, death
Vitamin D	Sunlight, saltwater fish	Rickets, bone deformity, muscular pain
Calcium	Dairy products, a few other foods	Decreased bone mineralization, osteoporosis (among adults)
Selenium	A wide range of food; selenium content is driven by the soil from which food is derived	Keshan disease, cardiac problems, cartilage tissue disorders, metabolic disorders
Fluoride	Fluorinated water	Dental caries

Data Source: WHO and FAO (2006)

communities are also arguably more prone to famine because they rely on a smaller diversity of foods, and so the failure of a crop can prove disastrous to nutritional health (Gardner and Halweil 2000). This was saliently illustrated by the Irish Potato Famine of 1846–51, when the spread of one disease—the potato blight—destroyed the key crop of Ireland, leading to an estimated one million deaths (one eighth of the Irish population) and the emigration of two million people (Donnelly 2001). While there were important political dimensions to this famine, overreliance on one crop was a key factor.

By the twentieth century, diet improved for many communities in the affluent world, with better agricultural technology and growing trade networks leading to greater food availability and diversity. By the early postwar period, many communities in Europe and America had excellent nutrition, with enough food to provide sufficient nutrition but, as yet, few processed foods on the shelves with their high salt, fat, and sugar content. In the Global South and poverty-stricken pockets of the Global North, however, many continued to struggle with malnutrition and even periodic famine. Conditions such as kwashiorkor (resulting from protein deficiency) and beriberi (vitamin B deficiency) remained common and may even have worsened in places as rapid population growth stretched resources (Table 3.2). In short, the inequalities of the mid-twentieth century left many malnourished, despite unprecedented wealth and surplus food (Gardner and Halweil 2000).

The **Green Revolution** was a deliberate effort on the part of the Global North to spread agricultural technologies (including high-yielding seed varieties, irrigation, mechanization, fertilizers, and pesticides) to the Global South, with the aim of improving nutrition for the world's poorer peoples. The effort began in the late 1960s and was very successful in increasing global crop yields over the next fifty years, although it has been criticized for its environmental impacts, as well as problems of increasing inequalities in many places. By the latter part of the twentieth century, agriculture could effectively feed the entire global population, and so famine increasingly began to be viewed as a political failure of redistribution rather than the environmental disaster that it had been in earlier eras. This growth in global calorie production was also accompanied by initiatives to address micronutrient and mineral deficiencies, such as efforts to iodize salt and fortify products with vitamins, leading to further reductions in malnutrition, even where populations were growing rapidly. By the year 2000, the number of malnourished people worldwide was approximately equivalent to the number showing patterns of overconsumption (approximately 1.1 billion in each case), although a further two to 3.5 billion people exhibited vitamin and mineral deficiencies (Gardner and Halweil 2000, 7).

Today, there are more *over*nourished than *under*nourished people in the world. Our efficient industrialized agricultural systems produce vast quantities of cheap food, meaning that obesity can afflict even the poor in the Western world. Indeed, the obesity epidemic is now largely concentrated among the poor in more affluent countries. We return to the issue of obesity in Chapter 11.

PATHOGEN AND VECTOR POPULATIONS

Population-scale processes are not just important in human populations; an appreciation of the dynamics of pathogen and vector populations is also integral to understanding the ways in which patterns of human disease change over time. A **virulent** pathogen that quickly kills its host may die with its host, which does not promote the pathogen's survival. A pathogen is more successful in evolutionary terms if it keeps its host alive to spread the pathogen widely. As a result, highly virulent pathogen strains tend to become outnumbered by less virulent strains over time, and so diseases tend to evolve to become more benign after a long relationship

with a host species. Syphilis, for instance, may have lost some of its virulence since the Middle Ages when, judging from accounts of the time, the disease's symptoms appear to have been far more severe than today. Given enough time, microorganisms may even evolve a symbiotic (mutually beneficial) relationship with their host, as probably occurred with many of the gut microbes in our bodies, which play an important role in the digestion of our food.

In other cases, pathogens may evolve strains with better mechanisms for spreading from one human host to another, providing an alternate route to reproductive success for the pathogen. For the human population this paves the way for an **epidemic** as the pathogen can now spread rapidly. Researchers have hypothesized that the 2015 Zika epidemic in Latin America may have been partly fueled by evolution of the Zika virus in ways that made it able to spread more efficiently in the human population (Gubler, Vasilakis, and Musso 2017). Microbes have short generation periods and mutations can occur rapidly, so pathogens are well positioned to develop rapidly into new strains. Viruses are particularly successful at adapting to new conditions—rather than relying solely on random mutations to adapt to new circumstances, viruses can also share genetic information, enabling them to pass favorable traits directly among individuals. Viral diseases of animal origin are the most likely source of **pandemic** disease (Morse et al. 2012), and influenza is widely believed to be the most likely contender for causing the next human pandemic (Gupta 2018).

Viral evolution: the threat of an influenza pandemic

Human influenza strains are continuously circulating in the human population and can rapidly mutate into different strains. Although a different array of human influenza strains circulates in the global population every year, these so-called seasonal flu strains are similar enough to strains from past years that our immune systems can usually recognize them and develop an effective immune response. Serious health complications can sometimes occur with seasonal flu infections, but usually only among the very young (as their immune systems have limited experience with flu pathogens) or the elderly (whose immune systems are weakened with age). We are also able to bolster individuals' immune responses with a flu vaccination. The flu virus mutates so rapidly that the previous year's vaccine may not be effective against the current year's flu strains and so the production of vaccine is an annual task, with public health officials having to predict which strains of flu are most likely to circulate in the upcoming flu season. Given these constraints, and the fact that mutations continue to occur in the vaccine strain even during the manufacturing process, flu immunizations are often less than 50 percent effective and have been as little as 10 percent effective in recent years (CDC 2018c; Moyer 2017). Despite these challenges, annual flu immunization programs are still widely regarded as successful public health interventions. One recent estimate suggests that approximately forty thousand deaths may have been averted between 2005 and 2014 in the US alone owing to flu vaccination, mostly among people over 65 years (Foppa et al. 2015).

Beyond seasonal outbreaks of flu, there is always the threat of new **pandemic** strains. Human flus probably evolved from flu strains in pigs or birds, and a variety of animal strains continue to circulate, providing a constant source of new viruses. When a new strain of flu emerges from an animal host, or a seasonal strain undergoes a dramatic mutation, the new strain may be sufficiently different that even healthy adults are susceptible to infection.

There is a well-recognized sequence of phases of influenza activity. Most flu strains initially develop in animals, particularly pigs, chickens, and waterfowl. Humans can sometimes pick these strains up from close contact with these animals. These animal viruses are rarely transmissible from person to person, however, and most strains never evolve beyond this stage. In a few cases, however, the virus develops the ability to pass directly between humans, which can lead to an epidemic. A grave concern with flu is that a person or animal may be co-infected with both an animal strain and a human flu strain, potentially enabling the two viruses to recombine into a "super-virus" that possesses the virulence of the animal flu with the contagiousness of a human flu.

The infamous post-war influenza pandemic of 1918–19 provides a cautionary tale of how serious the emergence of a new pandemic influenza strain can be. The 1918 pandemic, which probably emerged from an avian virus (Taubenberger and Morens 2006), sickened an estimated 20 to 40 percent of the global population (approximately 500 million people) and may have killed between 50 and 100 million people (Johnson and Mueller 2002). The **case fatality rate**—the proportion of people infected who subsequently died from the disease—was much higher than in seasonal flu outbreaks, probably over 2.5 percent (Taubenberger and Morens 2006). Nearly half of all deaths occurred in adults aged 20 to 40. The unusually small proportion of deaths in older age groups may be explained if elderly people had been exposed to a similar virus earlier in their lives, providing them with some immunity (Simonsen et al. 1998). The considerable physical toll of the First World War and widespread human migrations in its aftermath probably contributed to the timing and severity of the pandemic.

Strains of flu have continued to emerge from animal populations, with major, although less well-known, epidemics in 1957 and 1968, each causing more than one million deaths worldwide (CDC 2018b). More recently, flu hit the headlines in 2003 when an outbreak of influenza known as H5N1 caused disease in poultry in Southeast Asia, although this strain of flu had been circulating since at least 1997 when it caused an outbreak in Hong Kong that led to 18 human infections and six deaths (Walters 2014). Between 1997 and 2013, close contact with infected birds or their feces led to about six hundred human cases of H5N1 (Walters 2014). With a **case fatality rate** estimated at 60 percent, it was a particularly **virulent** flu strain (WHO 2011), creating widespread concern that a global pandemic would be devastating if the virus were to develop the ability to be transmitted directly from person to person. Most human cases occurred in young, healthy individuals, creating further concern. Young children may have been especially likely to contract the disease owing to the common cultural practice in parts of Southeast Asia of leaving the job of plucking birds for the table to children—an example of the role of **cultural ecology** in driving disease risk.

H5N1 is now considered endemic in several Asian countries and was the largest and most serious outbreak in poultry on record. Thankfully, H5N1 has stopped short of causing a global pandemic. There have been isolated cases where it appears that an infected individual may have passed the disease to someone through close contact (WHO 2005), but efficient human-to-human transmission has not yet developed. Nonetheless, the danger remains that H5N1 might combine with a human flu strain or mutate in ways that would give it the ability to spread efficiently from person to person.

The H5N1 outbreak closely followed the emergence of SARS (sudden acute respiratory syndrome) in 2002–2004 in China, focusing attention on South and East Asia's potential to generate disease epidemics. Rapid recent changes, including urbanization, population growth, migration, and new forms of land use and agriculture create particular vulnerabilities in the region (Coker et al. 2011). A long history of keeping domestic animals in or near the house puts many people in close contact with animals. Flu pathogens, in particular, have benefited from traditional agricultural systems that include domestic birds and pigs on farms under the flyways of migrating wild birds, providing an ideal opportunity for wild avian flu viruses to become integrated into agricultural systems (Walters 2014). Rising affluence in the region has led to increasing numbers of pigs and poultry over recent years, intensifying the threat. Crowded markets, in which both wild and domesticated species may be sold, provide another opportunity for close contact between humans and animal pathogens, enabling spillover events to occur. This is exemplified by the linking of the 2002–2004 outbreak of SARS to the sale of Himalayan palm civets (*Paguma larvata*) in Chinese markets (Li et al. 2005), where the animal is eaten for medical purposes in the belief that it helps people withstand cold weather (Walters 2004). High human population densities in the region also facilitate the spread of new pathogen strains by human-to-human transmission. Disease surveillance in this region is particularly critical.

A subsequent H1N1 flu outbreak that began in Mexico in March 2009 reinforced just how susceptible human populations may be to a serious flu pandemic. Related to certain swine flu viruses, H1N1 was a previously unidentified strain of flu that is believed to have undergone reassortment, whereby different strains recombine into a new strain (CDC 2009; Walters

2014). Pigs, which have the ability to host avian, bird, and swine flu strains, may have been co-infected with more than one strain of flu virus, allowing the virus to recombine into a new strain that was then able to cross into the human population (CDC 2009). In this case, H1N1 developed efficient human-to-human transmission, but fortunately did not prove as deadly as many had initially feared. Nonetheless, by the end of 2009, laboratory-confirmed cases had been reported from 208 countries/overseas territories/communities and caused at least 11,500 deaths (WHO 2009). Subsequent analysis has suggested that the actual death toll of the 2009 outbreak could have been over 200,000 (Roos 2012). Many of these deaths were from respiratory or cardiac complications of infection; 80 percent of deaths probably occurred in people under 65, as is typical of a pandemic strain. Africa and Southeast Asia accounted for a disproportionate number of deaths for their population sizes (Roos 2012), underscoring the role of economic vulnerability to pandemic disease.

A third potential pandemic flu strain, H7N9, emerged in 2013, but has so far been restricted to China (Walters 2014). Between 2013 and June 2017, only 1,533 laboratory-confirmed cases of the diseases had been reported, but approximately one third of them proved fatal (Silva, Das, and Izurieta 2017). H7N9 can infect birds and humans asymptomatically, making its progress difficult to track (Walters 2014). There is little evidence of sustained human-to-human transmission to date, but experts fear that the disease could initiate a pandemic if it were to develop efficient human transmission routes (Silva, Das, and Izurieta 2017).

Although many epidemiologists predict that another human pandemic is almost inevitable (Walters 2014; Gupta 2018), there are signs of progress in our efforts to avoid pandemic disease. Disease surveillance is improving, and early detection provides critical time to contain epidemics (Walters 2014). Scientists are also improving their ability to identify pathogens and develop vaccines (Gupta 2018). We have also increased our capacity to make vaccines—as recently as 2006, according to the WHO, we could only make 350 million doses of flu vaccine; today, given the right coordination of global facilities, we could theoretically make 5.4 billion doses (Gupta 2018). The production of vaccines still occurs more slowly than the evolution of the virus, and international coordination remains challenging, but we have come a long way since 1918.

Antibiotic and pesticide resistance

Rising resistance to antibiotics also illustrates evolutionary processes in populations. The development of resistance in a population is a form of **genetic adaptation**. When someone takes an antibiotic, the drug first destroys the most susceptible bacteria, leaving bacteria with higher levels of resistance. If any bacteria survive the course of antibiotics, the next generations will inherit these antibiotic-resistant traits, increasing the proportion of resistant bacteria in the population. The same process can occur in vector populations exposed to pesticides. In this way, resistance is a result of the rapid evolution of pathogen and vector populations, but through applying pesticides and antibiotics in large quantities, we are effectively increasing the selection pressure on these populations to develop into more virulent strains.

The accidental discovery that penicillin mold could kill bacteria in 1928 opened a new chapter in humankind's relationship with infectious disease. By 1944, penicillin had been introduced worldwide as an effective antibiotic and was soon followed by a range of others; by 1965, more than 25,000 antibiotic products had been created (Garrett 1995, 36). At the time, many people thought the era of bacterial disease was over. Unfortunately, as early as the 1950s, scientists identified strains of bacteria that were resistant to penicillin, and the number of cases of antibiotic resistance has multiplied alongside increased use of antibiotics. In 1952, penicillin was effective against almost all *Staphylococcus* infections; a mere thirty years later, it was effective in fewer than 10 percent of clinical cases (Garrett 1995, 411).

Since then, the number of diseases displaying antibiotic resistance has increased, including strains of pneumonia, tuberculosis, and foodborne pathogens (Figure 3.6). Antibiotic resistance is considered by the US Centers for Disease Control (CDC) to be "one of the biggest public health challenges of our time" (CDC

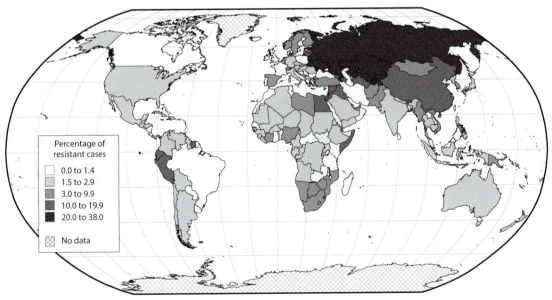

Figure 3.6 Estimated percentage of new TB cases resistant to rifampicin

Source: Data from WHO (2019)

Data are from the most recent year available.

2018d), with antibiotic resistant infections sickening more than two million people and killing 23,000 every year in the US alone (CDC 2013). Once antibiotic-resistant bacteria have developed, they can then spread directly among individuals. For instance, the sexually transmitted bacterial infection gonorrhea hit the headlines recently when a man returned to the UK from Southeast Asia with a multi-drug resistant form of the infection, generating fears that the disease would then spread in Britain (BBC 2018). The European Antimicrobial Resistance Surveillance Network (EARS-Net) closely monitors antibiotic resistance in Europe, including outbreaks such as this that might indicate the presence of a new drug-resistant pathogen (ECDC 2017). In the US, drug-resistant gonorrhea is also classified as an "urgent threat" by the CDC (CDC 2018d). A further "urgent threat" is growing antibiotic resistance in *Clostridium difficile* (CDC 2018d). *C. difficile* is the "most common microbial cause of healthcare-associated infections in US hospitals," causing 500,000 infections per year and 15,000 deaths in the US, as well as US$4.8 billion dollars of excess healthcare costs in acute facilities alone (CDC 2015). Antibiotic resistance has become such a critical issue that the US Department of Health and Human Services issued a global challenge to address antibiotic resistance at the 2018 United Nations General Assembly Meeting in New York (Kuehn 2018).

The more antibiotics we use as a society, the greater the selection pressure we place on bacterial communities. One estimate suggests that 51 tons of antibiotics are consumed daily in the US (Hollis and Ahmed 2013). Many public health practitioners are critical of what they perceive to be unnecessary use of antibiotics. For instance, antibiotics may be prescribed as a precautionary measure when the disease agent is not known, just in case it is a bacterial infection. Many of these infections turn out to be viral and so the antibiotic is ineffective against the disease, but nonetheless provides a selection pressure on any bacterial communities in the patient's body. Hospitals have begun to appoint special pharmacists to oversee prescriptions for infectious disease to stem this tide of antibiotic overprescription (Baker 2007). Public education campaigns have also been designed to discourage patients from demanding antibiotics (WHO 2017b). Some have argued that vaccination campaigns may

be a better long-term solution given the challenges of attempting to control antibiotic resistance (Andre et al. 2008; Lieberman 2003).

Failing to complete a course of antibiotics can exacerbate the problem. The shorter and less aggressive a course of antibiotics, the more likely that some bacteria will survive and pass their antibiotic resistance to future generations of the pathogen. Unfortunately, little empirical evidence is available to suggest exactly how long a course of antibiotics should be in order to be effective, leading to a recent editorial in the *British Medical Journal* suggesting that current antibiotic course lengths are based on precedent, rather than solid evidence (Llewelyn et al. 2017). More research is clearly needed. This problem is exacerbated by the widespread use of antibiotics in agriculture, where they are often given at low doses to prevent disease rather than to treat a particular infection. We explore the agricultural use of antibiotics in detail in the case study presented later.

These examples demonstrate some of the tensions between the well-being of individual patients and the interests of the population at large. For an individual patient, it may be advantageous to take an antibiotic for a disease of unknown origin, just in case the disease is bacterial, but this approach is problematic from the perspective of population health. Similarly, an individual patient may not understand why they should continue taking their course of an antibiotic if they feel better before the course is completed. Infections that require a long course of antibiotics are especially troublesome in this respect: tuberculosis (TB) is a prime example. Full treatment typically requires at least a six-month course of antibiotics, and yet patients often feel better sooner. Unsurprisingly, antibiotic resistance is a particularly grave problem for TB and some strains of tuberculosis are now resistant to multiple antibiotics. We return to TB in Chapter 4.

It is not just bacterial populations that are developing resistance. Pesticide resistance is a significant problem in many mosquito populations, as well as in organisms such as the plasmodium parasite that causes malaria. Some fungi are also starting to show resistance, such as *Candida*, a type of yeast. Although commonly found on the skin and usually benign, *Candida* infections can become a problem if the yeast gets into certain parts of the body such as the bloodstream or lungs. A drug-resistant form of *Candida*, dubbed *Candida auris* because it was first found in a patient's ear, was identified in Japan in 2009 (Casadevall, Kontoyiannis, and Robert 2019). *C. auris* infections have now been identified across all continents. Patients are particularly vulnerable to infection when antibiotics reduce the populations of benign bacteria that normally out-compete the fungus. Infection usually occurs in hospital settings, where more rigorous hygiene is currently our best defense against the spread of this pathogen (Ciric 2019).

Healthcare facilities provide good conditions for pathogens to evolve and spread by concentrating immuno-suppressed individuals in hospitals and then circulating people and equipment among them. Many common hospital activities, such as puncturing the skin for injections, facilitate the spread of pathogens. The US CDC now recognizes a variety of hospital-acquired infections associated with particular procedures, including "central-line associated bloodstream infections," "catheter-associated urinary tract infection," and "ventilator-associated pneumonia" (CDC 2014). Infections contracted in hospitals and hospital-like settings are called **nosocomial infections**. Many antibiotic-resistant strains of bacteria appear to have emerged as nosocomial infections, most notably MRSA (methicillin-resistant *Staphylococcus aureus*).

Nosocomial infections: the case of MRSA

Antibiotic-resistant infections remain concentrated in healthcare settings, where the elderly, very young, or immuno-suppressed are the most susceptible (EARS 2005). These **nosocomial infections** complicate surgical and other healthcare procedures and have even led some patients to seek medical treatment in other countries with lower risk. There is always a danger that these antibiotic-resistant strains may spread into the community. These "community-associated" infections have become a significant problem in the US.

Antibiotic resistance grew rapidly in many countries during the 1990s and 2000s. By 2005, approximately 70 percent of bacteria that cause infections in hospitals were resistant to at least one common antibiotic (EARS 2005). MRSA, first observed in the 1960s, is now one

of the most important causes of healthcare-acquired antibiotic-resistant infections worldwide (ECDC 2017). *Staphylococcus aureus* colonizes the skin of about 30 percent of healthy humans (EARS 2005). Although normally harmless, it can cause severe infections of the blood, skin, or soft tissues, and can sometimes be fatal. MRSA is most commonly transmitted in healthcare settings from the skin of one patient to other patients, particularly when healthcare workers do not practice sufficient hygiene (CDC 2007). Patients in intensive care units are at particularly high risk (Tiemersma et al. 2004). In the US, MRSA accounted for 2 percent of *Staphylococcus* infections in 1974, 22 percent in 1995, and 63 percent in 2004 (CDC 2007).

The European Antimicrobial Resistance Surveillance System (EARS-Net), founded in the late 1990s to provide comprehensive surveillance of antibiotic-resistant strains in Europe, provides some of the best data on MRSA. The proportion of bacteria resistant to methicillin varies widely in Europe, both among countries and among hospitals within particular countries (Tiemersma et al. 2004). In 2004, the proportion of *Staphylococcus aureus* showing resistance to methicillin ranged from less than 1 percent in Iceland to 73 percent in Romania (EARS 2005). By 2013, the proportion of cases showing resistance to methicillin had dropped somewhat across Europe, to an average of 18 percent of all cases, but MRSA remains a public health problem throughout Europe (ECDC 2017). Broadly speaking, rates tend to be lowest in Scandinavia and get progressively higher westwards and southwards across Europe (Figure 3.7). This may be a reflection of patterns of diffusion as strains disseminate from the area where they evolved. Regional variations in MRSA may also reflect differences in the types of care provided in different hospital settings, varying application or effectiveness of control strategies such as hygiene campaigns, or even differences in the way that data are collected (Tiemersma et al. 2004).

Increasingly, cases of MRSA are being identified among people without exposure to healthcare settings, suggesting that resistant bacteria are circulating more widely in the community. One study in 2003 suggested that about 12 percent of clinical MRSA infections in the US may be community associated (CDC 2008). Although these strains can usually be treated

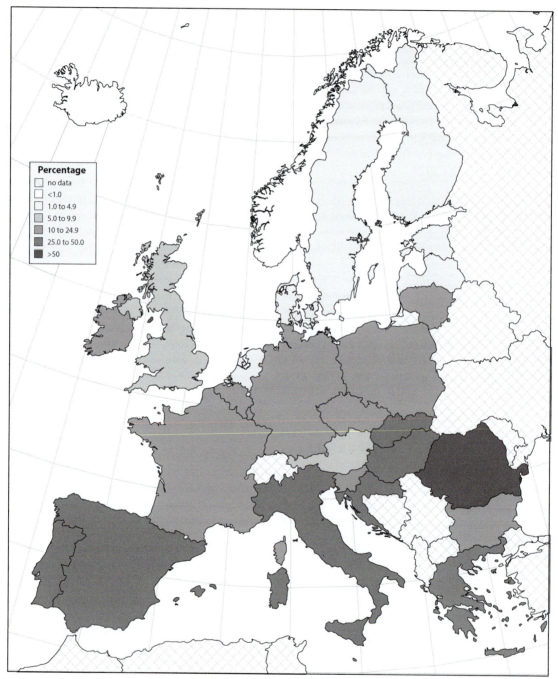

Figure 3.7 Geographic variations in MRSA prevalence in Europe (1999–2002)

Source: Data from ECDC (2017)

This map shows the percentage of methicillin-resistant Staphylococcus aureus *infections as a proportion of all* Staphylococcus aureus *infections.*

with second-line antibiotics (antibiotics used when initial antibiotic treatments prove ineffective), they are often more virulent than hospital-acquired infections and affect otherwise healthy individuals. Outbreaks of community-associated MRSA (CA-MRSA) have been identified in populations such as prisoners, athletes, and military trainees, tied to factors such as close personal contact, crowded living conditions, and poor hygiene (CDC 2008). It is hoped that improved hygiene can contain the problem. In September 2008, the US CDC launched a campaign to educate parents about how to protect their children from MRSA skin infections. The campaign aimed to inform parents about early signs of an MRSA infection and encourage good hygiene such as effective cleaning and covering of cuts.

Antibiotic resistance in foodborne disease and industrial agriculture

Another example of the significance of antibiotic resistance is the shift towards more intense forms of industrial agriculture, which has been held responsible for many recent outbreaks of foodborne disease in the West. While antibiotics clearly play an important role in treating infections in farm animals, antibiotics are often used **prophylactically** in animals simply to improve productivity (Economou and Gousia 2015). The economic benefits of administering low doses of antibiotics as growth promoters were first established in the 1950s when researchers noted that animals gained weight after receiving sub-therapeutic levels of penicillin and tetracycline (Stokstad and Jukes 1950; Economou and Gousia 2015). Farm animals fighting infections do not efficiently convert fodder into meat, milk, or eggs, and so antibiotics are used in low doses to allow animals to grow more rapidly (Walters 2004). The use of antibiotics in this way became common practice since this discovery, and agricultural use of antibiotics now far exceeds human usage in many countries. In the US, in a climate of cheap antibiotics and limited regulation, antibiotics have even been known to be sprayed on fruit trees or added to marine paints to inhibit barnacle growth (Hollis and Ahmed 2013). As we have already noted, there are compelling reasons to believe that widespread antibiotic application exerts a selection pressure on the

microbe population that is likely to lead to the evolution of antibiotic resistance. Indeed, every time we use an antibiotic, we affect its future efficacy, and so many have argued that we should avoid "wasting" antibiotics on non-essential agricultural applications (e.g., Hollis and Ahmed 2013). Donald Kennedy, a former US Food and Drug Administration (FDA) commissioner, argued that antibiotics must be viewed as a "kind of Commons" that is shared by all members of the public and which is threatened by agricultural misuse (Kennedy 2013).

Since the 1990s, aquaculture has also begun to rely on applications of antibiotics to enable fish to be stocked at high densities (Economou and Gousia 2015). In the US, 80 percent of all antibiotics consumed in 2013 were used for agriculture and aquaculture—a total of approximately 13 million kilograms (Hollis and Ahmed 2013). Despite aggressive agricultural lobbying, in 2017 the US FDA banned the use of medically important antibiotics in agriculture for growth promotion or feed efficiency in recognition of the problem (FDA 2017). The European Union (EU) has also attempted to regulate antibiotic use in agriculure—antimicrobial growth promoters have been banned in the EU since 2006—but the scale of usage remains a significant problem in many European countries as well.

These recent attempts to reduce veterinary use of antibiotics reflect rising public and scientific concern over antibiotic resistance. There are several pathways through which antibiotic resistance could be transferred from animal populations to people, including direct infection with resistant bacteria from farm animals, ingestion of infected meat or animal products, and the transfer of genes for resistance from animal to human pathogens (Chang et al. 2015). Several studies provide circumstantial evidence that human infections may be linked to agricultural use of antibiotics, although the issue remains highly controversial, particularly owing to the economic implications of reducing agricultural use of antibiotics. Avoparcin, for instance, has been widely used in agriculture but is part of the same group of drugs that includes vancomycin and teicoplanin, which have become drugs of last resort for human MRSA and drug-resistant *Enterococcus* infections. Resistance to vancomycin and teicoplanin emerged in human *Enterococcus* infections alongside widespread agricultural use of avoparcin. In countries that allow the use of avoparcin, vancomycin-resistant *Enterococcus* bacteria have been found in the bowel of between 2 and 17 percent of people who had not otherwise been exposed to these antibiotics, suggesting that the resistant bacteria may be finding their way into humans via consumption of tainted meat products (Baker 2007). A similar story has been reported for antibiotic-resistant strains of *Salmonella* contracted by humans from meat (Walters 2004), where there is growing scientific consensus on the role played by antibacterials in generating resistant strains (Economou and Gousia 2015).

Powerful agricultural and pharmaceutical lobbying groups have argued that there is not enough evidence to implicate the agricultural use of antibiotics and question particularly whether agricultural usage of antibiotics contributes to significant levels of clinical disease in people (Chang et al. 2015). In a summary of evidence from Denmark's ban on agricultural use of antibiotics for growth promotion, for instance, the Animal Health Institute, which represents veterinary drug manufacturers in the US, argues that "a ban on [antibiotics for agricultural growth promotion] in Denmark has not had the intended benefit of reducing antibiotic resistance patterns in humans; it has had the unintended consequence of increasing animal suffering, pain and death" (Animal Health Institute 2018, emphasis in original). Many stakeholders nonetheless argue that evidence is emerging that the potential consequences of inaction are so severe that legislation must be moved forward immediately, even in the absence of incontrovertible evidence.

Other aspects of the modern food system further complicate the story. Pollan (2006) argues that part of the reason that animals are prone to sickness on industrial farms is that they are kept at high densities and fed foods they have not evolved to eat, such as corn. These arguments have been extended to aquaculture, where fish are kept at much higher densities than under natural conditions and are often fed high-protein feed pellets. In some respects, therefore, the disturbed ecology associated with our modern industrial food system *necessitates* the use of antibiotics, resulting in a precarious dilemma. Without widespread reform of entire agricultural systems, which would have huge economic implications, agricultural lobbyists can make a compelling case that

the agricultural use of antibiotics is critical for animal welfare (as the Animal Health Institute argues earlier). This is reflected in the fact that bans on antibiotics have been difficult to enforce as there is a rather fuzzy distinction between antibiotics being used to cure infections and being used for growth promotion (Economou and Gousia 2015). Indeed, the rate at which antibiotics are prescribed therapeutically has generally risen in countries once their use for growth promotion has been banned (Hollis and Ahmed 2013).

Regardless of the precise role of antibiotics in outbreaks of foodborne disease, we know that foodborne pathogens remain a significant health burden worldwide, and there is mounting evidence that our industrial food systems may contribute to the problem. The problem first came to the fore in the UK, when several food scares hit the headlines in the 1980s, including concerns over *Salmonella* in eggs and *Listeria* in soft cheese and pâté (Freidberg 2004). The 1980s also heralded the beginning of Britain's "mad cow" crisis, with bovine spongiform encephalopathy (BSE) first detected in 1986, although it was not officially accepted as having any connection to the human neurological condition Creutzfeldt-Jakob Disease (CJD) until ten years later (Freidberg 2004) (Box 3.3). More recently, the US has found itself in a series of food-related crises with periodic outbreaks of *Salmonella*, *Escherichia coli*, and *Campylobacter* traced to the agricultural use of antibiotics (Maki 2006), and contaminated meat and cross-contaminated foods probably cause millions of cases of foodborne disease in the US (Scallan et al. 2011). Many commentators view these outbreaks as a consequence of the social and ecological changes that have occurred in our food production methods (Freidberg 2004; Walters 2004; Pollan 2006).

Although most cases of foodborne illness are from foods of animal origin, prepared fruits and vegetables and even processed foods are increasingly being implicated, suggesting that cross-contamination of foods is a problem. For example, an outbreak of a virulent strain of *Escherichia coli* associated with the consumption of fresh spinach in the US in 2006 led to 199 confirmed human infections and several deaths. This instance was at least the twenty-sixth reported outbreak of this strain of *E. coli* linked to leafy greens since 2003 (Maki 2006). Since then, other outbreaks in the US have been reported,

Box 3.3

BRITAIN'S BSE CRISIS

The outbreak of bovine spongiform encephalopathy (BSE) in the UK in 1986 shows how the dramatically altered ecology of industrial agriculture can have unforeseen consequences. At its peak in 1992, 36,680 cases of BSE were confirmed in cattle before aggressive control measures implemented by the British government (involving destroying cattle and limiting what animals could be fed) began to turn the tide (Baker 2007). BSE causes the brains of infected animals to become riddled with tiny holes, leading to impairments in behavior and movement, followed by death. A spongy-looking brain and mental disruption characterize a small group of diseases known as transmissible spongiform encephalopathies (TSEs). These include scrapie in sheep, chronic wasting in deer, and Kuru (associated with cannibalism) and Creutzfeldt-Jakob disease in humans.

Although there are competing hypotheses as to the origins of TSEs, the most widely accepted explanation is that these diseases are caused by changes in proteins in nerve tissue initiated by an infectious agent known as a **prion**. Prions have no genes of their own and appear to consist of little more than proteins, leading to questions about whether they can be considered to be living. Nonetheless, they are transmissible and appear to have the potential to leap over the species barrier in some circumstances. The fact that prions are not really "alive" makes them very hard to "kill"; notably, they are not destroyed by cooking, sterilization, or exposure to ultraviolet radiation.

In the case of BSE, prions from scrapie-infected sheep may have found their way into the food chain and infected cows. Although cows are naturally herbivorous, they are sometimes fed high-protein feeds that include animal protein to encourage rapid weight gain in modern factory farms. In the mid-twentieth century, industrial agriculture began rendering discarded by-products of butchered animals into feed for farm animals. The process of rendering uses heat, pressure, and chemical solvents to process these waste products, killing most pathogens. The peculiar resilience of prions, however, may have enabled some to survive the process and persist in animal feed (Walters 2004).

Shortly after the identification of BSE, strange new cases of the human TSE, Creutzfeldt-Jakob disease (CJD), began to appear, generating the fear that consumption of BSE-infected meat could lead to human cases of CJD. While CJD was recognized long before the BSE crisis and occurs all over the world, it was traditionally associated with the elderly. The new form of the disease that began to appear, "variant CJD" (vCJD), affects much younger individuals—the median age of onset being just 28 (Baker 2007). After the onset of symptoms such as memory loss and balance problems, the disease can progress to coma and death within a year.

Fear of a potential UK-wide epidemic of variant CJD prompted a massive country-wide campaign to cull cattle and reform animal feeding practices. Thankfully, this epidemic has not yet occurred. The UK's national CJD surveillance unit in Edinburgh reported 178 cases of probable or confirmed variant CJD by 2017, with the last reported case to date in 2014 (The National CJD Research & Surveillance Unit 2017). The number of reported variant CJD deaths peaked in the year 2000 with 28 deaths; since then, the number has steadily declined (Baker 2007) and it seems unlikely that a large-scale epidemic will occur. The CJD Surveillance Unit nonetheless warns that incidence could increase again if genetic subgroups of the disease with longer incubation periods emerge (*ibid.*). Even if mortality rates associated with the event remain low, the economic costs of the crisis were enormous, and many people were shaken by what was considered to be a near miss.

Cultural ecology played several key roles in the emergence of BSE. First, the practice of feeding

continued

Animal Health Institute. 2018. "The Antibiotic Ban in Denmark: A Case Study on Politically Driven Bans." Accessed January 10, 2019. www.ahi.org/issues-advocacy/animal-antibiotics/the-antibiotic-ban-in-denmark-a-case-study-on-politically-driven-bans/.

Armelagos, G. J., P. J. Brown, and B. Turner. 2005. "Evolutionary, Historical and Political Economic Perspectives on Health and Disease." *Social Science & Medicine* 61 (4): 755–65. doi:10.1016/j.socscimed.2004.08.066.

Baker, R. 2007. *Quiet Killers: The Fall and Rise of Deadly Diseases*. Sutton Publishing.

Barrett, R., C. W. Kuzawa, T. McDade, and G. J. Armelagos. 1998. "Emerging and Re-emerging Infectious Diseases: The Third Epidemiologic Transition." *Annual Review of Anthropology* 27: 247–71.

[BBC] British Broadcasting Company. 2018. " 'World's Worst' Super-Gonorrhoea Man Cured." Accessed January 10, 2019. www.bbc.com/news/health-43840505.

Borrell, B. 2013. "Faroe Islands Aim to Sequence Genes of Entire Country." *Discover*. Accessed January 10, 2019. http://discovermagazine.com/2013/julyaug/01-faroe-islands-aim-to-sequence-genomes-healthcare.

Bruni, Laia. 2017. "Global Vaccine Uptake and Projected Cervical Cancer Disease Reductions." *HPV World* 1 (19): 6–9.

Casadevall, Arturo, Dimitrios P. Kontoyiannis, and Vincent Robert. 2019. "On the Emergence of *Candida auris*: Climate Change, Azoles, Swamps, and Birds." *mBio* 10 (4): e01397-19. doi:10.1128/mBio.01397-19.

[CDC] Centers for Disease Control and Prevention. 2007. "MRSA in Healthcare Settings." Accessed June 8, 2009. www.cdc.gov/ncidod/dhqp/ar_MRSA_spotlight_2006.html.

[CDC] Centers for Disease Control and Prevention. 2008. "Community-Associated MRSA Information for the Public." Accessed June 9, 2009. www.cdc.gov/ncidod/dhqp/ar_MRSA_ca_public.html.

[CDC] Centers for Disease Control and Prevention. 2009. "Origin of 2009 H1N1 Flu (Swine Flu): Questions and Answers." Accessed February 5, 2019. www.cdc.gov/h1n1flu/information_h1n1_virus_qa.htm#a.

[CDC] Centers for Disease Control and Prevention. 2013. *Antibiotic Resistance Threats in the United States, 2013*. US Department of Health and Human Services.

[CDC] Centers for Disease Control and Prevention. 2014. "Types of Healthcare-associated Infections." Accessed January 10, 2019. www.cdc.gov/hai/infectiontypes.html.

[CDC] Centers for Disease Control and Prevention. 2015. "Nearly Half a Million Americans Suffered from *Clostridium difficile* Infections in a Single Year." Accessed January 10, 2019. www.cdc.gov/media/releases/2015/p0225-clostridium-difficile.html.

[CDC] Centers for Disease Control and Prevention. 2018a. "List of Selected Multistate Foodborne Outbreak Investigations." Accessed January 10, 2019. www.cdc.gov/foodsafety/outbreaks/multistate-outbreaks/outbreaks-list.html.

[CDC] Centers for Disease Control and Prevention. 2018b. "Past Pandemics." Accessed February 5, 2019. www.cdc.gov/flu/pandemic-resources/basics/past-pandemics.html.

[CDC] Centers for Disease Control and Prevention. 2018c. "Seasonal Influenza Vaccine Effectiveness, 2004–2018." Accessed February 5, 2019. www.cdc.gov/flu/professionals/vaccination/effectiveness-studies.htm.

[CDC] Centers for Disease, Control, and Prevention. 2018d. "Biggest Threats and Data." Accessed January 10, 2019. www.cdc.gov/drugresistance/biggest_threats.html.

Chang, Qiuzhi, Weike Wang, Gili Regev-Yochay, Marc Lipsitch, and William P. Hanage. 2015. "Antibiotics in Agriculture and the Risk to Human Health: How Worried Should We Be?" *Evolutionary Applications* 8 (3): 240–47. doi:10.1111/eva.12185.

Ciric, L. 2019. "Candida auris: The New Superbug on the Block." *BBC News*. Accessed August 29, 2019. www.bbc.com/news/health-49170866.

Coker, R. J., B. M. Hunter, J. W. Rudge, M. Liverani, and P. Hanvoravongchai. 2011. "Emerging Infectious Diseases in Southeast Asia: Regional Challenges to Control." *Lancet* 377 (9765): 599–609. doi:10.1016/s0140-6736(10)62004-1.

Diamond, J. M. 1987. "The Worst Mistake in the History of the Human Race." *Discover*. Accessed January 16, 2019. http://discovermagazine.com/1987/may/02-the-worst-mistake-in-the-history-of-the-human-race.

Donnelly, J. S. 2001. *The Great Irish Potato Famine*. Sutton Publishing.

Drewnowski, A., M. Maillot, and C. Rehm. 2012. "Reducing the Sodium-Potassium Ratio in the US Diet: A Challenge for Public Health." *American Journal of Clinical Nutrition* 96 (2): 439–44. doi:10.3945/ajcn.111.025353.

[EARS] European Antimicrobial Resistance Surveillance System. 2005. "Annual Report, 2004." www.rivm.nl/earss/Images/EARSS%20annual%20report%202004%20webversie_tcm61-25345.pdf.

[ECDC] European Centre for Disease Control. 2017. *Summary of the Latest Data on Antibiotic Resistance in the European Union*. Stockholm, Sweden: European Health Institute.

Economou, V., and P. Gousia. 2015. "Agriculture and Food Animals as a Source of Antimicrobial-Resistant Bacteria." *Infection and Drug Resistance* 8: 49–61. doi:10.2147/idr.S55778.

Ewald, Paul W. 2002. *Plague Time the New Germ Theory of Disease*. Anchor Books.

[FDA] US Food and Drug Administration. 2017. "FACT SHEET: Veterinary Feed Directive Final Rule and Next Steps." Accessed January 10. www.fda.gov/AnimalVeterinary/DevelopmentApprovalProcess/ucm449019.htm.

Foppa, Ivo M., Po-Yung Cheng, Sue B. Reynolds, David K. Shay, et al. 2015. "Deaths Averted by Influenza Vaccination in the U.S. During the Seasons 2005/06 Through 2013/14." *Vaccine* 33 (26): 3003–9. doi:10.1016/j.vaccine.2015.02.042.

Frassetto, L., R. C. Morris, Jr., D. E. Sellmeyer, K. Todd, and A. Sebastian. 2001. "Diet, Evolution and Aging: The Pathophysiologic Effects of the Post-Agricultural Inversion of the Potassium-to-Sodium and Base-to-Chloride Ratios in the Human Diet." *European Journal of Nutrition* 40 (5): 200–13.

Freidberg, Susanne. 2004. *French Beans and Food Scares: Culture and Commerce in an Anxious Age*. New York: Oxford University Press.

Frenk, Julio, and Octavio Gomez-Dantes. 2011. "The Triple Burden: Disease in Developing Nations." *Harvard International Review* 33: 36–40.

Gallagher, K. E., D. S. LaMontagne, and D. Watson-Jones. 2018. "Status of HPV Vaccine Introduction and Barriers to Country Uptake." *Vaccine* 36 (32 Pt A): 4761–67. doi:10.1016/j.vaccine.2018.02.003.

Gardner, Gary T., and Brian Halweil. 2000. *Underfed and Overfed: The Global Epidemic of Malnutrition, Worldwatch Paper, 150*. Washington, DC: Worldwatch Institute.

Garrett, Laurie. 1995. *The Coming Plague: Newly Emerging Diseases in a World out of Balance*. New York: Penguin.

Gilkey, M. B., W. A. Calo, M. W. Marciniak, and N. T. Brewer. 2017. "Parents Who Refuse or Delay HPV Vaccine: Differences in Vaccination Behavior, Beliefs, and Clinical Communication Preferences." *Human Vaccines and Immunotherapeutics* 13 (3): 680–86. doi:10.1080/21645515.2016.1247134.

Gubler, D. J., N. Vasilakis, and D. Musso. 2017. "History and Emergence of Zika Virus." *Journal of Infectious Diseases* 216 (Suppl 10): S860–67. doi:10.1093/infdis/jix451.

Guerrant, Richard L., Mark D. DeBoer, Sean R. Moore, Rebecca J. Scharf, and Aldo A. M. Lima. 2012. "The Impoverished Gut—A Triple Burden of Diarrhoea, Stunting and Chronic Disease." *Nature Reviews Gastroenterology & Hepatology* 10: 220. doi:10.1038/nrgastro.2012.239.

Gupta, A. 2018. "The Big One Is Coming, and It's Going to Be a Flu Pandemic." Accessed January 28, 2019. www.cnn.com/2017/04/07/health/flu-pandemic-sanjay-gupta/index.html.

Haub, C. 2008. "Tracking Trends in Low Fertility Countries: An Uptick in Europe?" Accessed January 6, 2010. www.prb.org/Articles/2008/tfrtrendsept08.aspx/.

Hollis, A., and Z. Ahmed. 2013. "Preserving Antibiotics, Rationally." *New England Journal of Medicine* 369 (26): 2474–76. doi:10.1056/NEJMp1311479.

Jabr, F. 2013. "How to Really Eat Like a Hunter-Gatherer: Why the Paleo Diet Is Half-Baked." *Scientific American*. Accessed January 10, 2019. www.scientific american.com/article/why-paleo-diet-half-baked-how-hunter-gatherer-really-eat/.

Jacobs, Gerald H., and Jeremy Nathans. 2009. "The Evolution of Primate Color Vision." *Scientific American* 300 (4): 56–63.

Johnson, N. P., and J. Mueller. 2002. "Updating the Accounts: Global Mortality of the 1918–1920 'Spanish' Influenza Pandemic." *Bulletin of the History of Medicine* 76 (1): 105–15.

Kalra, S. 2013. "Diabesity." *Journal of the Pakistan Medical Association* 63 (4): 532–34.

Kennedy, Donald. 2013. "Time to Deal with Antibiotics." *Science* 342 (6160): 777. doi:10.1126/science.1248056.

Kuehn, B. 2018. "Antibiotic Resistance Challenge." *JAMA* 320 (18): 1851. doi:10.1001/jama.2018.16587.

Ladusingh, L., S. K. Mohanty, and M. Thangjam. 2018. "Triple Burden of Disease and out of Pocket Healthcare Expenditure of Women in India." *PLoS One* 13 (5): e0196835.

Larson, H. J., R. Wilson, S. Hanley, A. Parys, and P. Paterson. 2014. "Tracking the Global Spread of Vaccine Sentiments: The Global Response to Japan's Suspension of Its HPV Vaccine Recommendation." *Human Vaccines Immunotherapeutics* 10 (9): 2543–50. doi:10.4161/216 45515.2014.969618.

Leichenko, R., and K. O'Brien. 2008. *Environmental Change and Globalization: Double Exposures*. Oxford University Press.

Li, Wendong, Zhengli Shi, Meng Yu, Wuze Ren, et al. 2005. "Bats Are Natural Reservoirs of SARS-Like Coronaviruses." *Science* 310 (5748): 676–79.

Lieberman, D. E. 2015. "Human Locomotion and Heat Loss: An Evolutionary Perspective." *Comprehensive Physiology* 5 (1): 99–117. doi:10.1002/cphy.c140011.

Lieberman, J. M. 2003. "Appropriate Antibiotic Use and Why It Is Important: The Challenges of Bacterial Resistance." *Pediatric Infectious Disease Journal* 22 (12): 1143–51. doi:10.1097/01.inf.0000101851.57263.63.

Little, C. L., and I. A. Gillespie. 2008. "Pre-pared Salads and Public Health." *Journal of Applied Microbiology* 105 (6): 1729–43. doi:10.1111/j.1365-2672.2008.03801.x.

Llewelyn, Martin J., Jennifer Fitzpatrick, Elizabeth Darwin, Sarah Tonkin-Crine, et al. 2017. "The Antibiotic Course Has Had Its Day." *BMJ* 358: j3418. doi:10.1136/bmj.j3418.

Maki, D. G. 2006. "Don't Eat the Spinach—Controlling Foodborne Infectious Disease." *New England Journal of Medicine* 355 (19): 1952–55.

Mardigan, T. 2014. "Back to the Stone Age." In *Harvard Medical School Commentaries on Health*. Harvard Health Publications and Credo Reference.

Marshall, S. J. 2004. "Developing Countries Face Double Burden of Disease." *Bulletin of the World Health Organization* 82 (7): 556.

Minster, R. L., N. L. Hawley, C. T. Su, G. Sun, et al. 2016. "A Thrifty Variant in CREBRF Strongly Influences Body Mass Index in Samoans." *Nature Genetics* 48 (9): 1049–54. doi:10.1038/ng.3620.

Morse, S. S., J. A. Mazet, M. Woolhouse, C. R. Parrish, et al. 2012. "Prediction and Prevention of the Next Pandemic Zoonosis." *Lancet* 380 (9857): 1956–65. doi:10.1016/s0140-6736(12)61684-5.

Moyer, M. 2017. "Flu Vaccine 'Factories' Create Errors That Reduce Protection." *Scientific American*. Accessed February 5, 2019. www.scientificamerican.com/article/flu-vaccine-ldquo-factories-rdquo-create-errors-that-reduce-protection/.

Murray, Christopher J. L., Charlton S. K. H. Callender, Xie Rachel Kulikoff, V. Srinivasan, et al. 2018. "Population and Fertility by Age and Sex for 195 Countries and Territories, 1950–2017: A Systematic Analysis for the Global Burden of Disease Study 2017." *Lancet* 392 (10159): 1995–2051. doi:10.1016/S0140-6736(18)32278-5.

National Cancer Institute. 2015. "Risk Factors: Infectious Agents." Accessed March 31, 2019. www.

cancer.gov/about-cancer/causes-prevention/risk/infectious-agents.

National Cancer Institute. 2019. "HPV and Cancer." Accessed August 29, 2019. www.cancer.gov/about-cancer/causes-prevention/risk/infectious-agents/hpv-and-cancer#what-is-hpv.

The National CJD Research and Surveillance Unit. 2017. *Creutzfeldt-Jakob Disease Surveillance in the UK.* Available at: https://www.cjd.ed.ac.uk/sites/default/files/report26.pdf. Accessed: 6.16.19.

Neel, J. V. 1962. "Diabetes Mellitus: A 'Thrifty' Genotype Rendered Detrimental by 'Progress'?" *American Journal of Human Genetics* 14 (4): 353–62.

Omran, A. R. 1971. "The Epidemiologic Transition: A Theory of the Epidemiology of Population Change." *Milbank Memorial Fund Quarterly* 49 (4): 509–38.

Pollan, Michael. 2006. *The Omnivore's Dilemma: A Natural History of Four Meals.* New York: Penguin Press.

Rasmussen, J., O. W. Nielsen, N. Janzen, M. Duno, et al. 2014. "Carnitine Levels in 26,462 Individuals from the Nationwide Screening Program for Primary Carnitine Deficiency in the Faroe Islands." *Journal of Inherited Metabolic Disease* 37 (2): 215–22. doi:10.1007/s10545-013-9606-2.

Roos, R. 2012. "CDC Estimate of Global H1N1 Pandemic Deaths: 284,000." Accessed February 4, 2019. www.cidrap.umn.edu/news-perspective/2012/06/cdc-estimate-global-h1n1-pandemic-deaths-284000.

Sabeena, Sasidharanpillai, Parvati V. Bhat, Veena Kamath, and Govindakarnavar Arunkumar. 2018. "Global Human Papilloma Virus Vaccine Implementation: An Update." *Journal of Obstetrics and Gynaecology Research* 44 (6): 989–97. doi:10.1111/jog.13634.

Scallan, Elaine, Robert M. Hoekstra, Frederick J. Angulo, Robert V. Tauxe, et al. 2011. "Foodborne Illness Acquired in the United States: Major Pathogens." *Emerging Infectious Diseases* 17 (1): 7–15. doi:10.3201/eid1701.P11101.

Schulz, Leslie O., and Lisa S. Chaudhari. 2015. "High-Risk Populations: The Pimas of Arizona and Mexico." *Current Obesity Reports* 4 (1): 92–98. doi:10.1007/s13679-014-0132-9.

Silva, W., T. K. Das, and R. Izurieta. 2017. "Estimating Disease Burden of a Potential A(H7N9) Pandemic Influenza Outbreak in the United States." *BMC Public Health* 17 (1): 898. doi:10.1186/s12889-017-4884-5.

Simonsen, L., M. J. Clarke, L. B. Schonberger, N. H. Arden, et al. 1998. "Pandemic Versus Epidemic Influenza Mortality: A Pattern of Changing Age Distribution." *Journal of Infectious Diseases* 178 (1): 53–60.

Singh, Ram B., Daniel Pella, Viola Mechirova, Kumar Kartikey, et al. 2007. "Prevalence of Obesity, Physical Inactivity and Undernutrition, a Triple Burden of Diseases During Transition in a Developing Economy. The Five City Study Group." *Acta Cardiologica* 62 (2): 119–27. doi:10.2143/AC.62.2.2020231.

Stokstad, E. L. R., and T. H. Jukes. 1950. "Further Observations on the 'Animal Protein Factor'." 73 (3): 523–28. doi:10.3181/00379727-73-17731.

Taubenberger, Jeffery K., and David M. Morens. 2006. "1918 Influenza: The Mother of All Pandemics." *Emerging Infectious Diseases* 12 (1): 15–22. doi:10.3201/eid1201.050979.

The National CJD Research and Surveillance Unit. 2017. *Creutzfeldt-Jakob Disease Surveillance in the UK.* Accessed June 16, 2019. https://www.cjd.ed.ac.uk/sites/default/files/report26.pdf.

Tiemersma, E. W., Slam Bronzwaer, O. Lyytikainen, J. E. Degener, et al. 2004. "Methicillin-Resistant Staphylococcus aureus in Europe, 1999–2002." *Emerging Infectious Diseases* 10 (9): 1627–34.

United Nations. 2017. *World Population Prospects: The 2017 Revision.* New York: United Nations.

United Nations, Department of Economic and Social Affairs, Population Division. 2019. "World Population Prospects 2019: Data Booklet." (ST/ESA/SER.A/424).

Walters, M. J. 2004. *Six Modern Plagues and How We Are Causing Them.* Island Press.

Walters, M. J. 2014. *Seven Modern Plagues: And How We Are Causing Them.* Island Press.

[WHO] World Health Organization. 2005. "Avian Influenza, Frequently Asked Questions." Accessed May 14, 2019. www.who.int/csr/disease/avian_influenza/avian_faqs/en/index.html.

[WHO] World Health Organization. 2009. "Pandemic (H1N1) 2009—Update 80." Accessed February 4, 2019. www.who.int/csr/don/2009_12_23/en/.

[WHO] World Health Organization. 2011. "FAQs: H5N1 Influenza." Accessed February 5, 2019. www.who.int/influenza/human_animal_interface/avian_influenza/h5n1_research/faqs/en/.

[WHO] World Health Organization. 2017a. *Comprehensive Cervical Cancer Control.* 2nd ed. Accessed January 16, 2019. www.searo.who.int/publications/bookstore/documents/Cancer/en/.

[WHO] World Health Organization. 2017b. "Creative Campaigns Spread Awareness on Antibiotic Resistance." Accessed January 10, 2019. www.who.int/news-room/feature-stories/detail/creative-campaigns-spread-awareness-on-antibiotic-resistance.

[WHO] World Health Organization. 2018a. "Noncommunicable Diseases." Accessed January 10, 2019. www.euro.who.int/en/health-topics/noncommunicable-diseases.

[WHO] World Health Organization. 2018b. "Obesity and Overweight." Accessed June 21, 2019. www.who.int/news-room/fact-sheets/detail/obesity-and-overweight.

[WHO] World Health Organization. 2018c. "The Top 10 Causes of Death." Accessed June 22, 2019. www.who.int/news-room/fact-sheets/detail/the-top-10-causes-of-death.

[WHO] World Health Organization. 2019. "Tuberculosis (TB)." Accessed September 1, 2019. www.who.int/tb/country/data/download/en/.

[WHO and FAO] World Health Organization, and Food and Agricultural Organization of the United Nations. 2006. *Guidelines on Food Fortification with Micronutrients.* France: World Health Organization.

4

ENVIRONMENTAL CHANGE AND EMERGING INFECTIOUS DISEASES

Human activities have had such a profound influence on the earth's ecosystems that in 2008 a group of geologists proposed the term **Anthropocene**—or human epoch—to convey the idea that humans are now one of the dominant forces influencing the earth's natural processes. Some of the key issues identified include climate change, biodiversity loss, the ubiquity of plastics in the environment, increased nitrogen and phosphorous in the soil, and particulate pollution in the atmosphere (Carrington 2016). In addition to potentially devastating consequences for the earth's natural ecosystems, all these factors affect human health and well-being. In this chapter, we discuss how human disruptions to natural ecosystems influence patterns of *infectious disease*, with a focus on the processes of climate change and ecosystem disruption. We then consider a series of case studies in the context of these changing environments. In the following chapter, we extend our discussion of environmental change by considering how human activities are also increasing our exposure to *non-living agents* of disease (geogens).

ENVIRONMENTAL CHANGE AND DISEASE PROLIFERATION

Many environmental changes have consequences for disease patterns. After the Indian Ocean tsunami of 2004,

for instance, tetanus infection may have caused hundreds of deaths (Baker 2007). Deaths from a cholera outbreak in the wake of Haiti's 2010 earthquake are thought to have numbered in the thousands (Domonoske 2016). Airborne, waterborne, and vector-borne disease outbreaks often follow disasters, although the threat of disease outbreaks related to environmental hazards may be overestimated (Watson, Gayer, and Connolly 2007). Far more significant to human health are subtler, longer-term changes in the disease landscape associated with ongoing processes of ecological change such as deforestation, climate change, and biodiversity loss. The term "ecodemic" reflects this association between epidemics and ecological change (Walters 2004).

The consequences of environmental change are not always negative—for instance, changes to water courses or vegetation patterns could reduce disease transmission if they make conditions less amenable for disease vectors. Many historic efforts to reduce disease transmission have involved deliberate environmental changes, including draining swamps to prevent malaria and building reservoirs to provide clean drinking water. However, there is strong evidence that the large-scale environmental changes we are experiencing today are likely to result in a net increase in human disease. Human populations have adapted to local conditions over the course of many centuries, and disrupting this stability is likely to have numerous unintended consequences. As Walters (2004, 11–12) states, "Over the past century or more, humans have so disrupted the global environment and its natural cycles that we risk evicting ourselves from our shelter of relative ecological stability. . . . If the upsurge in new diseases is any indication, microscopic predators are taking full advantage of the instability."

Pathogens and vectors possess a reproduction strategy that enables them to rapidly adapt genetically to changing conditions—far outpacing the evolution of more complex animals such as humans. These different approaches to reproduction are referred to as "r" and "K" strategies. Broadly speaking, "r" strategists have short generation times and large numbers of offspring, ensuring that at least some portion of the offspring can cope with a changed environment. "K" strategists, with much longer generation times, fare better in stable conditions because it takes many years for sufficient changes to build up in the gene pool for new adaptations to develop. The net result is that, as "K" strategists, we humans may be creating problems for ourselves if we ignore rapid environmental change.

Environmental change may also contribute to disease if it overwhelms infrastructure such as water treatment systems. In London, for example, the Victorian-era sewage system is under pressure from the rapid increase in London's population that has occurred over the past hundred years. It is now being strained to breaking point by periodic dramatic increases in winter rainfall associated with climate change. Excess storm water pours into storm drains and combines with sewage water bound for treatment plants. The system was originally designed with an overflow feature so that excess sewage and storm water would flow into the River Thames during extreme weather events, rather than seeping up under the streets of London. Unfortunately, "extreme" weather now happens so frequently that the Thames has been dubbed a public health risk because of its repeated tainting with microbe-rich sewage overflow. The city of London is currently investing billions of pounds in a super-sewer, the Tideway Tunnel, to address the problem.

Changes to climate, soils, and vegetation are also likely to cause stressful conditions in many communities. As soil erosion, changes to hydrologic systems, and climate change render traditional ways of life unsustainable, people are forced to move to find new ways to make a living. Many low-lying islands and coastal areas, for example, are becoming inundated as the sea level rises, leading to the abandonment of some coastal communities, such as in the South Pacific and the Southeast US. The **environmental migrants** that result from these ecological evictions may suffer problems with malnutrition (associated with the disruption of traditional agriculture), outbreaks of infectious disease in overcrowded migrant camps or urban areas, diseases of poverty and marginalization, as well as psychological conditions associated with the stress of dislocation.

Indeed, the mental health implications of a rapidly changing world should not be underestimated. The health impacts of increased stress from more frequent environmental disasters, erratic or failing harvests, environmentally induced migration, and disease epidemics are likely to be significant. The psychological strain of adapting to changing conditions may also contribute to conflict over resources. Even in the absence of conflict, research reveals hidden physical and psychological costs associated with human lifestyles that are increasingly divorced from the natural environment (Box 4.1).

Box 4.1

DIVORCE FROM THE NATURAL ENVIRONMENT

An intriguing area of research explores the ways in which human health is being affected by people's increased disconnection from the natural environment. Considerable evidence links the activities that we do outdoors versus indoors with health outcomes. The idea that exercise promotes health is well established, and there is evidence that our increasingly sedentary, indoor lifestyles limit opportunities for exercise. The notion of limiting "screen time" for children, for instance, is partly to encourage children to find other pastimes that include more physical activity.

Other subtler connections between health and the outdoors are being identified as well. For instance, rates of short-sightedness (myopia) have rapidly increased in recent years in affluent countries, doubling in the UK over the past fifty years (McCullough, O'Donoghue, and Saunders 2016). In parts of Asia, myopia is reaching epidemic proportions—80 percent of Singaporean young adults are short-sighted (Wilson 2015). Children are also developing myopia at younger ages (McCullough, O'Donoghue, and Saunders 2016). Conventional knowledge attributed this rise in short-sightedness to eye strain from reading books and screens as populations become more affluent and educated. However, recent work suggests that it may be better explained by a decrease in time spent outdoors. One theory suggests that the need to focus on distant objects when outside may be protective against short-sightedness, others that the brightness or other qualities of outdoor light might help prevent myopia (Landis et al. 2018; Wilson 2015). Regardless, the evidence suggests that getting more time outdoors—where humans evolved as a species—may be important for our bodies' normal functioning.

A further line of research describes mental health benefits of spending time outdoors. Being in or near greenspaces has been associated with lower levels of stress (Ward Thompson et al. 2012), reductions in depression and anxiety (Berman et al. 2012; Beyer et al. 2014), and improvements in cognition among children with attention deficit disorders (Faber Taylor and Kuo 2009), among other benefits (Pearson and Craig 2014). Feeling connected to nature has also been shown to have positive psychological impacts (e.g., Howell et al. 2011; Nisbet, Zelenski, and Murphy 2011). Many commentators have noted that mental health and behavioral disorders have increased, especially among children, as children spend less time outdoors. Although scientists know that "association does not imply causation" (just because two things occur at the same time, it does not mean that one event *caused* the other), causal links are nonetheless worth exploring in an era when wild and green spaces are rapidly diminishing. Kuo (2015) identifies several potential mechanisms that could link time outside with mental health outcomes, including: 1) direct chemical and biological impacts

of compounds found in nature, 2) physiological changes in the body that occur from exposure to natural environments (particularly reductions in stress hormones), and 3) changes in human behavior elicited by the experience of being in the natural environment (for instance, getting better sleep or more exercise). Although the individual impact of each of these factors is probably small, the combination of multiple influences may explain why we often find measurable responses to being in natural areas (Kuo 2015). We must be wary of overstating the role of the natural world in supporting mental health, however, particularly as there remain important gaps in the empirical evidence and some studies have generated conflicting results (Pearson and Craig 2014).

A final area of research suggests that a lack of exposure to microbe-rich environments such as soil and farms may be detrimental to our health if it compromises our ability to host a healthy community of microbes in our bodies. Known as the "**hygiene hypothesis**," the idea is that a lack of exposure to microbes and parasites early in life may lead the body's immune system to overreact to benign substances. As we ruthlessly eradicate microorganisms from our living spaces with antibacterial products and spend more time indoors away from significant microbe communities, many people today are probably exposed to fewer microorganisms than at any other point in our evolutionary history.

Noting a significant increase in allergic conditions such as hay fever in the twentieth century, researchers in the 1980s and '90s began to investigate why allergies appeared to occur more frequently in populations with limited exposure to microorganisms. Early evidence was provided by Strachan (1989), who reported lower rates of hay fever in children from larger families. Strachan argued that exposure to disease in early childhood from unhygienic contact with siblings had a protective effect against the development of hay fever. Further evidence emerged from farms, with studies suggesting that children exposed early in life to farm animals and the microbes they harbor had lower rates of asthma and allergies. This effect has now been observed in studies from all over

continued

Box 4.1 *continued*

the world (Ege 2017). Some researchers theorize that longstanding associations between humans and microorganisms mean that the human immune system has had to moderate its operation in order to co-exist with "friendly" microbes. It is possible, then, that lack of exposure to microbes early in life may lead to the over-stimulation of the immune system when exposed to benign substances such as pollen, leading to an allergic reaction (Alexandre-Silva et al. 2018). This can then cause autoimmune conditions such as Celiac disease, where the body exerts an immune response to a protein found in wheat, attacking the small intestine whenever that protein is eaten.

Today, a large number of variations of the "hygiene hypothesis" are being explored related to different environmental pressures, microbial exposures, and parasite infections (Figure 4.1).

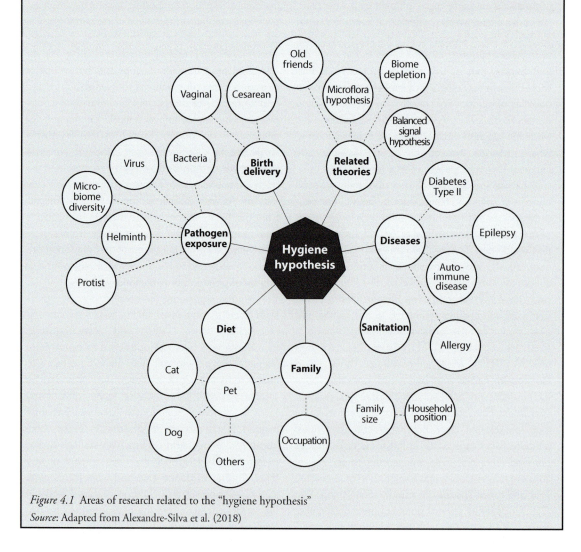

Figure 4.1 Areas of research related to the "hygiene hypothesis"
Source: Adapted from Alexandre-Silva et al. (2018)

For instance, some studies have shown success in treating autoimmune conditions with helminth (worm) infections, although helminth therapies are not yet standard practice (McSorley, Chaye, and Smits 2018). More broadly, the "hygiene hypothesis" remains controversial, perhaps because the basic idea that "being too clean can be harmful to your health" has become overgeneralized and overused. Nonetheless, with perhaps 20 to 40 percent of the global population sensitive to foreign proteins (Alexandre-Silva et al. 2018), the issue is worthy of further investigation.

Global climate change

Climate change is perhaps the greatest single threat to our planet's ecological stability. It is now widely accepted that temperatures are increasing globally, resulting in changes in rainfall patterns, melting ice caps and glaciers, lakes and rivers freezing later and thawing earlier seasonally, warming of the deep ocean, extension of growing seasons, and changes in plant and animal distributions (Epstein 2005; IPCC 2014; IPCC et al. 2007). These and other changes have a wide range of implications for human health, with negative climate-related health impacts likely to outweigh any health benefits during this century (IPCC 2014). Although the complexity of climate systems and climate-human interactions make it difficult to predict how climate change will affect human health in any particular region, the IPCC (IPCC 2014, 713) concludes that the major impacts at the global scale are likely to be:

- Greater risk of injury, disease, and death because of more intense heat waves and fires (very high confidence)
- Increased risk of undernutrition resulting from diminished food production in poor regions (high confidence)
- Consequences for health of lost work capacity and reduced labor productivity in vulnerable populations (high confidence)
- Increased risks of food- and waterborne diseases (very high confidence) and vector-borne diseases (medium confidence)
- Modest reductions in cold-related mortality and morbidity in some areas because of fewer cold extremes (low confidence), geographic shifts in food production, and reduced capacity of disease-carrying vectors resulting from exceedance of thermal thresholds (medium confidence)
- These positive effects will be increasingly outweighed, worldwide, by the magnitude and severity of the negative effects of climate change (high confidence).

The IPCC (2014) notes that the best way to reduce vulnerability to these changes is to implement basic health and sanitation measures and alleviate poverty. We return to these ideas of social vulnerability in the second part of the book. Here, we assess some of the physical impacts of climate change on human health.

At the most basic level, human bodies are stressed by climatic extremes. Heat stress can damage health through dehydration, acute conditions such as heat stroke, or by exacerbating chronic respiratory or cardiovascular conditions (Kovats and Hajat 2008). As a result, heat waves are a significant and often underestimated source of **excess mortality** (deaths above the number we would expect over a particular time period). A severe heat wave in Europe in 2003, with temperatures 10°C (18°F) above average seasonal values and little relief overnight, killed 21,000 to 35,000 people in five countries (Epstein 2005). In India, a 2010 heat wave claimed more than a thousand lives in the city of Ahmedabad, after temperatures soared to 47°C (Sen et al. 2017). Heat waves have been occurring with increasing frequency in subsequent years, with extreme temperatures once again hitting many parts of the Northern hemisphere as recently as July 2019. Although subsequent death tolls have not reached the devastating figures from 2003, partly owing to better planning and response mechanisms, people and animals continue to die in these events.

Most climate change models predict increases in other extreme weather events too, including tropical storms, flash floods, droughts, and wildfires, which may result in increased mortality and morbidity. Both the immediate risk of trauma associated with hazard events and the aftermath of disasters have important implications for health. For instance, flooding could lead to increases in waterborne and vector-borne diseases from disrupted water and sanitation systems and standing water. Wildfires may lead to poor air quality and increased rates of respiratory disease. Storm surges can inundate agricultural areas with saline water, leading to crop failure and malnutrition.

The greatest impact of climate change may not be direct effects from floodwaters or extreme temperatures, however, but the result of numerous indirect impacts of a changing climate (Figure 4.2). For instance, warmer temperatures have been associated with increased incidence of food poisoning (Bentham and Langford 2001) and diarrheal disease (Hashizume et al. 2008) because pathogens grow more quickly in the warmer conditions. Warmer temperatures may also increase the risk of fungal infections in humans. Fungi are adapted to thrive in the normal ambient temperatures of a region. One theory suggests that mammals have been protected to some degree from fungal infections because their bodies are considerably hotter than these environmental temperatures and therefore hostile to the growth of many fungi. Warmer temperatures associated with climate change are likely to select fungal lineages that can tolerate higher temperatures, increasing the number of fungi in the environment that can tolerate temperatures closer to those of the human body (Casadevall, Kontoyiannis, and Robert 2019).

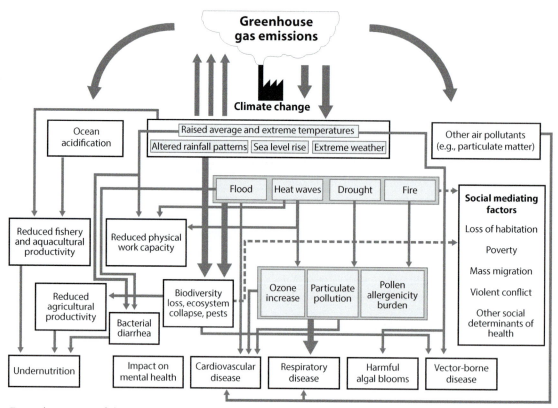

Figure 4.2 Impacts of climate change
Source: Adapted from Watts (2017)

Climate change is also likely to affect water and air quality. Water scarcity may lead to unsafe pathogen levels if sewage effluent contributes an increasing proportion of water to rivers, effectively concentrating the pathogen load (Rechenburg and Kistemann 2009). Changes in air quality are predicted to increase rates of respiratory and allergenic diseases associated with more particulate matter in the air and chemical changes in the atmosphere, such as an increase in ground-level ozone (IPCC et al. 2007). Climate change may increase concentrations of particulate matter through changes in wind patterns, dust from dry soils, and ash from wildfires (Epstein 2005; IPCC et al. 2007). Greater pollen production as plants grow more vigorously may also lead to an increase in airborne allergens (Beggs, Walczyk, and Health 2008). Some researchers believe climate-related factors may be partly responsible for dramatic increases in asthma over the past two decades, as well as outbreaks of thunderstorm asthma (see Box 2.1).

Researchers have also paid considerable attention to the impact of climate change on vector-borne diseases (e.g., Wimberly et al. 2008; Gubler et al. 2001). Warmer temperatures may increase the range of many vector species and prolong the season over which disease can be transmitted. Rising temperatures can also increase the reproduction rates of vectors and pathogens, shorten the time needed for vectors and pathogens to mature, and even increase the number of blood meals a vector is likely to take (Epstein 2005), although the precise impact of temperature will vary from vector to vector (Gubler et al. 2001). Climatic fluctuations associated with El Niño events have already been linked to malaria epidemics in parts of South America, providing an indication of the significance of climate to health. The relationship between El Niño and malaria varies across space (Gagñon, Smoyer-Tomic, and Bush 2002), demonstrating that climate-related changes will likely be environment specific. Broadly speaking, however, longer-term changes in temperature and precipitation patterns will alter vegetation patterns and other micro-environmental characteristics, changing the distribution of vectors, hosts, and reservoirs of disease. Human movements may exacerbate the situation by transferring pathogens and vectors into new habitats more rapidly than they would spread on their own. Fiji, for instance, experienced one

of its largest outbreaks of dengue fever in 2013–2014, which researchers linked to climate change and increased urbanization on the island, combined with increasingly mobile populations introducing new strains of disease (Moceituba and Tsang 2015). These cases illustrate the negative synergisms that may emerge when processes of social and environmental change occur together (see also Leichenko and O'Brien 2008).

Other factors associated with climate change are making human activities such as agriculture and collecting drinking water more challenging. The impact of crop failure or the loss of a local water source will be especially critical for people in poverty, who have less **resilience** to changing conditions. For instance, communities where agriculture is rain-fed are at a greater risk from erratic rainfall than farmers with access to irrigation. Worldwide, approximately 400 million farmers and their dependents rely on rain-fed agriculture, accounting for 60 percent of farmland in South Asia, 90 percent in Latin America, and 95 percent in sub-Saharan Africa (Ramanathan, Seddon, and Victor 2016). These communities will be on the front lines as climate changes.

Evidence such as this has made climate change a focal point in discussions of **environmental justice**, as scholars observe that those societies most responsible for climate change (industrialized nations) will be less affected by its impacts than poorer countries. Many communities in the Sahel, for instance, already experience erratic rainfall, villages in Bhutan are suffering an increased frequency of landslides and flash floods associated with warming temperatures, while receding glaciers and diminishing snowfall are leading to a scarcity of drinking water in Nepal (Sen et al. 2017). Increasing temperatures will also make manual work outdoors more dangerous in hot summer seasons (Smith et al. 2016), again disproportionately affecting poorer individuals.

Climate change may also have an impact on mental health. Cunsolo-Willox et al. (2013, 255) note that the social, economic, and environmental disruptions caused by climate change "may cause increased incidences and prevalence of mental health issues, emotional responses, and large-scale sociopsychological changes." The detrimental impact of more frequent environmental disasters on people's mental well-being has already been well documented (Page and Howard 2010), but there is evidence for other connections between environment and mental health too. For example, Obradovich et al. (2018) found that hotter temperatures, increased precipitation, multiyear warming, and exposure to tropical cyclones had negative mental health implications in the US. Changing climate may have particularly profound implications on the mental health of indigenous communities by disrupting traditional lifeways. In one study of an Inuit community in Northern Canada, for instance, people reported that environmental and climate-related changes were associated with increased family stress, a greater likelihood of using alcohol and drugs, and increased potential for suicidal ideation, as well as amplifying other mental health stressors (Cunsolo Willox et al. 2013). Poor mental health may also be associated with the *social* upheaval that is predicted to accompany climate change, such as mass migration and regional economic collapse (Page and Howard 2010).

Finally, climate change may lead to conflicts over resources. Water scarcity, food shortages, an increase in hazard events, and climate change–initiated migrations could all result in conflict. Fragile states and regions with ongoing conflicts are especially likely to be destabilized by climate change (Sida 2018). Yemen, for instance, is currently in the throes of a civil war. Yemenis already import more than 80 percent of their food, fuel, and water, and experts predict that Sanaa, Yemen, may be the world's first capital city to run out of water as climate change makes the region even more inhospitable (Erickson 2017), providing multiple potential flashpoints in the future.

Biodiversity and ecosystem disruption

Perhaps the most obvious way in which biodiversity affects human disease is the provision of an abundant pool of pathogens. In an era of growing human populations and widespread ecological disruption, **spillover events** (where a human contracts a pathogen that normally infects another animal) are likely to occur at higher rates than previously. **Zoonotic** pathogens accounted for the majority of emerging infectious disease events between 1940 and 2004 (Jones et al. 2008), and most pandemics originate from zoonotic viruses (Morse et al. 2012). Jones et al. (2008, 990) found that

60 percent of emerging infectious disease events were zoonotic, 72 percent of which originated from wildlife (as opposed to domestic animals). In recent years, SARS (sudden acute respiratory syndrome) was probably contracted by people from the Himalayan palm civet, although the natural reservoir of the pathogen is likely a bat (Li et al. 2005). Several bird flu strains have moved into the human population from wild birds, and a new virus emerging in the Middle East—Middle East respiratory syndrome coronavirus (MERS-CoV), may have been contracted from camels or bats (Walters 2014). As is common for diseases that have only recently entered the human population, most of these diseases have high **case fatality rates** as people have little resistance to them. For instance, MERS-CoV was fatal for about 35 percent of cases. Fortunately, it has not yet spread widely—most cases are restricted to Saudi Arabia—but the disease had caused more than two thousand laboratory-confirmed cases by 2018 (WHO 2018a).

The contact between people and wildlands that accompanies landscape conversion activities such as logging and agricultural expansion has led to more frequent interactions between people and zoonotic pathogens. As wildlands have been made more accessible by logging and mining roads, the trade in bushmeat (wild animals killed for meat) has also increased, providing abundant opportunities for humans to come into close contact with wild animals. The bushmeat trade generates contact with animal fluids through butchering, bites, and meat consumption (Walters 2014), and has been implicated in outbreaks of Ebola, anthrax, SARS, and HIV (Wolfe et al. 2005). HIV may have been contracted from a chimpanzee butchered for meat in the forests of West Africa, as chimps are known to carry a virus closely related to HIV (Walters 2014). Thankfully, most **spillover events** are limited in scope because zoonotic pathogens are generally not adapted to spread efficiently from person to person. In a few rare cases, however, zoonotic pathogens evolve in ways that allow direct human-to-human transmission, leading to the potential for a human epidemic, as occurred in the case of HIV. Figure 4.3 illustrates some of the key factors facilitating the spread of zoonotic diseases into the human population.

Nipah virus is a disease in the early stages of moving from animal to human populations. Fruit bats act

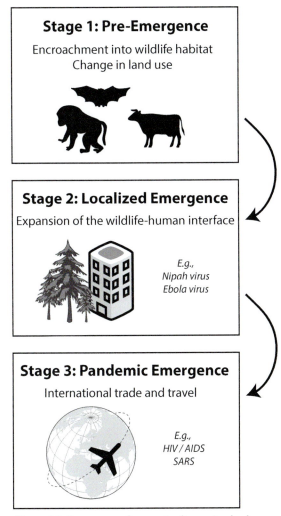

Stage 1: Pre-Emergence

Encroachment into wildlife habitat
Change in land use

Stage 2: Localized Emergence

Expansion of the wildlife-human interface

E.g.,
Nipah virus
Ebola virus

Stage 3: Pandemic Emergence

International trade and travel

E.g.,
HIV / AIDS
SARS

Figure 4.3 Emergence of zoonotic disease in the human population
Source: Adapted from Morse et al. (2012)

as a wild reservoir of the virus, and antibody analysis suggests that it is common among South and Southeast Asian bat populations. Humans are thought to become infected through close contact with bats or bat secretions. Although many human infections are mild or asymptomatic, the disease can lead to encephalitis and death—among symptomatic cases an average of 50 to 75 percent of cases are fatal (Epstein et al. 2006; Walters

2014; WHO 2018a). The first recognized outbreak occurred in Malaysia in 1998–1999, when domestic pigs became infected from wild fruit bats and then passed the disease on to humans. A combination of drought, wildfires, and deforestation caused the natural fruit crop to fail, forcing the bats into closer contact with people (Walters 2014). Subsequent outbreaks in Bangladesh may have been related to consumption of raw date palm sap contaminated with the pathogen by fruit bats (WHO 2018c). At least 15 outbreaks of Nipah virus have now occurred in the human population of Bangladesh and eastern India (WHO 2018c). Recent spillover events of Nipah virus in South and Southeast Asia have begun to show evidence of human-to-human transmission (WHO 2018c). Although the disease has not yet shown epidemic transmission patterns, evidence of human-to-human transmission makes this a possibility, and the disease is now one of several being carefully monitored for their potential to cause epidemic disease.

Evidence such as this has led scientists to promote the idea of "One Health," which suggests that human survival is so intricately interwoven with animal populations and the environment that we must tackle human-animal-environment systems as an integrated whole (Atlas and Maloy 2014). Organizations such as the World Health Organization (WHO) and US Centers for Disease Control (CDC) have leveraged the One Health initiative to encourage communication and collaboration among physicians, veterinarians, and ecologists.

The One Health perspective promotes the idea that any disruption to the complex of ecological relationships can have significant implications for human health and well-being. One pertinent meta-study concluded that loss of biodiversity was associated with outbreaks of infectious disease across a variety of ecosystems (Keesing et al. 2010). The authors cited multiple lines of evidence to support their argument, including evidence that low bird diversity leads to a higher incidence of West Nile encephalitis in humans (Swaddle and Calos 2008); that lower diversity among small mammals increases the prevalence of hantavirus among mammal hosts, producing a greater risk of human infections (e.g., Dizney and Ruedas 2009); and that rates of schistosomiasis infection in snails may be higher in less diverse aquatic communities (Johnson et al. 2009; Keesing et al. 2010).

Keesing et al. (2010) propose several mechanisms to explain this connection between biodiversity and infectious disease. First, decreasing biodiversity in a community can lead to greater success of a vector or reservoir species, increasing incidence of a disease. Second, encounter rates between pathogens and hosts may be increased by the loss of other species from an ecosystem. Third, those species that tend to thrive in circumstances of biodiversity loss—sometimes referred to as "weedy" species—may be the very ones most likely to host and spread disease. Such a connection has already been made in plants and may also occur in vertebrates. Finally, loss of biodiversity from a host's own **microbiome**—the microbial communities that we host in our own bodies—may allow disease-causing organisms to establish themselves, as has been noted with *Clostridium difficile* infections. Box 4.2 describes the potential role of the human microbiome with respect to infectious disease.

Box 4.2

THE HUMAN MICROBIOME

Since the advent of germ theory, we have been aware of the wide variety of microorganisms with which we share the earth. Our focus has traditionally been on organisms with the potential to do us harm—indeed, the very definition of a "germ" has developed over time from being simply a small organism to one that causes disease. The sanitation revolution that accompanied this understanding of microorganisms as agents of disease has undoubtedly saved millions of lives, but today there is increasing recognition of the large number of benign and beneficial microbes that share our earth and even our bodies. These ideas are consistent with the "**hygiene hypothesis**" discussed in Box 4.1, which suggests that people may benefit from increased exposure to benign microorganisms. This evolving understanding of the role of microorganisms raises questions about the unintended consequences of the rapid changes that we are bringing to our microbial environment.

The term "**human microbiome**" was coined to reflect the idea that we are host to trillions of microorganisms—including viruses, bacteria, and fungi—that live **symbiotically** with us. These microbes reside on the skin, in body cavities, and particularly in the gut. Scientists have found that these microorganisms help us metabolize food, create certain vitamins, and fill **ecological niches** that might otherwise support harmful microbes. The use of **probiotics** is based on the idea that we can boost certain beneficial microbe communities by artificially seeding them in our gut from a capsule or tablet, although the effectiveness of probiotics in healthy individuals, and even for the treatment of specific conditions, remains controversial (e.g., Crow et al. 2015).

Some of the best evidence for the importance of a healthy microbiome comes instead from what happens when we disrupt our microbial community with antibiotics. Although antibiotics are prescribed to kill a harmful strain of bacteria, they also kill many benign and beneficial strains. This is most problematic when a long and powerful course of antibiotics is prescribed for a serious infection, with potentially devastating consequences for the patient's communities of beneficial bacteria. The gut, with its dense and diverse microbial flora, is often particularly badly affected. As the gut is stripped of its benign bacteria, ecological niches open up, which can then be filled by harmful bacteria.

Clostridium difficile is one such opportunistic organism that is known to cause disease in patients after competing organisms have been removed. *C. difficile* causes diarrhea and stomach cramps and has traditionally required more courses of antibiotics to enable recovery. If these antibiotics fail, however, the infection can become chronic or recurrent. *C. difficile* has become a major problem in affluent regions with high antibiotic usage—particularly the

European Union and the US, but cases are now also increasingly occurring in East Asia, Australia, and New Zealand.

The significance of the human **microbiome** in the case of *C. difficile* is emphasized by the fact that an infection can be treated successfully by approaches that focus on restoring a healthy microbial community in the gut, rather than attempting to kill the *C. difficile* bacteria themselves. Fecal microbial transplants, in which fecal matter from a healthy patient is transplanted to a sick patient either directly into the colon or orally, have proved effective at curing otherwise challenging *C. difficile* infections (Gupta and Khanna 2017). The donor's mostly benign microbe species in the transplanted fecal matter quickly colonize the patient's gut where they can outcompete the pathogen, causing symptoms to recede, often within 24 to 48 hours. A recent **meta-analysis** found an average 89 percent cure rate from several recent **observational studies**, and an 80 to 90 percent cure rate from two recent **randomized controlled trials**, indicating that the method is very effective (Crow et al. 2015). Few adverse effects of the procedure have been reported, despite concerns that it could spread other infectious agents (*ibid.*). The success of this treatment has led some researchers to suggest that fecal microbial transplants may soon become a standard of care for recurrent *C. difficile* infections (Leffler and Lamont 2015).

This kind of evidence helps to explains why ecosystem disruption and biodiversity loss are often associated with disease outbreaks. Deforestation and agricultural expansion lead to dramatic changes in vegetation patterns, often with implications for health. In the Amazon, for instance, outbreaks of malaria are known to be associated with the clearing of land for agriculture and with logging and mining camps. Mosquitos of the *Anopholes* genus, the main vector for malaria, appear to benefit from the disturbance of newly cleared land, while the introduction of the pathogen is facilitated by high human mobility in these contexts. The situation is complex, however, as the incidence of malaria often rises after initial clearing of the land but then typically subsides again after agriculture has been established (Fraser 2010).

Lyme disease: lessons in ecological simplification

Lyme disease provides a good example of the impact of ecosystem disruption on disease. Lyme disease is the most common vector-borne disease in the US, with approximately thirty thousand cases reported to the CDC every year, although underreporting may mean that the actual figure is up to ten times higher (CDC 2018b). Most of these cases occur in the US Northeast and the northern Midwest, although incidence in these regions appears to be stabilizing and the disease spreading to neighboring states (Schwartz et al. 2017) (Figure 4.4). Thousands of new cases occur each year in Europe and Asia too, where underreporting may be significant because these regions have far less experience with the disease than the US. In the UK, for instance, Lyme disease is not a reportable disease, and many general practitioners have only a limited familiarity with the infection; one expert suggests that only one third of cases may be reported (Mills 2018). Despite problems with reporting, research suggests there has been a genuine increase in Lyme disease in the US and many other countries over the past twenty to thirty years. This rise in Lyme disease incidence is often attributed to the impacts of losing species from an ecosystem, or **ecosystem simplification**, as a result of agricultural development and urbanization.

The Lyme disease cycle involves different hosts and vectors in different parts of the world, and so here we focus on just one region. In North America, Lyme disease is caused by the bacterium *Borrelia burgdorferi*, transmitted to humans through the bite of infected ticks of the genus *Ixodes*. These tick species have specific habitat and micro-climate requirements (Kilpatrick et al. 2017), which explains the high variability in tick abundance that can be found over a small area. *Ixodes scapularis* is one of the two main North American vectors. It passes through four main developmental stages

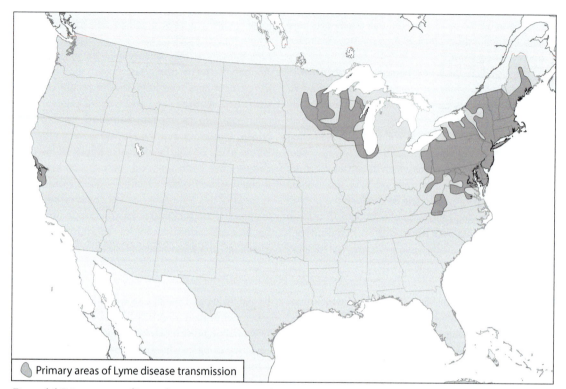

Figure 4.4 Primary areas of Lyme disease transmission in the US
Source: Adapted from CDC (2019)

in its two-year life cycle (van Buskirk and Ostfeld 1995). Adult females lay their eggs in the spring, and the resulting larvae seek a host, typically a bird or small mammal from which they take a single blood meal during the summer. The larvae then develop into nymphs and overwinter. The following late spring or early summer, the tick seeks a second blood meal before developing into its final adult form. The adult seeks a final host in the autumn, often a white-tailed deer. The adults mate on the deer and then the females drop off, laying their eggs the following spring (Ostfeld and Keesing 2000).

Larval ticks obtain the pathogen from one of their vertebrate hosts and remain infectious for the rest of their lives. If one of the tick's subsequent meals is a human, that individual may become infected and develop the disease. Humans are probably most commonly infected by *larval* ticks because the adult tick is large enough to

be noticed and removed before the 18 to 48 hours of feeding required for the infection to be transmitted (Kilpatrick et al. 2017). Symptoms of Lyme disease initially include fever, headache, fatigue, and a distinctive "bulls-eye" rash. If untreated, infection can affect the joints, heart, and nervous system, developing into conditions such as arthritis, meningitis, heart arrhythmia, and problems with memory and concentration (CDC no date).

Ecosystem simplification is associated with increased incidence of the disease for several reasons. Different animal species have different likelihoods of transmitting the pathogen to the ticks that feed on them, and so the combination of species in an ecosystem can influence the proportion of ticks infected with the pathogen. In the eastern US, for instance, the white-footed mouse (*Peromyscus leucopus*) is the principal natural reservoir of the pathogen and the most effective transmitter of the disease: 40 to 80 percent of the ticks that feed on infected white-footed mice acquire the bacterium. Several other hosts serve as moderately effective reservoirs (e.g., eastern chipmunks, *Tamias striatus*), but most other species are unable to transmit the disease effectively to ticks (Mather 1993).

Differences in the abundance of the disease reservoir species are therefore probably correlated with Lyme disease risk. The white-footed mouse is a generalist in terms of habitat and diet and, as such, often becomes abundant where other species have been removed. Because infected white-footed mice efficiently transmit the pathogen to ticks, Lyme disease risk has been theorized to increase in the simplified ecosystems where the mice thrive. Both computer simulations (van Buskirk and Ostfeld 1995) and studies of actual communities (Ostfeld and Keesing 2000) have contributed evidence that supports this idea. Other studies in eastern and central North America suggest that white-footed mouse populations are higher in small, isolated stands of forest embedded in agricultural or urban landscapes than in continuous forest (Lewellen and Vessey 1998; Nupp and Swihart 1998). This may result from limited predator populations and less competition from other small mammal species, meaning that there are more white-footed mice in these peri-urban settings, which in turn presents a higher risk of Lyme disease to humans.

Rapid suburbanization over the past fifty years may thereby have indirectly increased incidence of the disease. The significance of the white-footed mouse to Lyme disease transmission has now become so well established that they are targeted in some Lyme disease control programs. One group of researchers from MIT has proposed releasing genetically engineered, Lyme-resistant white-footed mice to control the disease. Nantucket and Martha's Vineyard—two islands in the US Northeast with some of the highest rates of Lyme disease in the world—have been proposed as experimental test sites for the project because the spread of the genetically engineered mice could be controlled on the islands. The hope is that a community of the genetically engineered mice will become established, passing immunity to subsequent generations until most of the islands' mice are unable to transmit the disease to ticks and ultimately to humans (MIT Media Lab no date).

Ecological relationships with respect to disease transmission are often very complicated, as illustrated by efforts to link Lyme disease incidence to habitat change. It has been hypothesized that increasing deer populations in the US Northeast, associated with expanding forest cover and reduced predator numbers, may be increasing incidence of the disease as deer provide blood meals to ticks. However, there appears to be little clear evidence of an association between Lyme disease incidence and deer abundance. One possible explanation is that the relationship between tick density and deer abundance is nonlinear, with tick abundance initially increasing with deer abundance but then reaching a saturation point at moderate deer densities (Kilpatrick et al. 2017). The role of forest cover and habitat fragmentation is equally complex. Fragmented landscapes are associated with high deer numbers, which might increase tick abundance, while intact forests may have more favorable microclimates and leaf litter conditions for ticks. As such, we might expect to find high tick numbers in both fragmented ecosystems and intact forests, complicating our efforts to draw robust conclusions about which ecosystems are conducive to the spread of the disease (Kilpatrick et al. 2017). In summary, declining biodiversity and human encroachment into forested environments may be responsible for increasing human exposure to Lyme disease, with forest fragmentation, reforestation, hunting, and climate change all factors in Lyme disease ecology. The precise ecological interactions that support or suppress the transmission of Lyme disease are incredibly complex, however, and highly place-specific.

EMERGENT AND RESURGENT INFECTIOUS DISEASES

Although the majority of deaths worldwide today are from non-communicable causes, infectious diseases such as malaria, HIV/AIDS, lower respiratory infections, and diarrheal diseases continue to be leading causes of death in children under age 5, and infectious diseases remain a significant disease burden for all age groups (Ritchie and Roser 2018). In some countries, infectious disease is still the primary cause of death; for instance, diarrheal disease is the leading cause of death in Kenya and HIV/

AIDS in Botswana and South Africa (*ibid.*). Globally, infectious disease may still cause one out of every three deaths (Walters 2014, xvii).

The late twentieth and early twenty-first centuries have also witnessed the emergence of a variety of *new* infectious diseases. It must be acknowledged that we may be more aware of emerging infectious disease events than we have been in the past owing to better surveillance, however. Indeed, the very notion of an **emerging infectious disease** is culturally situated as diseases newly recognized in the West may have circulated in remote or marginalized populations for years. However, many researchers have concluded that there has been a real increase in the number of new diseases over the past fifty years. In one global analysis of trends in emerging infectious diseases, Jones et al. (2008) report the emergence of 335 infectious diseases between 1940 and 2004. This list includes diseases caused by pathogens completely new to human populations (e.g., HIV-1), pathogens that have probably been present in human populations for some time but have recently increased in incidence (e.g., Lyme disease), and newly evolved pathogen strains (e.g., multi-drug resistant tuberculosis). Bacterial and rickettsial pathogens were responsible for 54.3 percent of emerging infectious disease events because of the large number of antibiotic-resistant strains that have emerged over this time (Jones et al. 2008, 990). As discussed earlier, zoonoses were also very significant in causing disease, with many diseases emerging from contact with wild or domestic animals. After controlling for the potential bias of improved surveillance and reporting of disease, the study confirmed a steady increase in the number of emerging infectious disease events reported per decade from the 1940s to a peak in the 1980s.

Geographic approaches can help us to understand the underlying causes behind emerging infectious diseases. We have already noted that environmental change has altered the range and incidence of many diseases. Additionally, increased contact between human and wild animal species through human encroachment into remote areas and trade in wild animals provides abundant opportunities for cross-species transfers of pathogens (Mayer 2000). Cultural changes can also alter disease patterns. For example, changing agricultural practices have contributed to emerging diseases, including bovine

spongiform encephalitis (BSE) ("mad-cow disease") and avian flu. The decline in funding and political will for vector control programs in the second half of the twentieth century has led to renewed outbreaks of malaria and dengue fever. Declining vaccination coverage in some countries has been significant in regional outbreaks of measles, pertussis, and diphtheria. The speed and volume of human movement around our globalizing world spreads pathogens more quickly than ever before.

As Farmer (1996) points out, the political and economic contexts in which these changes occur is critically important and yet frequently overlooked. From this perspective, factors such as poverty and political turmoil underlie many changing disease patterns. The HIV/AIDS epidemic has, for instance, caused far more deaths and disease in the low-income nations of sub-Saharan Africa than in the US, and is often found at higher rates in poor communities than affluent ones. Similarly, political and economic distress following the collapse of communism probably underlaid the resurgence of diphtheria in Eastern Europe in the 1990s. Consideration of these structural aspects of infectious diseases is therefore warranted, as we illustrate with the subsequent case studies.

The very notion of an "emerging" infectious disease is influenced by particular constellations of power, with the affluent Global North focusing attention and research funding on infections "emerging" from the Global South into the broader global community, while concurrently neglecting other diseases that have plagued the Global South for generations but rarely threaten communities in the North. Ebola and Zika, for instance—diseases with the potential to spread globally—have unlocked abundant research dollars and gained the attention of the popular media. In contrast, many tropical diseases that could easily be prevented with basic hygiene and sanitation, but pose little threat to the West, struggle for funding, even as thousands of people from poor communities suffer. These **neglected tropical diseases**—including Chagas disease, leishmaniasis, Buruli ulcer, and schistosomiasis—collectively affect more than one billion people and cost developing economies billions of dollars every year in lost productivity (WHO 2018b).

A spatial approach to health can offer additional critical insights. Figure 4.5 shows maps of the distribution of different types of emerging infectious disease events

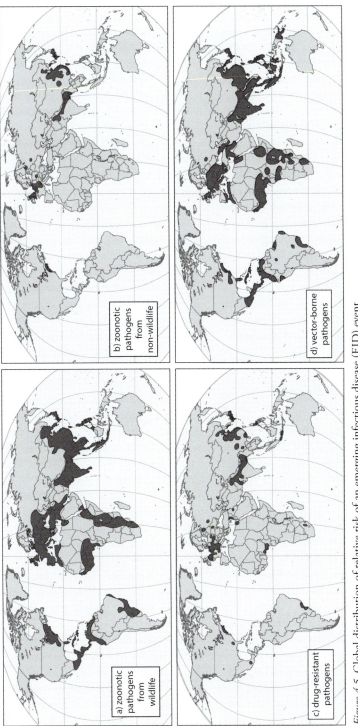

Figure 4.5 Global distribution of relative risk of an emerging infectious disease (EID) event

Source: Adapted with permission from Macmillan Publishers Ltd: [*Nature*] (Jones, K. et al. "Global trends in emerging infectious diseases." *Nature* 451(7181): 990–94) © 2008

"Maps are derived for [emerging infectious disease] events caused by *a,* zoonotic pathogens from wildlife, *b,* zoonotic pathogens from non-wildlife, *c,* drug-resistant pathogens and *d,* vector-borne pathogens. The relative risk is . . . categorized by standard deviations from the mean and mapped on a linear scale from green (lower values) to red (higher values)" (Jones et al. 2008, 993).

from Jones et al.'s (2008) study, revealing several spatial patterns. First, human population density significantly predicts events in all categories, even after controlling for **reporting bias**. This can be seen clearly in the dense populations of North India and East China, which show up as hotspots of disease emergence on all four maps. Second, Jones et al. (2008, 991) argue that "disease emergence is largely a product of anthropogenic and demographic changes, and is a hidden 'cost' of human economic development." For instance, zoonotic pathogens from non-wildlife sources tend to emerge in Europe and East Asia associated with intensive agricultural practices and dense human populations. By contrast, pathogens from wildlife sources and vector-borne diseases are likely to emerge either in regions with high-density human populations, or in areas where humans have significant contact with wild animal species such as in sub-Saharan Africa. Overall, Jones et al. (2008, 991) hypothesize that "socioeconomic drivers (such as human population density, antibiotic drug use and agricultural practices) are major determinants of the spatial distribution of [emerging infectious disease] events."

Modeling the spread of disease: diffusion

Spatial methods are useful in the study of EIDs because they can be used to model the spread of disease. **Diffusion** refers to the spread of a phenomenon across space. Geographers have built diffusion models to study a diverse range of geographic phenomena such as population, language, and cultural practices. Health geographers are especially interested in studying the spread of communicable disease and medical innovations.

There are two main types of diffusion: **expansion** and **relocation diffusion**. **Expansion diffusion** refers to the spread of a phenomenon to neighboring areas (Figure 4.6a). Any disease that spreads through direct transmission is likely to diffuse in this way. For instance, we could model the spread of bubonic plague across Europe in the Middle Ages using an expansion diffusion model. At a finer scale, we could use expansion diffusion to explain the spread of head lice among pupils at a school. These examples would also be considered **contagious diffusion**, which occurs when a phenomenon

spreads outwards in a uniform way. Expansion diffusion does not always exhibit a uniform pattern, however. **Hierarchical diffusion** refers to a type of expansion diffusion in which the phenomenon moves according to a hierarchy, such as a settlement hierarchy. For example, a disease might initially spread between large cities, before eventually moving to smaller towns, and finally to rural areas (Figure 4.6b). Many infectious diseases, as well as other cultural and social phenomena, spread in this hierarchical pattern because there is a greater exchange of people and goods between larger settlements than smaller ones. Additionally, cities often have conditions conducive to the spread of disease: dense human populations, numerous public spaces where people interact, and abundant waste that can act as a source of disease. The anonymity of city life has even been cited as a factor fostering disease in urban areas. For instance, sexually transmitted diseases and diseases associated with illicit practices such as intravenous drug use usually spread more rapidly in places where people are distanced from the strict social rules and informal surveillance that typify village life. These factors contribute to the so-called **urban penalty** that compromises health in cities.

The initial diffusion of HIV across the US exemplifies hierarchical diffusion patterns. In the early 1980s, most cases were identified in a few large cities such as New York, San Francisco, and Los Angeles. By the late 1980s, cases had spread to many urban areas and some smaller towns near them. By the late 1990s, cases were spread across settlements of a wide variety of sizes and across much of the country. Similar diffusion patterns were recorded across much of sub-Saharan Africa (Gould 1993) during the early part of the epidemic there. More recent research has identified situations that do not seem to follow this pattern, however, illustrating the place-specific nature of disease patterns. For instance, Oppong (1998) reports that HIV/AIDS rates in rural areas were actually higher than in urban areas in the early course of Ghana's epidemic.

Relocation diffusion refers to the introduction of a phenomenon to a location outside its current range (Figure 4.6c). This often occurs when a community migrates from one location to another, bringing specific cultural practices and diseases with them. The introduction of smallpox during the conquest of the

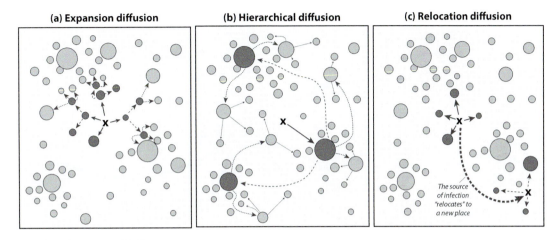

Figure 4.6 Three types of diffusion

The "x" symbolizes the original source of diffusion, and the circles symbolize population settlements; larger circles represent larger settlements. Arrows represent the movement of a pathogen or other phenomenon from one settlement to another.

New World is one example. Similarly, cholera outbreaks often began in port cities where ships introduced the cholera pathogen, either in a human host or in ballast water. An epidemic of cholera that devastated parts of Latin America in 1991 may have been initiated by the release of the cholera pathogen off the coast of Peru in ballast water from a ship arriving from Asia (Mayer 2000).

The potential for pathogens to spread by relocation diffusion has dramatically increased with the development of air transportation. Traditional modes of transport such as sailing boats and steam ships were slow enough that infected individuals on board had time to develop symptoms and be isolated before disembarking and releasing the pathogen into another community. By contrast, air transportation moves people so quickly that a pathogen can be transported around the world in a matter of hours, before an infected individual even feels sick. In 2003, sudden acute respiratory syndrome (SARS) quickly spread with air passengers from China, leading to isolated outbreaks as far away as Canada (Figure 4.7). In response to the threat from SARS, some airports set up thermal cameras to identify passengers arriving with a fever (Ng 2005), but this approach

cannot detect asymptomatic cases of disease and has not been widely adopted.

Vectors, including pesticide-resistant strains, can also be transported on airplanes, potentially introducing pathogens to a new area or leading to the establishment of new vector populations (Masterton and Green 1991). One study found that mosquitos could survive a nine-hour journey in the wheel bays of aircraft despite external temperatures plummeting as low as –54°C (–65°F) (Russell 1986). The phenomenon of "airport malaria," where individuals contract malaria at or near airports well outside the normal range of the disease, has been documented around airports in London, Geneva, and Detroit, among others (Mayer 2000). Multi-drug resistant strains of pathogens have also been transported with travelers, as occurred with the introduction of a strain of "super-gonorrhea" to the UK in 2018 (CDC 2018a; Gallagher 2018).

Pathogens and vectors can also be carried with transported goods, making the dramatic increase in global trade that has accompanied globalization another contributory factor to the spread of disease. The vector that causes dengue fever is renowned for traveling great distances, often surviving as larvae in

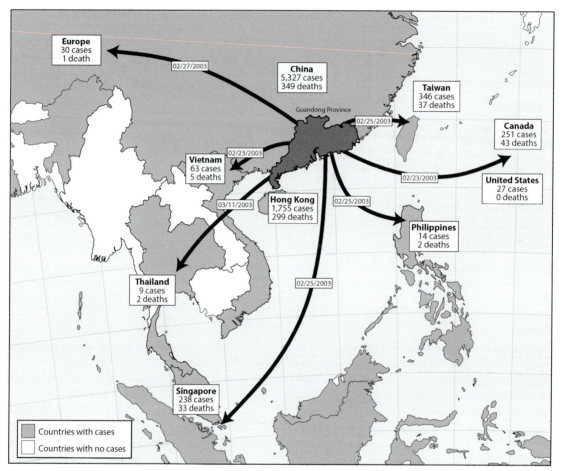

Figure 4.7 Spread of SARS, 2002–2003
Source: Data from WHO (2003a)

the small pools of water that collect inside tires during shipping (Pliego Pliego et al. 2018). Outbreaks of food poisoning have been traced to shipments of food arriving from abroad, particularly as consumer demand for a greater variety of food products and more out-of-season produce has increased food imports in many affluent countries. As Mayer (2000) points out, Westerners are often extremely cautious about what they eat when traveling and yet do not realize that many of the foods on their supermarket shelves are from regions of the world where they would avoid touching the salad! One recent CDC-affiliated assessment of foodborne disease in the US noted an increasing (although small) risk from imported foods, particularly seafood and produce, with the number of foodborne outbreaks associated with imported foods rising from an average of three per year during the period 1996–2000 to an average of 18 per year from 2009–2014 (Gould et al. 2017). Although food from Latin America, the Caribbean, and Asia was responsible for the largest number of outbreaks, Canadian foodstuffs were also responsible for multiple outbreaks, indicating that problems with the global food chain are not confined to the Global South (*ibid.*).

CASE STUDIES OF EMERGING AND RESURGENT INFECTIOUS DISEASES

Recognizing that most emerging and resurging infectious disease events develop from a combination of contributory factors, we now turn to two case studies, discussing the combination of ecological and social drivers that have led to increased incidence of the disease.

Zika virus

Zika is a flavivirus transmitted by mosquitos, primarily from the genus *Aedes*. Once a relatively obscure flavivirus (other flaviviruses cause dengue fever, yellow fever, and West Nile fever), Zika gained widespread media attention when it suddenly began to show epidemic activity and cause abnormal development in babies exposed in utero. However, most human infections are asymptomatic or cause only mild symptoms, including fever, rash, muscle and joint pain, and headache (WHO 2018e).

The virus was first identified in 1947 in monkeys in the Zika forest in Uganda during research on yellow fever (Weaver et al. 2016). A handful of isolated cases of human disease were reported across the tropics between the 1960s and 1980s, although **serosurveys** (tests for antibodies to particular pathogens in the blood) suggest that the pathogen may actually have been widespread in tropical Africa and Asia at this time (Musso and Gubler 2016; Weaver et al. 2016). The true extent of the disease is difficult to assess, however, because many Zika infections have probably been misdiagnosed as other tropical diseases, and even serological results are not completely clear, owing to the circulation of a variety of closely-related flaviviruses in the tropics (Gubler, Vasilakis, and Musso 2017; Weaver et al. 2016). Nonetheless, evidence suggests that Zika virus has circulated among humans, animals, and mosquitos for at least seventy years in Africa and Asia (Gubler, Vasilakis, and Musso 2017; Weaver et al. 2016). No epidemic activity was recorded for most of this period, however.

Despite evidence of human cases of Zika dating back to the 1960s, it was only in 2007 that epidemic activity was recorded for Zika (Figure 4.8). The first outbreak occurred on the island of Yap in the South Pacific, followed by sizeable outbreaks on other Pacific islands in subsequent years (WHO 2018e). Although most of these cases were mild, the disease began to be associated with neurological conditions in some patients, including Guillain-Barré syndrome (Gubler, Vasilakis, and Musso 2017), a rare condition in which the immune system attacks the nerves. It was only after an outbreak in Brazil in 2015 that Zika virus gained widespread public recognition. In the Brazilian outbreak, the disease was again associated with neurological symptoms, including a major epidemic of central nervous system malformations in babies, particularly **microcephaly** (a condition in which babies are born with smaller than normal heads and brains), attracting attention from the global media. The symptoms in newborns, collectively referred to as "**congenital** Zika syndrome," appeared in an estimated 5 to 15 percent of infants born to women infected with Zika during pregnancy, even when the mother's infection was asymptomatic (WHO 2018e). This quickly changed the perception of the virus from a mild (and largely ignored) tropical disease to an international public health concern as policymakers grappled with the emotive issue of congenitally impaired babies who would require years of specialist healthcare. Indeed, the massive surge of cases of microcephaly in Brazil initiated a WHO declaration of a "Public Health Emergency of International Concern" from February to November 2016 (Gubler, Vasilakis, and Musso 2017). Since the Brazilian outbreak, which may alone have caused more than one million infections (Brazilian Ministry of Health 2016), the disease has led to further outbreaks in the Americas and Asia. Sexual transmission of the virus—an unusual feature in a flavivirus (Gubler, Vasilakis, and Musso 2017)—was also first noted during these epidemic events. Between 2007 and 2017, eighty countries or territories reported active circulation of Zika virus, 23 reported an increase in Guillain-Barré syndrome potentially associated with the virus, and 31 an increase in congenital central nervous system abnormalities possibly associated with Zika infection (WHO 2018e; Gubler, Vasilakis, and Musso 2017).

Why the disease suddenly changed from a sporadic, mild infection to an epidemic disease with significant neurological symptoms has stimulated intensive investigation. Two main ideas have emerged so far. First, consistent with our emphasis on changing human

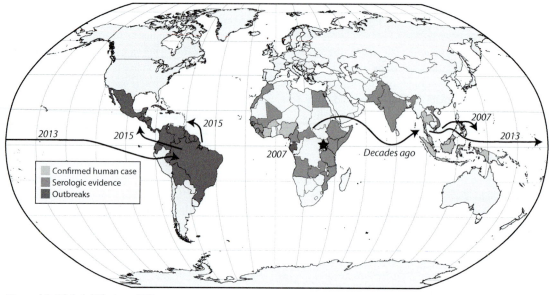

Figure 4.8 Global diffusion of Zika

Source: Adapted from Weaver (2016)

This map shows some of the key known dates around the spread of Zika. The star in Uganda shows the location of the Zika forest where the virus was discovered in 1947. Dates are based on evidence from phylogenetic studies.

environments, factors such as rapid human movements, urbanization, and poor vector control may facilitate the spread and survival of the virus and its anthropophilic mosquito vector, *Aedes aegypti* (WHO 2018e). Zika virus circulates both in a forest cycle with *Aedes* species and non-human primates, and a human cycle where humans contract the disease from mosquito species such as *Aedes aegypti* (Weaver et al. 2016). It seems likely that human migrations have introduced the virus to new vector species, leading to the development of new human cycles of the disease in different regions. Many different vector species have been implicated in Zika virus transmission, however, and so identifying the pathways through which infections have migrated from one place to another has proved challenging. Genetic analyses of Zika viruses indicate multiple strains of the virus, sometimes circulating simultaneously in the same population, suggesting multiple introductions of different viruses and supporting the significance of human movements in spreading the disease (Gubler, Vasilakis, and Musso

2017). Athletic competitions in Brazil during the period 2013–2014, including the World Cup and World Spirit Championships, may have been responsible for introducing the virus to Brazil with competitors arriving from the South Pacific, for instance (Gubler, Vasilakis, and Musso 2017; Weaver et al. 2016).

The sudden introduction of the disease to **immunologically naïve** populations in the South Pacific and the Americas may have initiated the recorded epidemics of disease (Weaver et al. 2016). Sudden epidemic activity may also explain the emergence of apparently "new" neurological symptoms. Neurological symptoms may always have occurred in a small proportion of Zika patients but associating these symptoms with Zika infection became easier with much larger outbreaks and so the connection was made only after the disease reached epidemic proportions (Weaver et al. 2016). In addition, **herd immunity** in Zika's traditional areas of circulation in Africa and Asia may have limited the number of in utero infections if most mothers had already been exposed to the virus in childhood, affording them immunity by the time they reached child-bearing age (Weaver et al. 2016).

Many infections in **immunologically naïve** populations may also have led to rapid genetic diversification of the virus as it spread into new territories and infected large numbers of people (Weaver et al. 2016), leading to a second hypothesis for Zika's sudden rapid spread. Genetic evolution of the virus may have enabled the disease's sudden transition to epidemic activity and prompted the evolution of other features of the disease that make it distinct from other flaviviruses, such as its ability to be transmitted sexually. One theory suggests that an Asian strain of the disease may have evolved to cause higher levels of **viremia** in the bloodstream, which could facilitate the transmission of the disease. Higher viremia could also encourage **vertical transmission** of the virus across the placenta, perhaps explaining the apparent surge in infants being infected in utero (Gubler, Vasilakis, and Musso 2017).

Reigning in the spread of Zika currently depends on vector control and careful surveillance of outbreaks because there is still no effective treatment or vaccine. Because of the devastating consequences of Zika infection for babies in utero, efforts to protect pregnant women from the disease are of greatest importance. Most interventions to date have relied on education, encouraging women to take precautions against mosquito bites, avoid travel to Zika-affected regions, and avoid sexual relations with men from Zika-affected regions during pregnancy. For people living in Zika-endemic areas, however, options are limited. *Aedes aegypti* is a day-biting species and so bed nets are ineffective. Many communities must therefore rely on vector-control programs for protection. Vector control for Zika coincides with efforts to control other diseases, notably dengue fever and yellow fever. Despite this, vector-control programs have been undermined in recent years by underfunding, rising pesticide resistance in vectors, and environmental concerns with pesticide spraying. Development of an effective vaccine might offer the greatest chance for success against the epidemic. Vaccines have been developed for other flaviviruses, generating hope that a Zika vaccine may be possible in the future (Weaver et al. 2016).

Tuberculosis (TB)

The tuberculosis bacterium may have existed since antiquity—there is evidence of infection in Egyptian mummies, and TB was probably one of the diseases labeled as "White Plague" in the seventeenth and eighteenth centuries. Today, the bacterium *Mycobacterium tuberculosis*—the main human cause of TB—is extremely widespread, infecting populations all around the world, and remains one of the top ten causes of death globally. In 2017, ten million people contracted the disease and 1.6 million died from it (WHO 2018d), although death rates have declined somewhat since the 1990s (Ritchie and Roser 2018). Most deaths from TB occur among the elderly, but it remains a significant cause of mortality and morbidity among people of working age, creating a major economic burden. Sub-Saharan Africa is at the center of the TB crisis (with mortality rates ranging from fifty to five hundred per 100,000), followed by South Asia (at 25 to fifty deaths per 100,000). In most other countries the TB death rate is under five per 100,000 (Ritchie and Roser 2018). HIV is the main reason for the resurgence of the disease in Africa, where death rates from TB almost doubled between 1990 and 2005 (WHO 2007) (Figure 4.9). Today, Southeast Asia and the Western Pacific are responsible for the majority of new cases (62 percent) (WHO 2018d).

In line with the Millennium Development Goal of reducing the burden of TB by 2015, the WHO launched a major initiative to identify and treat TB worldwide in 2006. The project aimed to reverse the increasing incidence of TB, improve detection of cases and cure rates, and reduce TB mortality (WHO 2006). Although TB incidence has begun to decline, TB was again noted in the United Nations' Sustainable Development Goals of 2015, which aimed to end the global TB epidemic by 2030. TB incidence is now declining at about 2 percent per annum, but we are not yet on target to end the TB epidemic by 2030 (WHO 2018d).

There is no known animal or environmental reservoir of TB, but the pathogen can lie dormant in human hosts for years. Approximately one quarter of the global population is thought to be infected, mostly as asymptomatic cases ("latent TB") (Raguenaud et al. 2008). Although people with latent TB cannot transmit the disease to

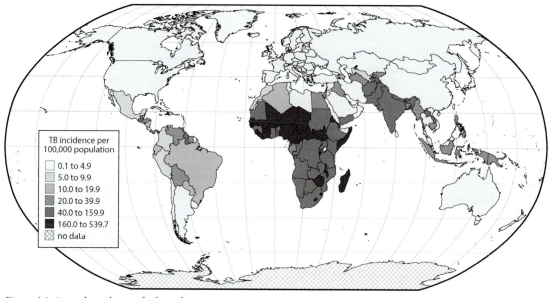

Figure 4.9 Annual incidence of tuberculosis, 2017
Source: World Bank (2019)

others, this huge disease reservoir makes combatting TB difficult. People infected with the TB pathogen have a 5 to 15 percent lifetime risk of developing an active infection, typically when the body becomes stressed by factors such as malnutrition, diabetes, alcoholism, tobacco use, or another infection (WHO 2018d). HIV infection is a major risk factor for developing active TB—people living with HIV have a twenty to thirty times increased risk of developing active TB, and just under one fifth of those who died from TB in 2017 were co-infected with HIV (WHO 2018d).

These risk factors mean that the social environment is integral to the story of tuberculosis. TB is often considered a disease of poverty because factors such as poor living conditions and inadequate diet can transform a latent infection into active disease, explaining why 95 percent of cases occur in the Global South (WHO 2018d). In high-income countries, risk factors such as alcoholism, smoking, and diabetes focus the disease among poor and marginalized populations. Because TB spreads by airborne transmission, people living in overcrowded conditions often also have the greatest exposure to the pathogen.

These sorts of social factors mean that tuberculosis outbreaks often occur during political unrest, war, or economic turmoil, as occurred in the 1990s following the dissolution of the Soviet Union. In this case, widespread economic and social distress, the decline of public health institutions, and high rates of incarceration led to a surge in tuberculosis cases. Incarceration provides the ideal conditions for a TB epidemic. Widespread economic and social malaise led to burgeoning prison populations when social budgets were tight, leading to overcrowding, poor diet, and inadequate healthcare in prisons. At the turn of the twenty-first century, incidence of TB among Russian prisoners was 40 to 50 percent higher than in civilian populations (Schulz and Robinson 2001). After infectious individuals were released from prisons, they exposed thousands of others to the disease.

New strains of the TB bacterium are also emerging as intensive use of antibiotics pose a significant selection pressure on the pathogen. With the development of an effective antibiotic therapy for the disease in 1951, there was great hope that the disease could be controlled (Murray 2004). Unfortunately, the long course of antibiotics required to cure the disease, combined with the side effects that many people experience from taking TB drugs, mean that many patients are reluctant to continue taking their medication once they feel better. Additionally, there are significant barriers to completing the complex drug regime required for effective treatment, particularly the cost and challenge of getting access to healthcare given that many infections occur in countries where people must pay out-of-pocket for healthcare. Even in high-income countries with good healthcare systems, problems such as homelessness, incarceration, and mental illness mean that individuals who begin a course of treatment may not complete it.

Evidence of antibiotic resistance in the TB pathogen emerged rapidly after drug therapies were introduced, and as early as the 1950s the standard treatment for TB had become a combination of antibiotics. This not only increased the potency of the treatment but also reduced the rate at which resistance built up in the pathogen population (Murray 2004). By the 1980s, the growing epidemic of HIV exacerbated problems of growing antibiotic resistance in the TB pathogen, and by the mid-1980s a resurgence of TB had been observed in some affluent countries, including the US (*ibid.*). Unfortunately, rising levels of antibiotic resistance coincided with the relaxation of traditional methods of TB control (such as isolation and careful surveillance) because public health budgets for TB were cut in the belief that the problem had been solved with new drug therapies. As a result, TB began to experience a major global resurgence in the 1980s and 1990s.

Strains of TB resistant to at least one antibiotic have now been found in every country surveyed (WHO 2007). Large numbers of TB infections are resistant to both isoniazid and rifampicin (the two most common TB antibiotics) and are considered to be "multi-drug resistant" (MDR-TB). MDR-TB can still be treated with **second-line drugs**, but the course of treatment takes longer (up to two years), and the drugs are more expensive and more toxic (WHO 2018d). Some TB strains are now resistant to many or all the antibiotic treatments available and are labeled "extensively drug-resistant" (XDR-TB), leaving patients with few treatment options. Many drug-resistant strains may have developed in the countries of the former Soviet Union, where almost half of all TB cases were resistant to at least one drug and 20 percent of cases considered MDR-TB as early as 2008 (WHO 2008). More recently, India, China, and Russia together accounted for nearly half of global cases of MDR-TB (Seaworth and Griffith 2017).

When drug-resistant strains began to spread in the 1990s, public health officials debated whether it was desirable or even possible to treat MDR-TB in poor countries, given already overstretched health systems and the high cost of treatment. The received wisdom for many years was that MDR-TB was simply too expensive to treat in low-income contexts. In addition, there were concerns that patients in low-income countries would lack the education or commitment to follow through their course of treatment correctly, allowing further antibiotic resistance to develop. Anthropologist and physician Paul Farmer was an early leader in the charge to counter these notions. Farmer (1996) argued that not only is TB treatment a human right, but also that there are practical concerns with leaving individuals untreated because MDR strains can spread so easily among countries. The tide began to turn as affluent countries eventually began to see that it was in their own best interests to treat MDR-TB infections before they could spread.

Many studies have explored reasons for non-compliance with TB drug regimens and found no consistent evidence that people in low-income countries are more prone to fail to complete their course of drugs. Instead, adherence to TB treatment programs is affected by a wide range of place-specific factors, including patient knowledge and beliefs about TB, cost of treatment, as well as the levels of supervision and attitudes of staff administering treatment (Skinner and Claassens 2016; Lei et al. 2016; Raguenaud et al. 2008). The treatment of MDR-TB among poor populations is now WHO policy, and the cost of second-line drugs has decreased significantly in recent years, putting the therapy in reach of more patients.

CONCLUSION

Emerging infectious diseases remain a considerable cause for concern, particularly in an era of rapid population growth, international travel, and widespread landscape conversion. These factors operate synergistically to increase the risk of a global pandemic. Many scholars argue that the pertinent question is not *if* we will see another global pandemic but *when*.

Rapid environmental change is implicated in changing disease patterns in a variety of ways, from upsetting traditional ecological balances between pathogens and people to increasing the likelihood that people will encounter novel pathogens. The role of environmental change to human health goes beyond infectious disease, however. Environmental change also leads to higher exposures to geogens and greater risk for *non-infectious* diseases, the topic of the next chapter.

DISCUSSION QUESTIONS

1 Discuss the direct and indirect links between climate change and health. What potential solutions can you think of that might address these issues—think about how you might address the problem at multiple scales, from the household to the global scale.

2 Using relevant case studies, discuss the many ways in which landscape change can lead to changes in the incidence of infectious diseases.

3 Think about recent experiences you have had of being in a natural environment. In what ways do you believe that this could have been beneficial to your overall health and well-being? In what ways might it have been harmful?

4 Given our evolving understandings of the beneficial role of microbes in our environment and bodies, how would you go about limiting, regulating, or encouraging exposure to microorganisms for a child in your care?

SUGGESTED READING

Curtis, S. E., and Oven, K. J. 2012. "Geographies of Health and Climate Change." *Progress in Human Geography* 36 (5): 654–66.

Jones, K. E., N. G. Patel, M. A. Levy, A. Storeygard, et al. 2008. "Global Trends in Emerging Infectious Diseases." *Nature* 451: 990–94.

Keesing, F., Lisa K. Belden, Peter Daszak, Andrew Dobson, et al. 2010. "Impacts of Biodiversity on the Emergence and Transmission of Infectious Diseases." *Nature* 468: 647–52.

Sen, B., Meghnath Dhimal, Aishath Thimna Latheef, and Upasona Ghosh. 2017. "Climate Change: Health Effects and Response in South Asia." *British Medical Journal* 359.

Walters, M. J. 2014. *Seven Modern Plagues and How We Are Causing Them*. Washington, DC: Island Press.

VIDEO RESOURCES

"Nova: What's Living in You." PBS. More information at: www.pbs.org/wgbh/nova/video/nova-wonders-whats-living-in-you/.

REFERENCES

Alexandre-Silva, G. M., P. A. Brito-Souza, A. C. S. Oliveira, F. A. Cerni, et al. 2018. "The Hygiene Hypothesis at a Glance: Early Exposures, Immune Mechanism and Novel Therapies." *Acta Tropica* 188: 16–26. doi:10.1016/j.actatropica.2018.08.032.

Atlas, R. M., and S. Maloy. 2014. *One Health: People, Animals, and the Environment*. ASM Press.

Baker, R. 2007. *Quiet Killers: The Fall and Rise of Deadly Diseases*. Sutton Publishing.

Beggs, Paul John, and Nicole Ewa. 2008. "Impacts of Climate Change on Plant Food Allergens: A Previously Unrecognized Threat to Human Health." *Air Quality, Atmosphere & Health* 1 (2): 119–23. doi:10.1007/s11869-008-0013-z.

Bentham, G., and I. H. Langford. 2001. "Environmental Temperatures and the Incidence of Food Poisoning in England and Wales." *International Journal of Biometeorology* 45 (1): 22–26.

Berman, M. G., E. Kross, K. M. Krpan, M. K. Askren, et al. 2012. "Interacting with Nature Improves Cognition and Affect for Individuals with Depression." *Journal of Affective Disorders* 140 (3): 300–5. doi:10.1016/j.jad.2012.03.012.

Beyer, K., A. Kaltenbach, A. Szabo, S. Bogar, F. J. Nieto, and K. Malecki. 2014. "Exposure to Neighborhood Green Space and Mental Health: Evidence from the Survey of the Health of Wisconsin. *International Journal of Environmental Research and Public Health* 11(3): 3453–72.

Brazilian Ministry of Health. 2016. "Protocolo de vigilância e resposta à ocorrência de microcefalia." Accessed February 12, 2019. http://portalarquivos2.saude.gov.br/images/pdf/2016/janeiro/22/microcefalia-protocolo-de-vigilancia-e-resposta-v1-3-22jan2016.pdf.

Carrington, D. 2016. "The Anthropocene Epoch: Scientists Declare Dawn of Human-Influenced Age." *The Guardian*, August 29. www.theguardian.com/environment/2016/aug/29/declare-anthropocene-epoch-experts-urge-geological-congress-human-impact-earth.

Casadevall, Arturo, Dimitrios P. Kontoyiannis, and Vincent Robert. 2019. "On the Emergence of Candida auris: Climate Change, Azoles, Swamps, and Birds." *mBio* 10 (4): e01397-19. doi:10.1128/mBio.01397-19.

[CDC] Centers for Disease Control. 2019. "Lyme Disease Maps: Historical Data." Accessed August 12, 2019. www.cdc.gov/lyme/stats/maps.html.

[CDC] Centers for Disease Control and Prevention. 2018a. "Biggest Threats and Data." Accessed February 4, 2019. www.cdc.gov/drugresistance/biggest_threats.html.

[CDC] Centers for Disease Control and Prevention. 2018b. "Lyme Disease Surveillance and Available Data." Accessed January 30, 2019. www.cdc.gov/lyme/stats/survfaq.html.

[CDC] Centers for Disease Control and Prevention. no date. "Lyme Disease: What You Need to Know." Accessed January 30, 2019. www.cdc.gov/lyme/resources/brochure/lymediseasebrochure.pdf.

Crow, J. R., S. L. Davis, D. M. Chaykosky, T. T. Smith, and J. M. Smith. 2015. "Probiotics and Fecal Microbiota Transplant for Primary and Secondary Prevention of *Clostridium difficile* Infection." *Pharmacotherapy* 35 (11): 1016–25. doi:10.1002/phar.1644.

Cunsolo Willox, Ashlee, Sherilee L. Harper, James D. Ford, Victoria L. Edge, et al. 2013. "Climate Change and Mental Health: An Exploratory Case Study from Rigolet, Nunatsiavut, Canada." *Climatic Change* 121 (2): 255–70. doi:10.1007/s10584-013-0875-4.

Dizney, Laurie J., and Luis A. Ruedas. 2009. "Increased Host Species Diversity and Decreased Prevalence of Sin Nombre Virus." *Emerging Infectious Diseases* 15 (7): 1012–18. doi:10.3201/eid1507.081621.

Domonoske, C. 2016. "U.N. Admits Role in Haiti Cholera Outbreak That Has Killed Thousands." Accessed January 27, 2019. www.npr.org/sections/thetwo-way/2016/08/18/490468640/u-n-admits-role-in-haiti-cholera-outbreak-that-has-killed-thousands.

Ege, M. J. 2017. "The Hygiene Hypothesis in the Age of the Microbiome." *Annals of the American Thoracic Society* 14: S348–53. doi:10.1513/AnnalsATS.201702-139AW.

Epstein, J. H., H. E. Field, S. Luby, J. R. Pulliam, and P. Daszak. 2006. "Nipah Virus: Impact, Origins, and Causes of Emergence." *Current Infectious Disease Reports* 8 (1): 59–65.

Epstein, P. R. 2005. "Climate Change and Human Health." *New England Journal of Medicine* 353 (14): 1433–36.

Erickson, A. 2017. "One Million People Have Contracted Cholera in Yemen. You Should Be Outraged." *The Washington Post*. www.washingtonpost.com/news/worldviews/wp/2017/12/21/one-million-people-have-caught-cholera-in-yemen-you-should-be-outraged/?noredirect=on&utm_term=.82ff4c87f080.

Faber Taylor, Andrea, and Frances E. Kuo. 2009. "Children with Attention Deficits Concentrate Better After Walk in the Park." *Journal of Attention Disorders* 12 (5): 402–9. doi:10.1177/1087054708323000.

Farmer, P. 1996. "Social Inequalities and Emerging Infectious Diseases." *Emerging Infectious Diseases* 2 (4): 259–69.

Fraser, B. 2010. "Taking on Malaria in the Amazon." *Lancet* 376 (9747): 1133–34.

Gagñon, A. S., K. E. Smoyer-Tomic, and A. B. Bush. 2002. "The El Nino Southern Oscillation and Malaria Epidemics in South America." *International Journal of Biometeorology* 46 (2): 81–89.

Gallagher, J. 2018. "Man Has 'World's Worst' Super-Gonorrhoea." *BBC News*. Accessed February 5, 2019. www.bbc.com/news/health-43571120.

Gould, L. H., J. Kline, C. Monahan, and K. Vierk. 2017. "Outbreaks of Disease Associated with Food Imported into the United States, 1996–2014(1)." *Emerging Infectious Diseases* 23 (3): 525–28. doi:10.3201/eid2303.161462.

Gould, Peter. 1993. *The Slow Plague: A Geography of the AIDS Pandemic*. Oxford, UK and Cambridge, MA: Blackwell Publishers.

Gubler, D. J., P. Reiter, K. L. Ebi, W. Yap, et al. 2001. "Climate Variability and Change in the United States: Potential Impacts on Vector- and Rodent-Borne Diseases." *Environmental Health Perspectives* 109 (Suppl 2): 223–33. doi:10.1289/ehp.109-1240669.

Gubler, D. J., N. Vasilakis, and D. Musso. 2017. "History and Emergence of Zika Virus." *Journal of Infectious Diseases* 216 (Suppl 10): S860–67. doi:10.1093/infdis/jix451.

Gupta, A., and S. Khanna. 2017. "Fecal Microbiota Transplantation." *JAMA* 318 (1): 102. doi:10.1001/jama.2017.6466.

Hashizume, M., Y. Wagatsuma, A. S. Faruque, T. Hayashi, et al. 2008. "Factors Determining Vulnerability to Diarrhoea During and After Severe Floods in Bangladesh." *Journal of Water and Health* 6 (3): 323–32.

Howell, Andrew J., Raelyne L. Dopko, Holli-Anne Passmore, and Karen Buro. 2011. "Nature Connectedness: Associations with Well-Being and Mindfulness." *Personality and Individual Differences* 51 (2): 166–71. doi:10.1016/j.paid.2011.03.037.

IPCC. 2014. *Climate Change 2014—Impacts, Adaptation and Vulnerability: Part A: Global and Sectoral Aspects:* *Working Group II Contribution to the IPCC Fifth Assessment Report. Volume 1: Global and Sectoral Aspects.* Vol. 1. Cambridge: Cambridge University Press.

IPCC, P. M. Parry, M. L. Parry, O. Canziani, et al. 2007. *Climate Change 2007—Impacts, Adaptation and Vulnerability: Working Group II Contribution to the Fourth Assessment Report of the IPCC.* Cambridge: Cambridge University Press.

Johnson, P. T., P. J. Lund, R. B. Hartson, and T. P. Yoshino. 2009. "Community Diversity Reduces Schistosoma mansoni Transmission, Host Pathology and Human Infection Risk." *Proceedings: Biological Sciences* 276 (1662): 1657–63. doi:10.1098/rspb.2008.1718.

Jones, K. E., N. G. Patel, M. A. Levy, A. Storeygard, et al. 2008. "Global Trends in Emerging Infectious Diseases." *Nature* 451 (7181): 990–94. doi:10.1038/nature06536.

Keesing, Felicia, Lisa K. Belden, Peter Daszak, Andrew Dobson, et al. 2010. "Impacts of Biodiversity on the Emergence and Transmission of Infectious Diseases." *Nature* 468: 647. doi:10.1038/nature09575. www.nature.com/articles/nature09575#supplementary-information.

Kilpatrick, A. M., A. D. M. Dobson, T. Levi, D. J. Salkeld, et al. 2017. "Lyme Disease Ecology in a Changing World: Consensus, Uncertainty and Critical Gaps for Improving Control." *Philosophical Transactions of the Royal Society of London. Series B, Biological Sciences* 372 (1722). doi:10.1098/rstb.2016.0117.

Kovats, R. S., and S. Hajat. 2008. "Heat Stress and Public Health: A Critical Review." *Annual Review of Public Health* 29: 41–55. doi:10.1146/annurev.publhealth.29.020907.090843.

Kuo, Ming. 2015. "How Might Contact with Nature Promote Human Health? Promising Mechanisms and a Possible Central Pathway." *Frontiers in Psychology* 6: 1093. doi:10.3389/fpsyg.2015.01093.

Landis, E. G., V. Yang, D. M. Brown, M. T. Pardue, and S. A. Read. 2018. "Dim Light Exposure and Myopia in Children." *Investigative Ophthalmology & Visual Science* 59 (12): 4804–11. doi:10.1167/iovs.18-24415.

Leffler, D. A., and J. T. Lamont. 2015. "Clostridium difficile Infection." *New England Journal of Medicine* 372 (16): 1539–48. doi:10.1056/NEJMra1403772.

Lei, Xun, Ke Huang, Qin Liu, Yong-Feng Jie, and Sheng-Lan Tang. 2016. "Are Tuberculosis Patients Adherent to Prescribed Treatments in China? Results of a Prospective Cohort Study." *Infectious Diseases of Poverty* 5: 38. doi:10.1186/s40249-016-0134-9.

Leichenko, R., and K. O'Brien. 2008. *Environmental Change and Globalization: Double Exposures*. Oxford University Press.

Lewellen, Ruth H., and Stephen H. Vessey. 1998. "The Effect of Density Dependence and Weather on Population Size of a Polyvoltine Species." *Ecological Monographs* 68 (4): 571–94. doi:10.1890/0012-9615(1998)068[0571:-TEODDA]2.0.CO;2.

Li, Wendong, Zhengli Shi, Meng Yu, Wuze Ren, et al. 2005. "Bats Are Natural Reservoirs of SARS-Like Coronaviruses." *Science* 310 (5748): 676–79.

Masterton, R. G., and A. D. Green. 1991. "Dissemination of Human Pathogens by Airline Travel." *Society for Applied Bacteriology Symposium Series* 20: 31S–38.

Mather, T. 1993. "The Dynamics of Spirochete Transmission Between Ticks and Vertebrates." In *Ecology and Environmental Management of Lyme Disease*, edited by H. S. Ginsberg. Rutgers University Press.

Mayer, J. D. 2000. "Geography, Ecology and Emerging Infectious Diseases." *Social Science & Medicine* 50 (7–8): 937–52.

McCullough, S. J., L. O'Donoghue, and K. J. Saunders. 2016. "Six Year Refractive Change Among White Children and Young Adults: Evidence for Significant Increase in Myopia Among White UK Children." *PLoS One* 11 (1): e0146332. doi:10.1371/journal.pone.0146332.

McSorley, H. J., M. A. M. Chaye, and H. H. Smits. 2018. "Worms: Pernicious Parasites or Allies Against Allergies?" *Parasite Immunology* e12574. doi:10.1111/pim.12574.

Mills, G. 2018. "Lyme Disease Cases on the Rise in the UK." *Veterinary Record* 183 (2): 44–45. doi:10.1136/vr.k3072.

[MIT] Massachusetts Institute of Technology Media Lab. no date. "Preventing Tick-Borne Disease by Permanently Immunizing Mice." Accessed January 30, 2019. www.media.mit.edu/projects/preventing-tick-borne-disease-by-permanently-immunizing-mice/overview/.

Moceituba, Atasa, and Monique Tsang. 2015. "Averting Climate Change's Health Effects in Fiji." *Bulletin of the World Health Organization* 93 (11): 746.

Morse, S. S., J. A. Mazet, M. Woolhouse, C. R. Parrish, et al. 2012. "Prediction and Prevention of the Next Pandemic Zoonosis." *Lancet* 380 (9857): 1956–65. doi:10.1016/s0140-6736(12)61684-5.

Murray, J. F. 2004. "A Century of Tuberculosis." *American Journal of Respiratory and Critical Care Medicine* 169 (11): 1181–86. doi:10.1164/rccm.200402-140OE.

Musso, D., and D. J. Gubler. 2016. "Zika Virus." *Clinical Microbiology Reviews* 29 (3): 487–524. doi:10.1128/cmr.00072-15.

Ng, E. Y. 2005. "Is Thermal Scanner Losing Its Bite in Mass Screening of Fever Due to SARS?" *Medical Physics* 32 (1): 93–97. doi:10.1118/1.1819532.

Nisbet, Elizabeth K., John M. Zelenski, and Steven A. Murphy. 2011. "Happiness Is in Our Nature: Exploring Nature Relatedness as a Contributor to Subjective Well-Being." *Journal of Happiness Studies* 12 (2): 303–22. doi:10.1007/s10902-010-9197-7.

Nupp, T. E., and R. K. Swihart. 1998. "Effects of Forest Fragmentation on Population Attributes of White-Footed Mice and Eastern Chipmunks." *Journal of Mammalogy* 79 (4): 1234–43.

Obradovich, Nick, Robyn Migliorini, Martin P. Paulus, and Iyad Rahwan. 2018. "Empirical Evidence of Mental Health Risks Posed by Climate Change." *Proceedings of the National Academy of Sciences* 115 (43): 10953–58. doi:10.1073/pnas.1801528115.

Oppong, Joseph R. 1998. "A Vulnerability Interpretation of the Geography of HIV/AIDS in Ghana, 1986–1995." *The Professional Geographer* 50 (4): 437–48. doi:10.1111/0033-0124.00131.

Ostfeld, R. S., and F. Keesing. 2000. "Lyme Disease Cases on the Rise in the UK." *Veterinary Record* 183 (2): 44. doi:10.1136/vr.k3072.

Page, L. A., and L. M. Howard. 2010. "The Impact of Climate Change on Mental Health (but Will Mental Health Be Discussed at Copenhagen?)." *Psychological Medicine* 40 (2): 177–80.

Pearson, David G., and Tony Craig. 2014. "The Great Outdoors? Exploring the Mental Health Benefits of Natural Environments." *Frontiers in Psychology* 5: 1178. doi:10.3389/fpsyg.2014.01178.

Pliego Pliego, Emilene, Jorge Velázquez-Castro, Markus P. Eichhorn, and Andrés Fraguela Collar. 2018. "Increased Efficiency in the Second-Hand Tire Trade Provides Opportunity for Dengue Control." *Journal of Theoretical Biology* 437: 126–36. doi:10.1016/j.jtbi.2017.10.025.

Raguenaud, M., R. Zachariah, M. Massaquoi, V. Ombeka, H. Ritter, and J. M. Chakaya. 2008. "High Adherence to Anti-Tuberculosis Treatment Among Patients Attending a Hospital and Slum Health Centre in Nairobi, Kenya." *Global Public Health* 3 (4): 433–39. doi:10.1080/17441690802063205.

Ramanathan, V., J. Seddon, and D. Victor. 2016. "The Next Frontier on Climate Change." *Foreign Affairs* 95(2): 134–42.

Rechenburg, Andrea, and Thomas Kistemann. 2009. "Sewage Effluent as a Source of Campylobacter sp. in a Surface Water Catchment." *International Journal of Environmental Health Research* 19 (4): 239–49. doi:10.1080/09603120802460376.

Ritchie, H., and M. Roser. 2018. "Causes of Death." *OurWorldInData.Org.* Accessed February 4, 2019. https://ourworldindata.org/causes-of-death.

Russell, R. C. 1986. "Transportation of Insects of Public Health Importance on International Aircraft." *Travel Medicine International* 7: 26–31.

Schulz, W. F., and M. Robinson. 2001. *In Our Own Best Interest: How Defending Human Rights Benefits Us All.* Beacon Press.

Schwartz, A., A. Hinckley, P. S. Mead, S. Hook, and K. J. Kugeler. 2017. "Surveillance for Lyme Disease— United States, 2008–2015." *Morbidity and Mortality Weekly Report* 66 (22): 1–12.

Seaworth, B. J., and D. E. Griffith. 2017. "Therapy of Multidrug-Resistant and Extensively Drug-Resistant Tuberculosis." *Microbiology Spectrum* 5 (2). doi:10.1128/microbiolspec.TNMI7-0042-2017.

Sen, Banalata, Meghnath Dhimal, Aishath Thimna Latheef, and Upasona Ghosh. 2017. "Climate Change: Health Effects and Response in South Asia." *British Medical Journal* 359. doi:10.1136/bmj.j5117.

Sida. 2018. *The Relationship Between Climate Change and Violent Conflict.* International Organisations and Policy Support.

Skinner, Donald, and Mareli Claassens. 2016. "It's Complicated: Why Do Tuberculosis Patients Not Initiate or Stay Adherent to Treatment? A Qualitative Study from South Africa." *BMC Infectious Diseases* 16 (1): 712. doi:10.1186/s12879-016-2054-5.

Smith, K. R., A. Woodward, B. Lemke, M. Otto, et al. 2016. "The Last Summer Olympics? Climate Change, Health, and Work Outdoors." *Lancet* 388 (10045): 642–44. doi:10.1016/s0140-6736(16)31335-6.

Strachan, D. P. 1989. "Hay-Fever, Hygiene, and Household Size." *British Medical Journal* 299 (6710): 1259–60. doi:10.1136/bmj.299.6710.1259.

Swaddle, J. P., and S. E. Calos. 2008. "Increased Avian Diversity Is Associated with Lower Incidence of Human West Nile Infection: Observation of the Dilution Effect." *PLoS One* 3 (6): e2488. doi:10.1371/journal.pone.0002488.

van Buskirk, Josh, and Richard S. Ostfeld. 1995. "Controlling Lyme Disease by Modifying the Density and Species Composition of Tick Hosts." 5 (4): 1133–40. doi:10.2307/2269360.

Walters, M. J. 2004. *Six Modern Plagues and How We Are Causing Them.* Island Press.

Walters, M. J. 2014. *Seven Modern Plagues: And How We Are Causing Them.* Island Press.

Ward Thompson, Catharine, Jenny Roe, Peter Aspinall, Richard Mitchell, et al. 2012. "More Green Space Is Linked to Less Stress in Deprived Communities: Evidence from Salivary Cortisol Patterns." *Landscape and Urban Planning* 105 (3): 221–29. doi:10.1016/j.landurbplan.2011.12.015.

Watson, J. T., M. Gayer, and M. A. Connolly. 2007. "Epidemics After Natural Disasters." *Emerging Infectious Diseases* 13 (1): 1–5.

Watts, Nick, W. Neil Adger, Sonja Ayeb-Karlsson, Yuqi Bai, et al. 2017. "The Lancet Countdown: Tracking Progress on Health and Climate Change." *The Lancet* 389 (10074): 1151–64. doi:10.1016/S0140-6736(16)32124-9.

Weaver, S. C., F. Costa, M. A. Garcia-Blanco, A. I. Ko, et al. 2016. "Zika Virus: History, Emergence, Biology, and Prospects for Control." *Antiviral Research* 130: 69–80. doi:10.1016/j.antiviral.2016.03.010.

[WHO] World Health Organization. 2003. "Update 95—SARS: Chronology of a Serial Killer." Accessed June 21, 2019. www.who.int/csr/don/2003_07_04/en/index.html.

[WHO] World Health Organization. 2006. "The Stop TB Strategy." Accessed June 9, 2019. http://whqlibdoc.who.int/hq/2006/WHO_HTM_STB_2006.368_eng.pdf.

[WHO] World Health Organization. 2007. "Tuberculosis' Factsheet No. 104." Accessed June 2, 2009. www.who.int/mediacentre/factsheets/fs104/en/index.html.

[WHO] World Health Organization. 2008. "Anti-Tuberculosis Drug Resistance in the World." Accessed June 9, 2009. www.who.int/tb/publications/2008/drs_report4_26feb08.pdf.

[WHO] World Health Organization. 2018a. "Middle East Respiratory Syndrome Coronavirus (MERS-CoV)." Accessed December 22, 2018. www.who.int/emergencies/mers-cov/en/.

[WHO] World Health Organization. 2018b. "Neglected Tropical Diseases." Accessed February 5, 2019. www.who.int/neglected_diseases/en/.

[WHO] World Health Organization. 2018c. "Nipah Virus Outbreaks in the WHO South-East Asia Region." Accessed January 27, 2019. www.searo.who.int/entity/emerging_diseases/links/nipah_virus_outbreaks_sear/en.

[WHO] World Health Organization. 2018d. "Tuberculosis." Accessed February 12, 2019. www.who.int/en/news-room/fact-sheets/detail/tuberculosis.

[WHO] World Health Organization. 2018e. "Zika Virus." Accessed February 12, 2019. www.who.int/news-room/fact-sheets/detail/zika-virus.

Wilson, C. 2015. "Looking Good." *New Scientist* 226 (3022): 33–37. doi:10.1016/s0262-4079(15)30414-0.

Wimberly, Michael C., Michael J. Yabsley, Adam D. Baer, Vivien G. Dugan, and William R. Davidson. 2008. "Spatial Heterogeneity of Climate and Land-Cover Constraints on Distributions of Tick-Borne Pathogens." *Global Ecology and Biogeography* 17 (2): 189–202. doi:10.1111/j.1466-8238.2007.00353.x.

Wolfe, N., P. Daszak, A. Kilpatrick, and D. Burke. 2005. "Bushmeat Hunting, Deforestation, and Prediction of Zoonotic Disease." *Emerging Infectious Diseases* 11 (12): 1822–7..

World Bank. 2019. "World Bank Open Data." Accessed August 22. https://data.worldbank.org/.

5

ENVIRONMENTAL EXPOSURES

Since the mid-twentieth century, the primary causes of death and morbidity in many countries have shifted from infectious to non-infectious disease as innovations such as antibiotics, vaccination, and sanitation have controlled infections. As a result, lifespans have extended and **degenerative** conditions and **chronic** exposures to non-living agents of disease, **geogens**, have become more significant.

A wide variety of substances can damage health, including airborne and waterborne chemicals, and solid wastes. Some geogens occur naturally, but **synthetic** (human-created) chemicals are particularly likely to have health consequences as the human body has had little time to adapt to these substances. **Pollutants**—waste products of human activity—therefore often have significant consequences for health. While pollution has always been a part of human civilization, there have been precipitous increases in the concentrations of synthetic chemicals in the air, water, and soil since the Industrial Revolution, many of which are now known to have toxic effects (Figure 5.1). While few would argue that the benefits of modernization are not worth some cost, pursuing economic advances with little regard for their effects on human health and the environment has serious consequences. The health impacts of geogens is the focus of this chapter.

The study of pollution requires both a solid understanding of interactions between humans and their environments and careful spatial analysis, making it a key focus for health geographers. Analyzing the impact of geogens on health often employs statistical analysis at the population scale, where small effects can be empirically observed. While it is difficult to conclude that a particular health outcome results from exposure to a specific substance at the scale of an individual, observing a consistent association across an entire population can make a strong case. Analyzing associations between non-living agents of disease and health outcomes to identify causative relationships is a major focus of the field of **epidemiology**.

We begin by discussing the nature and sources of non-living agents of disease, briefly considering the importance of naturally occurring geogens before turning to synthetic pollutants. After discussing some of the major types of pollution, we explore ways in which public health researchers gather evidence to verify causal links between **exposures** and health impacts. The study of environmental exposures also elicits questions of social equity because certain communities or socioeconomic groups often bear an unfair share of the health

Figure 5.1 Tanning industry, Fes, Morocco
Credit: Heike Alberts

Toxic by-products are produced by many traditional industries, such as this tanning industry in Morocco where chromium—a known carcinogen—has been integral to the industry since the 1800s.

burden associated with pollutants—an area of inquiry known as **environmental justice**. We address the concept of environmental justice at the end of the chapter, directing our attention towards the social issues that we address in the second half of this book.

EXPOSURE TO GEOGENS

All living creatures process materials such as food and oxygen for survival; as they do, they continually absorb other substances from their environment that can include harmful materials. Toxic substances can be ingested via the gastrointestinal tract, breathed in through the lungs, or absorbed through the skin during physical contact. Animals have evolved a variety of mechanisms to keep out or remove harmful substances. The skin acts as a major barrier between the vital organs and the external environment. Most ingested toxic substances are filtered out by the liver or kidneys and then excreted. In **acute** exposures, substances can be rapidly expelled by vomiting, diarrhea, or coughing. If toxins are not effectively expelled, they can damage the body. Ingestion of a sufficient quantity of a toxic substance may cause an acute physiological reaction, leading to a rapid breakdown of the body's systems and potentially death, as occurs during a drug overdose. More commonly, chronic exposures to low levels of substances, over months, years, or decades, can damage health.

Humans have invented hundreds of thousands of synthetic products that we now use in everyday life. Regulatory agencies are unable to keep up with the sheer volume of synthetic chemicals, and many products have come to market with no testing or understanding of their effects on health. We are all effectively participating in a massive natural experiment on the toxic effects of many of the substances we are exposed to, leading to legislation only when members of the public or scientific community notice and publicize harmful effects. Only after decades of using lead in paint and gasoline, for instance, were regulations implemented to minimize people's exposure to its toxic effects.

As we now understand that chronic exposures to even small amounts of some substances can be harmful to health, scientists are questioning the potential impacts of some widely used products. The impact of plastics raises significant concerns, for example, because plastic products and wastes have become ubiquitous in our environment. Sunscreen provides another example: although the application of sunscreen has clear health benefits in terms of protecting the skin, some researchers are now exploring whether the absorption of chemicals from sunscreen has long-term health implications (Matta et al. 2019).

The health impacts of exposure to geogens depend not only on the type, duration, and intensity of exposure, but also on the biology of the exposed individual. Susceptible populations, such as the young, old, ill, or pregnant, often experience the most severe health consequences. Children and fetuses are particularly vulnerable, owing to their small size and developing bodies. The placenta was once believed to provide a barrier that protected the developing fetus from harmful substances, but a growing body of evidence has shown that numerous substances can cross the placenta. Pregnant women are now advised to avoid exposure to a variety of substances to protect their unborn child, including

many prescription drugs, cigarette smoke, and mercury (a common contaminant of fish).

NATURALLY OCCURRING GEOGENS

Almost any substance can damage health if consumed in sufficient quantities—even seemingly innocuous substances like water or salt can cause acute physiological problems if consumed in excess. Most organisms have evolved behavioral or physiological adaptations to substances they have encountered over many generations and have mechanisms to regulate consumption or eliminate excess. Our bodies are designed to carefully excrete regulated quantities of water and salts in urine, for instance, to maintain a healthy concentration of salts in body tissues. Nonetheless, there are examples of naturally occurring substances that are dangerous to human bodies, including radon (a radioactive gas originating from uranium and found in certain rocks), aflatoxin (one of the most carcinogenic substances known; produced by various species of fungus), and arsenic (sometimes found in groundwater). Arsenic in groundwater in Bangladesh has caused one of the most devastating cases of poisoning in history, and yet the substance occurs completely naturally. We discuss this case before turning our attention to synthetic pollutants for the rest of the chapter.

Arsenic in Bangladesh

Arsenic is a naturally occurring metallic element found in a variety of minerals widely distributed throughout the earth's **lithosphere**. Exposure to arsenic in the air can cause throat irritation and acute pulmonary symptoms, contact with the skin can lead to redness and swelling, and acute arsenic ingestion can cause vomiting, diarrhea, and even death. Chronic, low-level exposure can also lead to a range of potentially deadly symptoms, including nausea, decreased production of red and white blood cells, damage to the heart and circulatory system, and the appearance of small warts on the palms, soles, and torso (EPHA 2007). Children are particularly susceptible to long-term exposure, which can result in stunted development. Arsenic compounds are classified as "known **carcinogens**" because there is substantial evidence that

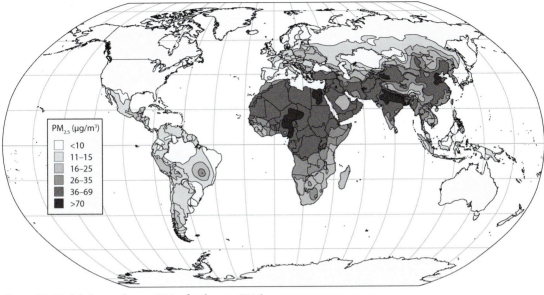

Figure 5.3 Modeled annual mean PM$_{2.5}$ for the year 2016

Source: Data from WHO (2019)

This map shows the estimated concentration of PM$_{2.5}$, particulate matter smaller than 2.5 microns in diameter.

three billion people cook on open fires fueled by substances such as kerosene and biomass, leading to almost four million premature deaths per year worldwide associated with pollution from cooking practices (WHO 2018c). In poorly ventilated houses where biomass is burned for fuel, indoor air pollution may reach concentrations one hundred times levels considered acceptable for outdoor air pollution, increasing individuals' susceptibility to respiratory problems, including chronic obstructive pulmonary disease (COPD), asthma, and lung cancer. In addition, stroke, ischemic heart disease, and pneumonia are associated with household air pollution (WHO 2018c). Women and children are especially vulnerable as they often spend a lot of time indoors. Almost half of all pneumonia deaths in children under age 5 are believed to be associated with particulate matter inhaled from household sources (*ibid.*).

In high-income countries, indoor air quality is usually better, although tightly sealing buildings to reduce energy loss can have detrimental consequences on indoor air quality (Fuentes-Leonarte, Ballester, and Tenías 2009).

Indoor air quality may also be compromised through a wide variety of chemicals used in the home, including paints, wall coverings, air fresheners, and cleaning products. Buildings themselves can also contain harmful substances such as asbestos (in insulation, floor coverings, and ceiling materials) and formaldehyde (in some furniture). The term "sick building syndrome" refers to situations in which chemicals from the fabric of a building or its furnishings lead to adverse health impacts. For example, some of the trailers provided by the US government to victims of Hurricane Katrina caused respiratory problems for susceptible individuals because they contained high concentrations of formaldehyde (Box 5.1).

Because many industrial processes produce airborne toxins, workplace air quality is an additional indoor health concern. Air quality in factories and other workplaces is carefully regulated in many countries to ensure the safety of workers. The Occupational Safety and Health Administration (OSHA) in the US and the Health and Safety Executive (HSE) in the UK, for example, monitor and regulate air quality in workplaces.

Box 5.1

ENVIRONMENTAL EXPOSURES AND HURRICANE KATRINA

On August 23, 2005, Hurricane Katrina formed over the Bahamas and landed on the Gulf coast of Louisiana within a week. Katrina flooded approximately 80 percent of the city of New Orleans, resulting in the immediate deaths of more than 1,800 people. At least 200,000 people were displaced from their homes and sent to evacuee shelters (CDC 2005). Alongside the many health problems resulting from the stress of displacement and living in crowded shelter conditions, the US Coast Guard and Environmental Protection Agency (EPA) reported 575 Katrina-related petroleum or chemical spills. In the aftermath of the hurricane, there were regular reports of illness as a result of contact with polluted water, including asthma and respiratory problems, sores, rashes, and nausea. The EPA also warned residents of New Orleans to take precautionary measures to prevent illness from exposure to contaminated food and water, mold resulting from the flood water, and mosquitos carrying West Nile virus.

The Federal Emergency Management Agency (FEMA) provided trailers for people rendered homeless by the hurricane in an effort to assist victims of Katrina. After a few months, some trailer residents began to complain of headaches, skin rashes, nosebleeds, and respiratory problems (Lohr 2007), leading to concerns that trailer residents might be exposed to formaldehyde, a chemical used in a diverse array of products. Acute exposure to formaldehyde can cause the health effects reported by the trailer residents, while long-term exposure can result in cancer (Environmental Protection Agency 2018a).

As a consequence of these concerns, FEMA and the CDC tested 519 trailers and found that they contained an average level of 77 parts per billion (ppb) of formaldehyde in the air, over three times the concentration found in most households and a level high enough to cause cancer with long-term exposure (CDC 2008). Consequently, FEMA began to move people out of the 38,297 trailers and into safer accommodation. After a recreational vehicle lobby unsuccessfully advocated sending the contaminated trailers to Haiti to assist with reconstruction following the 2010 earthquake (Associated Press 2010), presumably to avoid flooding the US market with cheap vehicles, many of the trailers were sold at low prices on the open market in the US (Hsu 2010).

In many poorer regions of the world, by contrast, legislation related to air quality may not exist or laws may not be enforced.

Although pollution of indoor air is still a significant risk to health, *outdoor* air pollution, or **ambient** air pollution, is progressively becoming the greater concern as industrialization creates pollutants to such an extent that they have altered the composition of the atmosphere. Major emitters include power plants, industry, vehicles, and biomass burning. Among the harmful substances produced by these activities, particulate matter (PM) and ground-level ozone are especially important as causes of ill health, with nitrogen dioxide (NO_2) and sulfur dioxide (SO_2) also significantly increasing the global disease burden (WHO 2018b, no date). **Primary air pollution** refers to pollutants that are emitted directly into the air; **secondary pollution** refers to a pollutant resulting from a chemical process after the initial chemical is emitted. For instance, ground-level ozone is a secondary pollutant because it forms when nitrous oxides (NO_x) and volatile organic compounds interact with sunlight. Ozone is beneficial in the upper atmosphere, where it shields the surface of the Earth from ultraviolet light but is considered a pollutant at ground level. Ground-level ozone can aggravate a variety of respiratory conditions and make the lungs more prone to infection, particularly in individuals with underlying respiratory problems such as asthma (EPA 2018b).

Recent estimates suggest that more than 90 percent of the world's population reside in places where outdoor air pollution exceeds WHO-recommended concentrations and that outdoor air pollution may cause 4.2 million premature deaths every year. Most of these deaths (91 percent) occur in low-income countries (WHO 2018b). Research since the early 1990s has consistently demonstrated a link between mortality and levels of particulate matter in the air, even at relatively low concentrations. The Harvard Six Cities Study, for instance, found that residents of cities with the lowest levels of airborne particulate matter in the study had life expectancies on average two years longer than people living in the cities with the highest levels (Dockery et al. 1993).

Urban residents are most likely to experience adverse health impacts from air pollutants caused by high concentrations of traffic and industrial and domestic pollutants, which can generate the heavy concentrations of pollutants that we commonly refer to as "smog." Globally, more than 80 percent of people living in urban areas are exposed to pollution levels exceeding WHO guidelines, including nearly all urban residents in the Global South (WHO no date). Vehicle pollution is a significant contributory factor. In the US, transportation contributed more than half of the carbon monoxide and nitrogen oxides, as well as nearly one quarter of hydrocarbons to the atmosphere in 2013 (Union of Concerned Scientists no date). In addition to their significant contribution to greenhouse emissions, vehicle exhausts produce particulate matter, volatile organic compounds, nitrous oxides, carbon monoxide, and sulfur dioxide—all with human health implications (Union of Concerned Scientists 2018). Efforts to address vehicular pollution often focus on reducing the number of vehicles on the roads. Several cities, including Paris, Delhi, and Beijing, have experimented with policies that allow only cars with odd or even number plates into the city on specific days. Other cities such as London have low emission zones where high-emitting vehicles, particularly diesel vehicles, are charged a fee for entry.

Vehicles vary widely in the volume of pollutants they produce, with many new technologies (such as high efficiency and hybrid cars) significantly reducing pollutant emission, so mandating the use of these new types of vehicles provides a potential policy approach for addressing air pollution. The UK government, for instance, has recently published its "Road to Zero" plan, which aims to end the sale of conventional gasoline and diesel cars by 2040. Facing pressure from vehicle manufacturers, the government ultimately backed down from its original plan to ban *all* new cars powered by fossil fuels by conceding that the sale of hybrid vehicles will still be permitted (Topham 2018). As this case illustrates, our exposure to vehicle emissions is influenced by political and economic decisions. This social framework around air pollution policy is evident in the recent history of diesel-fueled cars, which have transitioned from being seen as an environmental asset to a public health nemesis, following the revelation that the human health implications of diesel exhausts are far more devastating than was once realized (Box 5.2).

Box 5.2

THE RISE AND FALL OF THE DIESEL CAR

Diesel cars have traditionally been touted as more efficient and less polluting than petroleum-fueled cars, owing to their greater efficiency. Recent evidence, however, suggests that the risks associated with exposure to pollution from diesel cars may have been underestimated and that reducing diesel pollution should be a priority (Union of Concerned Scientists 2019).

The proportion of diesel cars in Western Europe increased from 13 percent in 1990 to more than half by 2015 (European Automobile Manufacturers Association 2018). Diesel cars were popular in Europe because high fuel taxes in many countries provided an economic incentive for consumers to switch to the more efficient diesel vehicles. After the Kyoto Protocol of 1997, European governments

continued

Box 5.2   *continued*

began to realize that a switch to diesel cars could also assist with meeting carbon emissions targets (Petzinger 2018), and so diesel cars were given further incentives in many countries.

Unfortunately, recent evidence has begun to suggest that the air pollution impacts of diesel may be far more significant than was originally thought, leading many European policymakers to reverse their stance. As early as 2012, the WHO increased its estimate of risk of diesel exhaust fumes from "probable carcinogen" to "major cancer risk" in response to research that concluded that diesel fumes could lead to lung cancer and perhaps also bladder cancer in high-risk workers such as truck drivers (Gallagher 2012; NHS 2012). Despite this, diesel-fueled cars continued to dominate the market in Europe over the next few years, with 52 percent of new cars produced in 2015 running on diesel, until a public scandal drew attention to the polluting effects of diesel (European Automobile Manufacturers Association 2017; Petzinger 2018).

In 2015, the German car manufacturer Volkswagen was found to have cheated on regulatory tests for nitrous oxides, making its cars appear to pollute less than they actually did. The US Environmental Protection Agency found that diesel Volkswagens in the US emitted up to forty times more toxic fumes than were permitted by US standards (Topham, Clark et al. 2015). The cars' software allowed the vehicle to operate in a low-emissions mode when being driven under test conditions but produce far greater emissions under normal driving conditions (*ibid.*). In the European Union, almost 8.5 million cars were sold with the fraudulent software, with an additional 482,000 vehicles implicated in the US (Oldenkamp, van Zelm, and Huijbregts 2016). Volkswagen eventually admitted that the problem applied to eleven million cars globally (Topham, Clark et al. 2015). One study estimated that nine million fraudulent vehicles sold in Europe and the US between 2009 and 2015 emitted 526 kilotons of nitrogen oxides beyond what they were permitted to emit. These additional emissions were calculated to have cost 46,000 **disability adjusted life years** (DALYs) (Oldenkamp, van Zelm, and Huijbregts 2016). Volkswagen subsequently received a hefty fine and recalled many of the affected cars. Other diesel car brands were subsequently subjected to more rigorous testing, and many were also found to emit higher levels of pollution than was previously believed.

Diesel exhaust fumes affect human health via three routes. First, particulate matter is created from the incomplete combustion of diesel fuel and often contains toxic chemical elements such as arsenic and cadmium, causing respiratory and cardiovascular problems (Oldenkamp, van Zelm, and Huijbregts 2016; Union of Concerned Scientists 2019). Second, diesel fumes contain large quantities of nitrous oxides; on-road diesel vehicles are thought to account for 20 percent of anthropogenic emissions of nitrous oxides globally (Anenberg et al. 2017). High concentrations of NO_2 can inflame airways, affect fetal lung development, reduce lung capacity, increase risk of asthma in childhood, and lead to increased mortality risk in adulthood from heart disease, lung disease, and stroke (Mills et al. 2015; Munro 2019). Finally, nitrous oxides can also transform into **secondary pollutants**, such as ozone, which cause additional health problems. As a result, diesel emissions exceeding national certification limits probably contributed to 38,000 premature deaths globally in 2015 (Anenberg et al. 2017).

By 2017, in response to "Dieselgate" and the associated spotlight that had been cast on the health impacts of diesel, the sale of diesel cars decreased across Europe (Petzinger 2018). Diesel bans are on the horizon in many European cities struggling with air pollution as citizen and physician advocacy groups have begun to lobby for stricter control of diesel emissions (e.g., Munro 2019).

History shows that pollution levels must reach concentrations with significantly detrimental impacts before the general public is willing to bear the cost of measures to reduce pollution. Consequently, countries in the middle stages of industrialization often experience the highest levels of pollution, as regulatory measures have not yet been introduced. This idea can be illustrated with an **environmental Kuznets curve** (Figure 5.4). The environmental Kuznets curve is an adaptation of work by economist Simon Kuznets that suggests that economic development initially leads to environmental degradation until society becomes affluent enough to be willing to invest in environmental protection measures. The current global distribution of air pollution is consistent with this theory, with the highest levels of pollution in countries such as India and China that have begun to industrialize but have not yet instituted widespread environmental protection measures. Many air pollutants, such as sulfur dioxide and nitrogen oxides, appear to be consistent with this model, although not all phenomena follow this pattern.

India, which currently has some of the worst air quality globally (Balakrishnan et al. 2019), provides a powerful example of the impact of air pollution on health. Recently, India's capital city Delhi hit the news as particulate matter from fireworks during the festival of Diwali exacerbated its already desperate air quality situation, causing Delhi's Air Quality Index to reach 1,000 (levels above 300 are deemed hazardous to health) in the early morning hours after the festival. A pall of smoke drifting away from Delhi was recorded by NASA satellites (Cuppucci 2018).

While fireworks and a temperature inversion made this event exceptional, air quality in Delhi and across India is regularly poor enough to have a substantial impact on health. More than half of all households in India burn solid fuels for heating and cooking. Additionally, high levels of ambient air pollution are generated by emissions from coal-burning power plants, industrial activities, construction and brick kilns, transportation, dust from roads, agricultural waste burning, and diesel generators (Balakrishnan et al. 2019).

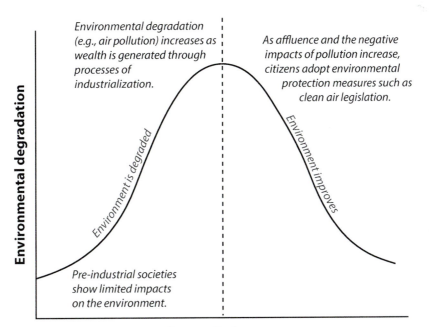

Figure 5.4 Environmental Kuznets curve showing the relationship between wealth and environmental degradation

As such, India suffers a double exposure to pollution associated with both its rural, agricultural history and its industrializing future, although urban smog is increasingly the dominant cause of poor health. Fully 77 percent of India's population is exposed to levels of particulate matter in ambient air that does not meet current Indian air quality standards (Balakrishnan et al. 2019). The city of Delhi has the highest levels of particulate matter among the heavily polluted group of megacities worldwide (WHO, Department of Public Health 2016). In 2017, 1.24 million deaths (12.5% of deaths in India) were attributed to air pollution (Balakrishnan et al. 2019). India's life expectancy would increase by an estimated 1.7 years if air quality could be improved to healthy levels (*ibid.*).

China has similarly been coping with high levels of air pollution in its major industrial cities, associated with its "economic miracle." Driven by extraordinary industrial growth over the past two decades, and particularly a heavy reliance on coal as an energy source (Millman, Tang, and Perera 2008), air pollution in China reached some of the highest levels ever recorded in the 1990s (WRI et al. 1998), initiating a public health crisis (Figure 5.5). One study found that outdoor air pollution in China in the early 2000s was associated with more than 300,000 deaths and 20 million cases of respiratory disease annually, at a cost of more than 3 percent of the country's gross domestic product (GDP) (Millman, Tang, and Perera 2008, 260). More than 70 percent of Chinese households burned coal or biomass for heating and cooking at the time, exacerbating problems of indoor air quality, and causing problems such as reduced fetal and child growth, pulmonary disease, developmental impairment, and elevated cancer risk (*ibid.*).

Figure 5.5 Air pollution, Beijing, China, 2009
Credit: Photo by Shane Houdek

China began instituting pollution control policies in earnest in 2005, when controls on fine particulate pollution were implemented on power plants, industry, and vehicle emissions (Zhao et al. 2018). Nonetheless, China's air quality was still so poor in 2008 that many people were concerned about the health risk to international athletes during the Beijing Olympics. The Chinese government undertook a massive effort to reduce airborne pollutants for the period of the Games by restricting vehicle use, closing some polluting facilities, and decreasing coal combustion (Friedrich 2011). One study found that the risk of lung cancer associated with exposure to one pollutant (polycyclic aromatic hydrocarbons) would have been halved if these mediation efforts had been continued after the Olympics (Jia et al. 2011).

Although efforts to control pollution were relaxed after the Olympic Games and China remains a global hotspot for pollution, renewed investment in pollution control measures over the past few years may finally be improving conditions. Chinese Premier Li Keqiang announced that China was waging a war against smog in March 2014, reversing long-standing policies that had prioritized economic development over the environment. Perhaps China has begun the path down the descending limb of the environmental Kuznets curve? One policy forced villages to turn from burning solid fuel to natural gas. The Chinese government claims that four million homes have now made the change (Yu 2018). A recent study suggests that this measure has reduced the most damaging forms of particulate pollution, preventing an estimated 400,000 premature deaths annually (Zhao et al. 2018). Indeed, cleaning up household fuels appears to have had a more significant impact on exposure to particulate matter than efforts to control industrial sources of pollution (*ibid.*). Another study that used remote sensing to quantify particulate pollution transported eastwards towards South Korea and Japan observed that particulates in the air increased until 2007, followed by a 10 to 20 percent decrease over the following decade (Zhang et al. 2017). Readings from a network of pollution monitoring stations that has been set up across China suggest that air pollution has finally begun to decline (Silver et al. 2018).

We should not forget that many affluent countries experienced similarly devastating pollution problems while they were industrializing. For instance, the Great Smog of December 1952 in London may have led to twelve thousand deaths (Hunt et al. 2003). The normal load of industrial pollutants in the air was exacerbated by diesel fumes (London had just phased out electric trams in favor of diesel buses) and coal smoke from domestic heat sources. A stagnant air mass trapped by a temperature inversion prevented the dissipation of these pollutants (*ibid.*). Although the death toll of this event was exceptional, morbidity and mortality from poor air quality was probably significant in London from the 1700s well into the twentieth century. Since the mid-twentieth century, pollution legislation, such as removing lead from gasoline, has greatly reduced environmental exposures in many affluent countries, again illustrative of the descending limb of the environmental Kuznets curve.

In some cases, the physical setting of a city can make it particularly vulnerable to air pollution. For instance, Mexico City is located in a basin encircled by higher ground, which results in pollution being trapped over the city (Figure 5.6). The main opening to this basin in the north funnels the prevailing northeast winds into the basin, pushing pollutants back over the city. To make matters worse, a major industrial zone situated just north of the city pumps pollutants into this moving air mass and over the city (Yip and Madl 2002). Mexico City's high elevation and intense sunlight also encourage the formation of ground-level ozone.

Climate change provides another example of how environmental context can affect air pollution, in what the WHO refers to as the "air pollution and climate nexus" (WHO no date). Many facets of global climate change, particularly pertaining to air temperature, humidity, and wind, could affect the production or dispersal of pollutants. In particular, the chemical processes associated with the creation of secondary pollutants are often influenced by atmospheric conditions, making them susceptible to the effects of climate change. For instance, production of ground-level ozone increases on hot, sunny days because it is created by photo-chemical processes. One study in Britain found that an increase in ground-level ozone concentrations associated with the unusually warm summers of 2003, 2005, and 2006

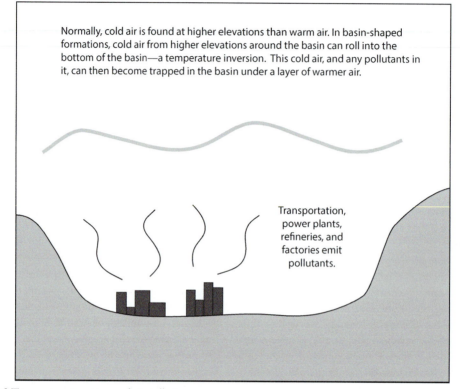

Figure 5.6 Temperature inversion and air pollution

may have led to approximately five thousand deaths (Doherty et al. 2009).

The destruction of stratospheric ozone by synthetic pollutants since the 1980s provides a cautionary tale of the complexity of our atmospheric systems and unforeseen human health effects. While ground-level ozone is a powerful pollutant, with detrimental impacts on human health, ozone in the stratosphere is critical for protecting health by filtering out some of the ultraviolet light from the sun. The depletion of ozone from human-produced chlorofluorocarbons has caused higher levels of exposure to dangerous forms of ultraviolet light for people in the mid- to high-latitudes, thereby increasing the incidence of skin cancer. Australia currently has the highest rate of skin cancer in the world, after experiencing a 60 percent increase in the incidence of melanoma between 1982 and 2010. An estimated 95 percent of skin cancers in Australia are caused by skin damage associated with

exposure to ultraviolet radiation from the sun (Alliance 2019). Climate change may compound these problems if warmer temperatures lead to behavioral changes such as spending more time at the beach that could lead to even greater exposures (*ibid.*).

Although chronic exposure to pollutants probably places the greatest burden on health in most societies today, acute exposures periodically occur and can be devastating. One of the most infamous industrial accidents in human history occurred on December 3, 1984, at the Union Carbide pesticide plant in the city of Bhopal, India. Methyl isocyanate leaked into the air from a gas tank at the plant, exposing hundreds of thousands of people in the surrounding urban area. Although the precise toll of the accident will never be known, the regional government of Madhya Pradesh confirmed approximately 3,800 deaths in the immediate aftermath of the event; other estimates suggest that as many as ten thousand deaths may have occurred within a few days of the accident (Sharma 2005; MacKenzie 2002), mostly in urban slums surrounding the plant. An additional fifteen to twenty thousand premature deaths may have occurred over the next two decades (Sharma 2005), and hundreds of thousands became ill as a consequence of the disaster (Trotter, Day, and Love 1989).

The US-owned Union Carbide factory was run by an Indian subsidiary, leading to some contention over who should be held responsible for the disaster (MacKenzie 2002). Problems with safety equipment and procedures at the plant had been identified two years before the leak, but the plant was nonetheless operating well below the safety standards of its sister plant in the US state of West Virginia (Broughton 2005). In 1989, Union Carbide paid US$470 million to a trust to compensate victims of the disaster, but survivors claim they received only about US$500 each and that clean-up efforts were inadequate. Recent evidence suggests that the plant was contaminating the environment before the leak, with waste left to evaporate in open lagoons and localized contamination of groundwater and soil with heavy metals, leading to long-term exposure of local residents (MacKenzie 2002). As recently as 2018, the area around the old Union Carbide factory was still considered to be heavily contaminated, with many of the poorest local people, who were not able to move away, still exposed to toxins that have leached into the soil and water supply (Mandavilli 2018).

Water quality

People are exposed to toxins in the water through ingestion or contact via activities such as bathing. Many industrial processes create water-soluble acids, heavy metals, and other hazardous wastes that may poison aquatic life and cause disease in humans (Figure 5.7). Additionally, household products such as detergents, hygiene products, cosmetics, antidepressants, birth control medication, and chemotherapeutic medication can enter the water through municipal waste systems. Scientists have detected the presence of chemicals originating from these products in waste slurries, potato fields, groundwater, and even aquifers that feed municipal water systems (Kolpin et al.

2002). The idea of mining sewage for valuable metals, including gold and silver, is now being explored as it is believed that millions of dollars' worth of metals could be found in sewage sludge, particularly in run-off from manufacturing regions. The city of Suwa in Japan has attempted this and retrieved more gold from sewage sludge than is found in many commercially mined ores (Corwall 2015).

Some chemical contaminants, such as lead, are known to pose a risk to human health and are stringently regulated in most places. Abundant evidence suggests that exposure to even low levels of lead can be harmful, particularly for children and fetuses. Many water pipes used to be constructed with lead and so public health agencies often carefully monitor levels of lead in drinking water.

The health impacts of a host of other chemicals found in water sources are not well understood. Even chemicals that are known to be toxic at high concentrations may have unknown effects from low-dose chronic exposures. Some chemicals may have toxic effects if they transform into new compounds with other substances. The cost of studying the human health impacts of all these substances and their potential combinations is prohibitive, and so many of these exposures currently present an unknown risk.

Water pollution is further complicated by the ecology of freshwater systems. Changes to freshwater environments can alter their ability to support disease-causing organisms or influence chemical reactions in the water. Processes such as *sedimentation* (the buildup of particles in the water), *eutrophication* (the loading of nutrients into an aquatic ecosystem), and *thermal pollution* (changes in water temperature caused by industrial processes or from reactions among pollutants) may indirectly influence human health. Climate change also plays a role by increasing water temperatures, altering flow regimes, and causing flooding. These processes stress aquatic systems that are already strained by human activities.

There is also a tradeoff between effectively treating water for pathogens and maintaining treatment chemicals at safe levels. In 1974, researchers discovered chloroform in drinking water, produced when chlorine reacts with organic matter, leading to a debate over the safety of using disinfectants in drinking water

Figure 5.7 Oil leak, Pearl Harbor, Hawaii
Credit: Melissa Gould

(Rosario-Ortiz et al. 2016). Disinfectants may also cause lead to leach from water pipes in older water systems, generating an additional potential exposure (Edwards, Triantafyllidou, and Best 2009). In response, some European countries (e.g., the Netherlands, Germany, and Austria) have moved towards alternative cleaning processes such as filtration, ozone, and ultraviolet treatments, to reduce chemical disinfectants in drinking water. In the US, by contrast, the use of chlorination to clean water remains common; critics argue that this is because this requires less investment in updating old infrastructure (Rosario-Ortiz et al. 2016). Although there is considerable disagreement over the best way to balance microbial and chemical risks in treating drinking water (*ibid.*), devastating events such as the lead contamination crisis in Flint, Michigan, bring these questions to the fore.

Water contamination in Flint, Michigan, US

The town of Flint, Michigan, traditionally drew its water from Lake Huron, under an arrangement with the Detroit Water and Sewage Department that had been in place for half a century. While awaiting the construction of a new pipeline to Lake Huron in 2015, the town began to draw water from the Flint River as a cost-saving measure (Hanna-Attisha et al. 2016). Soon after the switch, residents became concerned over changes in the color, taste, and smell of their drinking water and began to complain of skin rashes. Water testing showed bacterial contamination, including *Escherichia coli*, resulting in violations of the US's Safe Drinking Water Act (*ibid.*). The city responded by applying additional disinfectants to the water, only to result in further violations of the Safe Drinking Water Act—this time with respect to chemical contamination from trihalomethane (*ibid.*). The addition of extra chlorine to the water had even more devastating health consequences because it triggered the release of lead into the water from pipes. Flint's traditional water supply had low corrosivity for lead and included a corrosion inhibitor added by the Detroit water authority. By contrast, the new water supply from the Flint River was more corrosive, and Flint city officials decided not to invest in adding corrosion control chemicals to the water. This more corrosive water stripped away the rusty build-up that had lined the city's water pipes, exposing the lead pipe underneath and delivering lead straight into the water supply (Panko 2017).

Repeated complaints from the local community initially gained little attention, but eventually researchers from Virginia Tech University were brought in to investigate the water supply. The researchers recorded high levels of lead in the water, and subsequent research found elevated lead in the blood of local residents (Hanna-Attisha et al. 2016). Flint's community already had elevated blood lead levels before the crisis, owing to pre-existing risk factors such as poor nutrition, poverty, and older housing stock (which is more likely to contain lead in paints and pipes), but the water crisis doubled the proportion of Flint's children with elevated blood lead levels from 2.4 percent to 4.9 percent (*ibid.* p. 285). Results varied from neighborhood to neighborhood, however, with elevated readings concentrated in the poorest neighborhoods. Indeed, in some affluent neighborhoods, blood lead levels actually *dropped* from before the crisis, probably because middle-class residents switched to bottled water as the event unfolded (*ibid.*)

In October 2015, after thousands of people had been exposed to dangerous levels of lead for several months, Flint returned to its original water supply, and the city began adding additional phosphates to the water to reduce lead levels (Panko 2017). The long-term health effects of lead exposure include developmental delay and learning difficulties in children, as well as pre-term birth and low birthrate in babies exposed prenatally, and so the crisis will likely have an impact for years to come.

The water crisis in Flint is widely regarded as a case of **environmental injustice** (e.g., Balazs and Ray 2014;

Greenberg 2016). As a cash-strapped post-industrial city with a large proportion of poor residents, lack of funds and limited political power have been blamed for the constellation of events that led to the disaster. There were lead pipes in the aging infrastructure and a declining tax base, which contributed to the unfortunate decisions to cut corners with water safety. The marginalized status of Flint's population meant that it took several months for the problem to gain public attention and several more months before a solution was finally implemented.

Solid waste

Solid waste is not only unsightly but also poses significant threats to human health. Food waste can help support populations of disease-carrying animals such as rats, dogs, and flies. Impermeable packaging can trap water that can provide breeding grounds for vectors, a significant problem with respect to dengue fever. Geogens are also common in the wastes we produce, including substances such as lead, mercury, and asbestos, which can get into the broader environment through shoddy waste disposal practices. The primary route of human exposure to toxins from solid waste is through ingestion after toxins have leached into the soil or water supply and then into the food chain. Other forms of solid waste, such as asbestos, can lead to health problems through inhalation or exposure to tainted soil or dust (Table 5.3).

Until the 1950s, scant attention was given to waste disposal. Industry disposed of solid waste with little regard for its potential environmental impacts, and domestic garbage was dumped in landfills, where it was sometimes burned. In many cities of the Global South, refuse disposal remains chaotic and inadequate for the rapidly growing populations (Figure 5.8). In the Global North, waste disposal tends to be better regulated, but concerns remain over the potential toxicity of many of the products that we throw away, given evidence that living near a landfill has been associated with an increased risk of a variety of adverse health effects (Palmeira Wanderley et al. 2017). Many countries are also struggling to find enough appropriate sites for solid waste disposal, given the large quantity of non-biodegradable materials that are thrown away. Recycling waste would appear to

Table 5.3 Primary hazardous wastes

Substance	Source	Acute health effects	Chronic health effects
Arsenic	Natural, industrial effluents, combustion	Nosebleeds, headaches, loss of hair	Cancer, hyperkeratosis, cardiovascular disease, reproduce effects
Lead	Natural, lead-based paint, contaminated soil and dust	Convulsions, coma, death	Impairment of mental and physical development, damage to nervous system, kidneys, and blood cells
Mercury	Found in the earth's crust, combustion of coal, industrial processes	High blood pressure, impaired cognitive function, damage to kidneys	Impaired neurological development, impaired nervous system, birth defects
Vinyl chloride	Used to make pipes, wiring, upholstery	Coughing, shortness of breath, dizziness, nausea, death	Liver damage, weakened immune system, damage to nervous system
Polychlorinated biphenyls	Used as coolants and lubricants in electrical equipment, fluorescent lights	Skin conditions such as acne and rashes	Liver damage, immune system damage
Benzene	Naturally occurring, vehicle exhaust, production of many synthetic materials	Drowsiness, dizziness, headache, tremors, confusion, death	Tissue damage, anemia, cancer (especially leukemia)
Cadmium	Natural, used in batteries, pigments, metal coatings, and plastic	Vomiting and diarrhea	Cancer, kidney, lung, and bone damage
Polycyclic aromatic hydrocarbons	Combustion of coal, oil, gas, garbage, tobacco; also found in charbroiled meat	Skin and tissue damage, weakened immune system	Skin and tissue damage, weakened immune system, cancer

Note: Hazardous wastes are listed in approximate order of the quantity emitted.

Source: Agency for Toxic Substances Disease Registry (2009)

be an ideal solution but has traditionally relied on large numbers of low-paid workers in countries like China to process a lot of the waste created by high-income countries such as the US—another case of **environmental injustice**. As China becomes less willing to take America's waste, the US is currently struggling to find solid waste solutions.

The explosion in plastics production that has occurred since the 1950s means that plastic is now ubiquitous in our terrestrial and marine environments, with unknown health consequences. In many ways a miracle product—waterproof, hygienic, cheap to produce, and incredibly diverse—plastic production has escalated rapidly, with numerous different types of plastic invented, many of which are designed to be disposable and few of which

have been explicitly designed to be recyclable. Research has only just begun to reveal the health impacts of this plastics revolution.

Plastics play an important role in society, with some undeniably positive impacts, revolutionizing food storage, vehicle manufacture, and surgery, among many other facets of modern life. Globally, plastic production increased from 1.7 million tons in 1950 to 322 million tons in 2016 (Balazs and Ray 2014). The sheer scale of plastic production and the extremely slow rate at which plastics degrade have left the world with a significant environmental problem with potential health implications, however. Here, we focus on pollution from microplastics as one illustration of the growing problem of solid waste.

Figure 5.8 Refuse near self-built housing, Ko Lanta, Thailand
Credit: Melissa Gould

Microplastics pollution

All plastic items present a waste disposal challenge. Large plastic items provide an unsightly source of solid waste, both on land and in water courses, as well as a direct threat to wildlife. For human health, however, it is often the ingestion of tiny particles of plastic and exposure to toxic chemicals in plastics that are the greatest concerns. In the early 2000s, Thompson et al. (2004) began to investigate the impact of tiny particles of plastic in ocean environments, coining the term "microplastics" to refer to the minute plastic particles and fibers that they found. As large plastic items slowly degrade, their surfaces release numerous micro- to nano-sized fragments (Song et al. 2017), particularly in the presence of high temperatures and ultraviolet light (Andrady 2015). Industrial and domestic effluent are additional sources of

microplastics (Karbalaei et al. 2018), with laundry water alone contributing significant amounts, as synthetic fabrics shed fibers into wash water (Browne et al. 2011). Even personal care products have been implicated in adding microbeads of plastic to the environment (Lei et al. 2017).

Subsequent research illustrates that the problem has become significantly worse in recent years. An estimated 12.7 million tons of plastic waste entered the oceans in 2010 (Jambeck et al. 2015), and microplastics are now ubiquitous in marine environments (Shim, Hong, and Eo 2018). Microplastics can float on seawater and travel widely with winds and marine currents (Maximenko, Hafner, and Niiler 2012), with hundreds of publications documenting microplastic contamination from pole to pole and from the water's surface to the ocean's depths

(Rochman 2018). Microplastics are also widely found in freshwater systems. Soil may act as a long-term sink for microplastics, although less research has explored microplastics in terrestrial ecosystems (*ibid.*). Microplastic contamination may also occur through atmospheric deposition, explaining how even very remote locations can become contaminated (Dris et al. 2016).

Microplastics are rapidly ingested by aquatic life (Thompson et al. 2004) and have been linked to morbidity and mortality in numerous aquatic organisms (Karbalaei et al. 2018). The impacts of microplastics on human health is unclear, however, largely because of the limited period of research in this field. There are several mechanisms through which microplastics could affect health, however: microplastics could cause physical damage from the particles themselves, biological stresses could develop from exposure to the particles, or the chemical additives in plastic may have toxic effects (Revel, Châtel, and Mouneyrac 2018).

Humans may be exposed via foods, with plastic contamination recorded in a wide variety of seafood (see Revel, Châtel, and Mouneyrac 2018; Karbalaei et al. 2018). The minute size of many plastic particles means that even tiny ocean filter feeders may ingest plastic from the water and the particles can then accumulate up the food chain. Dust, groundwater, agricultural soils, sea salt, honey, and beer have been suggested as additional sources of exposure (Revel, Châtel, and Mouneyrac 2018; Rochman 2018; Karbalaei et al. 2018). Inhalation of nanoparticles of plastic may also be detrimental to health, perhaps leading to lung disease (Karbalaei et al. 2018). Even tap water has been implicated as a potential source of microplastic contamination, with one study finding particles in 83 percent of samples analyzed from five separate continents (Kosuth et al. 2017). Again, however, the human health implications are poorly understood.

Many microplastics contain chemical components used as stabilizers, flame retardants, pigments, plasticizers, and antioxidants, with a range of toxicities (Shim, Hong, and Eo 2018). Some plastics, such as polyethylene terephthalate (PET; used in drinking bottles, plastic film, and microwave packaging) and polyvinyl chloride (PVC), have been identified as suspected carcinogens (Li, Tse, and Fok 2016; Wang et al. 2016). Bisphenol A (BPA) is one of the additives to plastics that has recently attracted attention. BPA is used in the lining of many food and beverage cans and has been shown to leach into food under certain circumstances. Exposure to BPA has been associated with altered liver, reproductive, and brain function; changes in insulin resistance; changes in fetal development in pregnant women; as well as possible links to obesity and cardiovascular disease (Srivastava and Godara 2017). BPA is an example of an endocrine disruptor—a group of chemicals that mimic hormones in the human endocrine system (Box 5.3).

Box 5.3

ENDOCRINE-DISRUPTING CHEMICALS

Endocrine-disrupting chemicals (EDCs) are substances that can disrupt the endocrine system by mimicking or interfering with the hormones that regulate biological processes. The endocrine system plays a critical role in sex differentiation and reproduction via hormones such as estrogen, which causes feminization, and androgen, which causes masculine characteristics to develop. Another major hormone group, the thyroids, are integral to brain development (WHO and UNEP 2012). The interruption of endocrine processes can therefore have significant effects on development, particularly on sexual characteristics and the functioning of the reproductive system (*ibid.*). Some EDCs may also affect metabolic processes and so diabetes and obesity may also be linked to exposure to EDCs (Gioiosa et al. 2015). Currently, approximately eight hundred chemicals are suspected to interfere with the endocrine system (WHO and UNEP 2012), including the synthetic estrogen DES (diethylstilbestrol), PCBs (polychlorinated biphenols), and the pesticide DDT (dichlorodiphenyltrichloroethane) (National Institute of Environmental Health Sciences 2015).

Humans are exposed to EDCs by ingestion, inhalation, or through the skin. Many of these substances can also be transferred transplacentally to a developing fetus or to an infant via breast milk. Concerns over the influence of chemicals on the endocrine system surfaced in the 1960s when malformations were observed in infants after exposure to certain drugs taken by the mother during pregnancy. Thalidomide, a drug for alleviating morning sickness that caused malformed limbs in infants exposed in-utero, is the most notorious example.

Subsequent research has identified several lines of evidence that support the idea that EDCs may have implications for human health. First, many endocrine-related disorders have increased in human populations over the past fifty years. Second, research has revealed significant endocrine-related health effects in wildlife. Finally, laboratory studies have begun to link certain EDCs with specific disease outcomes (WHO and UNEP 2012). Nonetheless, EDCs remain a contentious topic, with many arguing that there is insufficient evidence to conclusively link EDCs with negative health outcomes.

Nonetheless, circumstantial evidence is mounting. Over the past few decades, there has been a rise in rates of genital malformations, poor semen quality, endocrine-related cancers, obesity, and Type II diabetes, all with likely endocrine associations (WHO and UNEP 2012). Up to 40 percent of young men now have low semen quality in some countries, with attendant fertility concerns (*ibid.*). Since male development relies on multiple hormones, males are typically more sensitive to endocrine disruption than females, with problems such as defects of the penis, interrupted testicular descent, and testicular cancer all associated with exposure to EDCs (*ibid.*). For females, early onset of breast development—a risk factor for breast cancer—is now common across the world (*ibid.*), and exposure to androgens has been associated with polycystic ovarian syndrome later in life (Pasquali et al. 2011).

Because of ethical concerns related to conducting research on human subjects, much of our evidence of health impacts of EDCs comes from animal studies.

A variety of studies have found connections between environmental exposures to EDCs and adverse effects in wildlife. For example, compounds excreted as a result of taking female contraceptive pills have been documented to cause feminization in fish (Kidd et al. 2007). There is also evidence that PCB pollution during the 1970s and '80s caused high rates of reproductive failure and abnormal female reproductive development among seals in the heavily polluted Baltic Sea (WHO and UNEP 2012). Other evidence from wildlife shows that the anti-fouling agent tributyltin (used in ship paints) causes abnormalities in mollusk sexual development, and that there are high rates of testicular non-descent in black-tailed deer in Alaska and Montana (*ibid.*).

Unfortunately, animal studies are not always good predictors of human disease. Additionally, traditional testing methods for chemicals, whereby a high dose is tested and then used to predict the potential impacts of lower doses of the chemical, is insufficient in studies of EDCs as endocrinologists have long known that responses to hormones do not occur in a linear, dose-dependent fashion. As such, testing of chemicals for endocrine-disrupting effects needs new study protocols (Gioiosa et al. 2015).

Removing EDCs from the water supply may seem like an obvious solution, but it is a difficult and complex process. Wastewater treatment plants are already designed to remove many chemicals from the water before discharge but always leave some trace contamination. Endocrine-disrupting chemicals are therefore often detected in rivers downstream of treatment plants (Barber et al. 2015). In addition, our knowledge of EDCs is still incomplete and so scientists do not know all potential sources of EDCs. A vast range of chemicals are discharged into water courses, including industrial chemicals, pesticides, cleaning products, and prescription drugs, any number of which could have endocrine-disrupting effects. Working out which of these chemicals is most critical to remove from the water supply is a key challenge. For instance, the common pain-relieving drug paracetamol (acetaminophen) has been shown to have endocrine-disrupting

continued

Box 5.3 *continued*

effects, even at the very low concentrations found in rivers after water treatment processes have removed almost all of the substance. One study showed that exposure to the low concentrations of paracetamol found downstream of a water treatment plant reduced testosterone levels in fish by more than half after exposure for just three weeks (Guiloski et al. 2017).

Bisphenol A is one chemical that has come to the public's attention for its potential endocrine-disrupting properties, although the nature of the threat remains controversial. Used in clear plastic bottles and as a coating on the lining of beverage and food cans, humans have significant potential for exposure by ingestion. BPA can also be passed transplacentally and to a lesser degree via breast milk

(Vandenberg et al. 2007). The CDC has reported finding BPA in the urine of almost all people tested in the US, indicating widespread exposure (CDC 2018). A number of studies have found BPA to have endocrine-disrupting and metabolic effects after chronic exposure at low doses (e.g., Gioiosa et al. 2015), although other studies have not recorded a significant influence on thyroid or hormone levels (WHO and UNEP 2012). The US Centers for Disease Control (2017) conclude that more research is needed to determine the true impacts on human health. Nonetheless, several countries have banned BPA from baby bottles and toys as a precautionary measure (WHO and UNEP 2012), and BPA is gradually being phased out of many other products in response to consumer concerns.

While the study of the health impacts of plastics is an active area of research and considerable circumstantial evidence suggests that exposure to microplastics *may* damage health, there is little clear cause-and-effect evidence. Important questions persist about the dose of microplastics needed to affect health, as well as the mechanisms through which different plastics could affect human health. As is common with efforts to assess exposure risk, demonstrating a clear causative relationship requires extensive research and careful statistical analysis. It is to this field of exposure assessment that we turn next.

ASSESSING ENVIRONMENTAL EXPOSURES

Owing to the grave potential health consequences of exposure to geogens, most governments attempt to regulate the substances to which their populations are exposed. This includes legislation on the disposal of hazardous materials, exposures in the workplace, and acceptable levels of contaminants in drinking water and food. To formulate policy rationally, we must first understand the costs and benefits associated with particular activities. If it is suspected, for example, that exposure

to benzene puts people at risk for leukemia, then there is an obvious incentive for governments to regulate the level of benzene in workplaces, and yet such regulation will inevitably come at some cost to industry. This leads to difficult questions: How much risk is acceptable? How much money are we willing to spend to protect health? How sure must we be of the impacts of a substance before we regulate its use? The process of collecting and assessing data to address such questions has led to a field of inquiry in public health called **risk analysis**.

Identifying and verifying a causal link between an exposure and a health effect is difficult. Rarely can the precise dose and timing of an exposure be verified. Establishing causality is often also complicated by a lag between the time of exposure and the onset of observable health effects. Furthermore, everyone is exposed to a huge number of toxins over their lifetime, making it difficult to determine which exposure is responsible for any particular outcome. Most health effects result from a complex combination of causal factors, including exposures, genetics, and behaviors, making it even more difficult to pinpoint a direct causal link between an exposure and a disease. A variety of analytical techniques can address these challenges in an attempt to generate objective conclusions about connections between exposures and disease (Figure 5.9).

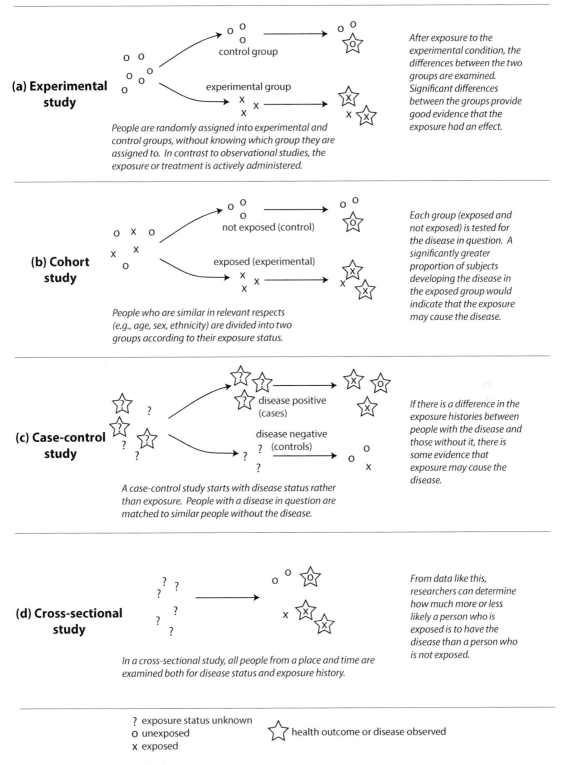

(a) Experimental study

control group

experimental group

People are randomly assigned into experimental and control groups, without knowing which group they are assigned to. In contrast to observational studies, the exposure or treatment is actively administered.

After exposure to the experimental condition, the differences between the two groups are examined. Significant differences between the groups provide good evidence that the exposure had an effect.

(b) Cohort study

not exposed (control)

exposed (experimental)

People who are similar in relevant respects (e.g., age, sex, ethnicity) are divided into two groups according to their exposure status.

Each group (exposed and not exposed) is tested for the disease in question. A significantly greater proportion of subjects developing the disease in the exposed group would indicate that the exposure may cause the disease.

(c) Case-control study

disease positive (cases)

disease negative (controls)

A case-control study starts with disease status rather than exposure. People with a disease in question are matched to similar people without the disease.

If there is a difference in the exposure histories between people with the disease and those without it, there is some evidence that exposure may cause the disease.

(d) Cross-sectional study

In a cross-sectional study, all people from a place and time are examined both for disease status and exposure history.

From data like this, researchers can determine how much more or less likely a person who is exposed is to have the disease than a person who is not exposed.

? exposure status unknown
o unexposed
x exposed

☆ health outcome or disease observed

Figure 5.9 Epidemiologic study designs

EPIDEMIOLOGIC APPROACHES TO ASSESSING EXPOSURES

Much of our understanding of the relationship between exposures and disease relies on work from the field of **epidemiology**. While physicians focus on the health of individuals, epidemiologists study health and disease at the population level. For epidemiologists, the notion of an **exposure** can be generalized to include not only environmental contaminants but also behaviors (such as how much exercise an individual normally gets or what types of food they eat) in order to examine the impacts of everyday activities on health. Epidemiology relies heavily on **positivist** and **quantitative** approaches for gathering information. A positivist perspective maintains that we can only know what we can observe and measure and relies heavily on scientific methods that test hypotheses through controlled experimentation. The field of **statistics** provides objective rules for how to interpret quantitative data.

One of the most rigorous forms of the scientific method is **experimental studies**, also known as "clinical trials" in the case of medical research on human subjects

(Figure 5.9a). In an experimental study, people are randomly assigned to two different groups. A treatment or exposure is administered to one group, the treatment group, for a specified period. A **control group** is subjected to conditions as similar as possible to the treatment group, except for the treatment or exposure in question. After the study period, the health outcomes of each group are carefully measured and compared. If there is a significant difference between the two groups, we can conclude that the exposure or treatment has had an effect.

The obvious problem with experimental studies is that it is not ethical to subject people to exposures that may damage their health. Historical experience with unethical clinical trials has generated great concern over the use of human subjects in scientific studies. One of the most infamous cases of unethical treatment of research subjects was the "Tuskegee Study of Untreated Syphilis in the Negro Male" (CDC 2015), in which a treatment was withheld from study participants long after it had been shown to be effective (Box 5.4). Many countries and institutions now have ethics committees to oversee and regulate research using human subjects, although

---

Box 5.4

RESEARCH ETHICS AND THE TUSKEGEE STUDY

The "Tuskegee Study of Untreated Syphilis in the Negro Male" that began in 1932 was backed by the US Public Health Service. It justified a treatment program for syphilis among African Americans to generate a greater understanding of the natural history of the disease, but the study has been widely criticized for compromising the rights of its research subjects.

Investigators recruited approximately six hundred African American men for the study in Tuskegee, Alabama, many of whom were living in poverty. Approximately two thirds of the participants had syphilis. In return for their participation in the project, participants received free medical exams, free meals, and burial insurance (CDC 2015).

A key criticism of the project was that research subjects were not adequately informed about the goals of the project and never provided informed consent of their participation. Furthermore, study participants suffering from syphilis were not properly treated for it, even though an effective treatment for the disease was developed during the study (CDC 2015). What began as a six-month-long study in 1932 ended up lasting forty years and costing the lives of many participants.

In October 1972, the study was stopped after being found to be "ethically unjustified" by an advisory panel's investigations. A class-action lawsuit and out-of-court settlement with the US government left the government paying medical and burial expenses for all the study's participants and subsequently for their wives and children (CDC 2015).

questions continue to be raised over research carried out in low-income countries, which may lack rigorous oversight.

Owing to these concerns, clinical trials are used largely for testing drugs and other therapies designed to *improve* health. Where test exposures may harm people, animals are often substituted for human subjects. Unfortunately, even if we overlook ethical objections to using animals as experimental subjects, animals are often a poor substitute for estimating the effects of exposures on humans. To avoid the ethical dilemma of deliberately exposing living subjects to harmful substances, a separate group of studies—observational studies—take advantage of the fact that people are already exposed to many toxins in their everyday lives.

Observational studies examine large populations within which some proportion of individuals are exposed to a hypothesized threat to health without intervention by the researcher. For instance, a research team could divide a population into smokers and non-smokers and then follow the health outcomes of the two groups. This is usually considered ethical because the researchers have not persuaded anyone to alter their behavior.

For example, Guillette et al. (1998) used an observational study to study the impact of agricultural chemicals on children's development in the Yaqui Valley of Mexico. One group of children was selected from the lowlands, where they were exposed to high levels of agricultural chemicals. A second group was selected from a community in the foothills that lacked exposure to the chemicals but was otherwise similar to the first group with respect to factors such as diet, ethnicity, and culture. After careful study, the researchers concluded that the lowland children exhibited significantly higher rates of neuromuscular and mental deficits than the foothills children, supporting the hypothesis that agricultural chemicals may cause developmental problems in children.

It should be noted that such observational studies rarely provide conclusive evidence of a causal link but instead simply indicate that there is some connection between an exposure and a health outcome. The phrase "*association does not imply causation*" reminds us that discovering that two things occur together in time and space does not mean that one necessarily *caused* the other. Observational studies are useful for building evidence about cause-and-effect relationships but should be followed up with additional research. To add support to the findings of Guillette et al.'s study, for instance, further research could investigate biochemical pathways through which pesticides could cause the observed neurological damage, or additional observational studies might examine whether agrochemicals are associated with developmental delay in other places.

There are multiple methods for assessing the relationship between exposures and health via observational studies. In a **cohort study**, a group of people with a suspected risk factor is compared with a second group of people who lack exposure to that factor, as exemplified by Guillette et al.'s study (Figure 5.9b). Because there is often a significant time lag between an exposure and a health outcome, decades of study may be required to establish a relationship. In some cases, it may be possible to use medical records to examine changes from an earlier point in time to avoid having to wait for health outcomes. Nonetheless, the expense and organization required to run a cohort study is significant.

A second type of observational study is a **case-control study** (Figure 5.9c). In a case-control study, a researcher assigns study participants to groups based on their disease status and then compares the exposure histories of the two groups for systematic differences. For instance, if we want to test which exposures are associated with the incidence of leukemia, the experimental group would consist of individuals who have leukemia (the "cases"). The control group would be identified from the same population but would consist of individuals who do not have leukemia. The researcher could then look at each group's exposure history to see if any patterns emerge in factors such as diet, lifestyle, or workplace exposures. Controls are selected to be as similar as possible to cases with respect to characteristics that are often associated with health, such as age, gender, and socioeconomic status. Because case-control studies typically use smaller sample sizes than cohort studies and do not require follow-up over extended periods of time, they are often less expensive than cohort studies.

A **cross-sectional study** is a type of observational study that examines associations between the prevalence of a disease and an exposure among individuals in a specified population at a *particular point in time* (Figure 5.9d). The main distinction from cohort or case-control studies is that the study population is chosen without regard for exposure or disease status (Aschengrau and Seage 2003). Instead a representative "cross-section" of a population is selected and examined. For instance, Maantay (2007) studied the relationship between asthma and air pollution by collecting clinical data from individuals to determine whether and how much they experience asthma attacks. Maantay estimated individuals' exposure to particulate matter from the location of their residence and other data on ambient levels of particulate matter. The study identified a positive association —that is, that people who lived in areas with high levels of particulate matter were more likely to suffer from asthma attacks—providing support for a causal link.

Once again, however, cross-sectional studies do not provide conclusive evidence of a cause-and-effect relationship, just support for an association. A further disadvantage of cross-sectional studies is that they only report how *current* disease status is related to *current* exposures, yielding little information about whether the exposure preceded the disease (which is critical information if we are to conclude that the exposure *led to* the disease), or how previous exposures might influence observed health patterns. Nonetheless, many government surveys use cross-sectional studies to get a picture of a nation or region's health at relatively low cost.

A final study design employs **aggregated data**, such as data for census tracts or counties, rather than data for individuals. Studies that examine aggregated data are called **ecological studies**. Ecological studies are often used in geographic studies of health because they tie health outcomes to groups of people aggregated by districts. Suppose, for example, that we suspect that exposure to traffic exhausts is related to a certain kind of cancer. We could compare rates of that cancer at the county level with ambient air quality at that same scale. Ecological studies are comparatively cheap and straightforward and can often be undertaken using existing data. A major drawback of ecological studies is the phenomenon of **ecological fallacy**. Stated simply, this means that an association found at an aggregated level is assumed to hold also at the individual level or

at another level of aggregation. For example, we may conclude from a national-scale ecological study that cities with high population densities are more likely to have higher rates of asthma than rural areas. This does not mean that this relationship will hold at other scales though. For instance, we cannot assume from this work that an individual living in a densely populated part of the city is more likely to have asthma than a resident of a less densely settled part of the city.

An additional and persistent problem across all types of study is the possibility of **confounding**. Confounding occurs when a factor that is not being examined in the study is associated with both the exposure and health outcome. Suppose that a researcher hypothesizes that exposure to chemical X causes pancreatic cancer and so she constructs a cohort study at a factory with high exposure to the chemical that compares the cancer rate among factory workers with a control group from the general population. The researcher is unaware of previous research that has identified a strong link between smoking and pancreatic cancer and neglects to control for smoking in her study. Unfortunately, the factory workers were more likely to smoke than members of the community, and so smoking behavior becomes a confounding factor in the study. After finding higher levels of cancer in the factory workers, the researcher erroneously concludes that occupational exposure to chemical X is responsible for the cancer, but these higher rates may in fact be due to smoking behavior. There are several methods for controlling for confounding factors. A simple solution in this example would be to ensure that an equal proportion of smokers and non-smokers were in both the treatment and control groups.

Problems with confounding make the analysis of links between exposure and disease extremely challenging, particularly as it may be hard for researchers to identify every variable that may potentially confound their results. Over time, however, using different study designs, different study populations, and different sets of assumption, researchers can build scientific consensus about the relative risk of exposures. Even if general agreement is achieved within the scientific community, however, political and economic factors can have a major influence on how this information is used and which substances are regulated, highlighting the influence of social and economic factors, a topic to which we turn in subsequent chapters.

An additional problem with epidemiologic methods that focus on populations (such as case-control and cohort studies) is that they can miss causes of disease that are evenly distributed across the whole population, as Rose (1985) articulated in a seminal paper. Rose demonstrated that the presence of soft water across the whole of Scotland means that it can never be identified as a causal factor in any case-control or cohort study in Scotland because everyone in the country has been exposed, although it is possible that it may have detrimental health impacts. Similarly, "if everyone smoked twenty cigarettes a day, then clinical, case-control, and cohort studies alike would lead us to conclude that lung cancer was a genetic disease; and in one sense that would be true, since if everyone is exposed to the necessary agent, then the distribution of cases is wholly determined by individual susceptibility" (Rose 1985, 32). Rose concludes that identifying high-risk groups via epidemiologic methods is a reasonable interim measure and may be of great benefit to susceptible individuals, but that the long-term goal should be uncovering "underlying causes of incidence," as the removal of these will mean that even susceptible individuals are no longer at risk. This search for underlying causes of disease invites the use of geographic methods that can investigate both physical and social environments at a variety of scales.

ENVIRONMENTAL JUSTICE

In previous chapters, we discussed how different groups of people are more or less likely to contract diseases based on the particular suite of environmental exposures they experience. We have also discussed situations in which some groups bear a greater burden of environmental exposures—slum dwellers suffering the worst consequences of methyl isocyanate exposure in Bhopal, for instance. The study of **environmental justice** investigates such inequalities by analyzing the social dimensions of exposures. Central to the concept of environmental justice is the idea that communities should have equal access to a clean, healthy environment, irrespective of factors such as race, income, and class

(Cutter 1995). The environmental justice movement highlights several issues: 1) inequalities in who bears the environmental health burden associated with industrial development, 2) how we can develop policy to ensure that people are equitably protected from environmental exposures, and 3) how we can give all groups an equal voice in the development of environmental policy.

The roots of the environmental justice movement lie in the US civil rights and environmental movements of the 1960s. By the early 1980s, African American leaders had begun to organize efforts to protest decisions to locate environmental waste disposal sites in predominately minority neighborhoods. In 1982, more than five hundred people were arrested for attempting to block trucks carrying toxic waste to a landfill in an African American community in North Carolina (Bullard 2008). In what may be the earliest official acknowledgement of **environmental racism**, a 1983 government report found that three quarters of the off-site toxic waste sites in the southeast US were located in predominately minority and impoverished communities (United States General Accounting Office 1983). Four years later, a report from a commission on racial justice indicated that race was the most significant factor in explaining the location of waste facility sites nationwide and that three out of five African Americans and Latinos lived in communities near toxic waste sites (Bullard 2008).

The environmental justice movement did not pick up steam in much of the rest of the world until the late 1990s. Since then, researchers have devoted significant effort to investigating empirically whether deprived and minority communities are more likely to be exposed to environmental hazards. Although results have been mixed, there is sufficient evidence to warrant concern and the need for further investigation (Cutter 1995). For example, Figure 5.10 shows the location of regulated toxic release sites in Salt Lake City and per capita income by census district. The map provides some evidence to support the idea that affluence may influence a community's exposure to toxic waste sites, although further investigation would be needed to establish a correlation. It is difficult to say whether toxic facilities may be deliberately located in areas with poor and minority populations—perhaps as a consequence of the limited economic and political power of these communities—or

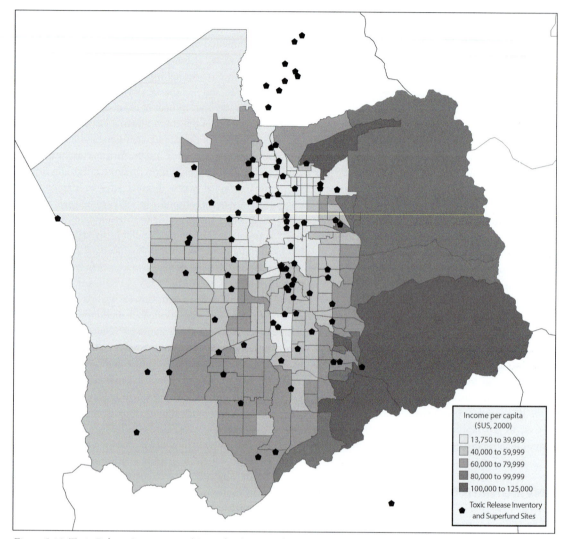

Figure 5.10 Toxic Release Inventory and Superfund sites and income per capita in Salt Lake City, Utah, US

Sources: United States National Library of Medicine (2009), US Census Bureau (2009)

The Toxic Release Inventory program (TRI) records and reports toxic chemical releases produced by the federal government and some private industries. Superfund is the US government's program to clean up controlled hazardous waste sites.

whether poor and minority populations move into neighborhoods with polluting facilities owing to lower land values. Nonetheless, it poses important questions about the financial and political ability of these communities to choose clean and healthy areas in which to live.

While the theory of environmental justice originated from local-scale analyses of waste facilities in the United States, its central ideas have been applied at different geographic scales and to a variety of subjects, including water quality (McDonald and Jones 2018), heat waves (Harlan et al. 2006), air pollution (Pearce and Kingham 2008), climate change (Vanderheiden 2008), and nutrition (Mawela and van den Berg 2018). For example, while wealthy regions of the world continue to reap many of the benefits of oil consumption, many studies have shown that the environmental and health burdens of oil production are disproportionately borne by poor and indigenous populations (O'Rourke and Connolly 2003). Owing to their significant but often localized environmental impacts, extractive industries have become a key focus for the environmental justice movement.

Environmental justice and mining

Mining has become a common way for low-income countries to develop exports and encourage investment. Pressure for low-income countries to develop their mineral resources often originates from outside powers, particularly institutions such as the International Monetary Fund (IMF) and World Bank, which have traditionally viewed extractive industries as a way for poor countries to develop exports and improve their balance of payments. Even in high-income countries, critics argue that poorer, rural populations bear most of the health impacts of mining, while wealthier urbanites reap many of the benefits. On a global scale, such inequalities are often magnified because lax environmental regulations in low-income countries fail to protect people from the worst consequences of mining, while profits leave the country with the controlling mining operation. As Hilson and Haselip (2004, 26) summarize, "The majority of developing countries have piecemeal legislation, evolving policy agendas, and under-resourced government departments, which gives a mining company little incentive to perform at a high level environmentally." Indeed, mining corporations actively exploit lax environmental legislation abroad in some cases (Roberts 1997).

With more than seventy active conflicts around mining operations, Peru illustrates many of these tensions. Peru privatized and deregulated its state-owned mining system in the 1990s with support from the IMF and World Bank (Slack 2009). Transnational mining companies were attracted to Peru's mineral wealth, investing in projects such as the huge Yanacocha gold mine in Cajamarca. Today, the mine is operated by a partnership between a US company, a Peruvian company, and the International Finance Corporation (IFC; a private sector arm of the World Bank). Local villagers' organizations have staged protests on the grounds of being unfairly compensated for land taken by the mine, and the contamination of local water bodies with mining by-products, including cyanide. Events came to a head in 2000 when a mine-operated tanker spilled 295 kilograms (650 pounds) of liquid mercury along the road through local communities (Slack 2009, 3). Unaware of the dangers of mercury, many villagers picked up the mercury and even brought it into their homes, leading to cases of mercury poisoning, including symptoms such as nausea, headaches, and skin lesions. Oxfam reports that nearly one thousand villagers experienced mercury poisoning (Slack 2009, 3), although difficulty establishing connections between mercury exposure and health impacts has led other authors to provide much lower figures. The event led to violent protests in 2001–2002 and conflicts over compensation (*ibid.*). As recently as 2016, a report was published investigating a conflict between the mine operators and local farmers over land rights, suggesting that the situation is not yet resolved (Slack 2016).

Electronic waste (e-waste)

Another critical contemporary environmental justice problem is the disposal of electronic waste. "E-waste" refers to high-tech computer products and other electrical equipment such as refrigerators and televisions that have reached the end of their useful life. Electronic waste contains a complex and toxic mix of hundreds of substances, including mercury, lead, arsenic, cadmium, chromium, polybrominated biphenyls (PCBs), and polybrominated diphenylethers (PBDEs), all of which present exposure risks. While electronics only represent a fraction of solid waste globally, they are responsible for a large proportion of the heavy metal content in landfills.

The volume of e-waste generated by industrialized societies is daunting. It has been estimated that about twenty million personal computers became obsolete in 1994; by 2004 this figure had risen to one hundred million (Puckett et al. 2002). More waste is generated by other computer products such as smart phones, computer games, and MP3 players. The problem is only set to get worse: younger consumers tend to have the most electronic devices, with more than half of all British 16- to 24-year-olds reported to have ten or more electronic devices in their home (Royal Chemistry Society 2019). As citizens of low-income countries begin to use more electronic goods, global volumes of e-waste will inevitably rise. In 2016, the planet generated an estimated 44.7 million metric tons of e-waste (Baldé et al. 2017). Growth in the volume of e-waste is estimated to be at least 3 percent per year (ibid. 2016).

Because of the rapid rate at which electronics technology has advanced and limited public understanding of the problem, few countries have well-established systems for labeling, handling, or disposing of e-waste. In the US, for example, even though much discarded electronic material contains chemicals classified as "hazardous waste" by the EPA, household electronics are not legally considered to be hazardous material, allowing them to be discarded with few precautions.

Not all e-waste is immediately discarded, however. A significant number of used devices are stored indefinitely: three quarters of all computers in the US may be sitting in garages or closets (Schwarzer et al. 2005), presenting significant disposal problems for the future. Without recycling these devices, we also risk running out of new sources of rare elements such as gallium, indium, and tantalum—natural sources of six of the elements in mobile phones are predicted to run out within the next hundred years. Furthermore, several important elements, including gold and tungsten, are classified as "conflict elements" because conflict and child labor are common in areas where they are mined (Royal Chemistry Society 2019). Other "obsolete" but usable devices are sent for use in poor countries where there is significant demand for cheap electronic equipment. About a quarter of all computers in use in India, a major recipient of used electronics, are older than 8 years (Schwarzer et al. 2005). While the environmental merit of reusing devices is laudable, sending old equipment to poor countries allows Western nations to displace the task of disposing of these materials, raising more questions of environmental justice.

In what is one of the most blatant modern examples of environmental injustice, huge quantities of e-waste have been deliberately exported from industrialized countries to the Global South for disposal for many years. An estimated 93 percent of exported used electronic equipment from the US went to Asia in 2016 (Baldé et al. 2017). Centers for e-waste processing have emerged in China, in particular, where some villages depend on disposal to sustain their economies. The Chinese government is beginning to reject shipments of waste from high-income countries, although this is likely to simply displace the problem to other, even poorer, countries.

The disposal of imported waste has significant negative health implications for many communities of the Global South (LaDou and Lovegrove 2008). Common disposal practices include stripping electronics in open-pit acid baths to recover metals, removing electronic components from printed circuit boards by heating them over a grill, chipping and melting plastics without sufficient ventilation, and disposing of unsalvageable materials in local rivers or lakes (Wong et al. 2007). In China, despite government efforts to discourage hazardous practices, much hazardous material is manually dissembled or illegally burned, often by workers who have little or no knowledge of the health impacts of the chemicals they are handling and wear no protective clothing (Chatterjee 2007). Such practices are not only dangerous for workers, but can also contaminate local air, water, soil, and food.

An array of epidemiologic studies in electronic disposal centers in China and elsewhere have demonstrated that the health impacts are grave. Children residing in e-waste–processing regions have elevated levels of lead in their blood, which can lead to problems in cognitive development (e.g., Huo et al. 2007). A cohort study comparing the population of Jinghai (the primary disposal center in north China) to unexposed populations suggests that e-waste also has the potential to cause mutations in chromosomes, resulting in cancer, infertility, spontaneous abortion, birth defects, and growth and development problems (Liu et al. 2009).

As the impending enormity of the problem of e-waste becomes apparent, policy is being developed to address the problem. By 2017, approximately two thirds of the world's population lived in countries with e-waste legislation, a significant improvement from three years earlier, when the figure was around two-fifths (Baldé et al. 2017). The European Union and its member countries have been particularly aggressive in developing e-waste policies by, for example, requiring producers to phase out the most toxic components of their products, designing new products in ways that allow for easier recycling, and guaranteeing products for at least two years in order to reduce the production of "disposable" electronic products. In addition, 186 governments have now ratified the "Basel Convention on the Control of Transboundary Movements of Hazardous Wastes and Their Disposal," which came into force in 1992 and restricts the export of e-waste (Widmer et al. 2005). The US, which is not a signatory to the Basel convention, now requires that producers recycle their own products in some states.

CONCLUSION

As the toll of infectious disease has been reduced and life expectancies have risen, so the risk to health from exposure to environmental contaminants has increased. At the same time, industrialization has begun generating synthetic substances at an unprecedented rate, many of which have unknown health impacts. Geographic approaches that consider connections between people and their environment are critical to exploring and analyzing relationships between potentially risky exposures and human health to inform decisions about regulating them.

The health impacts from environmental exposures are uneven, with some communities—often those with little political or economic power—affected to a greater degree than others. Environmental exposures are therefore not only a critical area of ecological study, but also generate important questions of equity, invoking consideration of the ways in which societal structures influence who is subjected to the greatest risk from health hazards. The task of teasing apart such social influences on health forms a second major theme within the geography of health—it is to this topic that we turn in the second part of the book.

DISCUSSION QUESTIONS

1 In what ways do geographic approaches lend themselves to the study of geogens? How might geographic approaches be combined with other approaches and techniques that address environmental exposures?

2 What is the difference between an experimental and an observational study? What are some of the pros and cons of these two methods?

3 Explain how e-waste can be considered an issue of environmental justice. What other issues of environmental justice are currently in the news?

4 Do you perceive any issues of environmental injustice in your own city? Do these local issues relate to national-scale or global-scale processes in any way?

5 Some Western pharmaceutical companies undertake drug trials in low-income countries despite the fact that the drugs they are testing will largely be sold in US or European markets. Is it ethical to undertake clinical trials in one part of the world, even if that community is unlikely to benefit soon from the drug under test?

SUGGESTED READING

Broughton, E. 2005. "The Bhopal Disaster and Its Aftermath: A Review." *Environmental Health: A Global Access Science Source* 4 (1): 6.

Cutter, S. L. 1995. "Race, Class, and Environmental Justice." *Progress in Human Geography* 19: 111–22.

Hanna-Attisha, M., Jenny LaChance, Richard Casey Sadler, and Allison Champney Schnepp. 2016. "Elevated Blood Lead Levels in Children Associated with the Flint Drinking Water Crisis: A Spatial Analysis of Risk and Public Health Response." *American Journal of Public Health* 106 (2): 283–90.

Hatton, T. 2017. "Air Pollution in Victorian-Era Britain—Its Effects on Health Now Revealed." *The Conversation.* https://theconversation.com/air-pollution-in-victorian-era-britain-its-effects-on-health-now-revealed-87208.

Human Rights Watch. 2016. *Nepotism and Neglect the Failing Response to Arsenic in the Drinking Water of Bangladesh's Rural Poor* [online]. www.hrw.org/report/2016/04/06/nepotism-and-neglect/failing-response-arsenic-drinking-water-bangladeshs-rural.

Mandavilli, A. 2018. "The World's Worst Industrial Disaster Is Still Unfolding." www.theatlantic.com/science/archive/2018/07/the-worlds-worst-industrial-disaster-is-still-unfolding/560726/.

Pearce, J., and Kingham, S. 2008. "Environmental Inequalities in New Zealand: A National Study of Air Pollution and Environmental Justice." *Geoforum* 39: 980–93.

Smith, R., B. Lourie, and S. Dopp. 2009. *Slow Death by Rubber Duck: How the Toxic Chemistry of Everyday Life Affects Our Health*. A.A. Knopf Canada.

VIDEO RESOURCES

"Poisoned Water." 2017. Frontline, PBS. More information at: www.pbs.org/wgbh/nova/video/poisoned-water/.

REFERENCES

Agency for Toxic Substances and Disease Registry. 2019. "Agency for Toxic Substances and Disease Registry." Accessed August 30, 2019. www.atsdr.cdc.gov/.

Andrady, Anthony L. 2015. "Persistence of Plastic Litter in the Oceans." In *Marine Anthropogenic Litter*, edited by Melanie Bergmann, Lars Gutow, and Michael Klages, 57–72. Cham: Springer International Publishing.

Anenberg, Susan C., Joshua Miller, Ray Minjares, Li Du, et al. 2017. "Impacts and Mitigation of Excess Diesel-Related NOx Emissions in 11 Major Vehicle Markets." *Nature* 545: 467. doi:10.1038/nature22086. www.nature.com/articles/nature22086#supplementary-information.

Aschengrau, Ann, and George R. Seage. 2003. *Essentials of Epidemiology in Public Health*. Sudbury, MA: Jones and Bartlett.

Balakrishnan, Kalpana, Sagnik Dey, Tarun Gupta, R. S. Dhaliwal, et al. 2019. "The Impact of Air Pollution on Deaths, Disease Burden, and Life Expectancy Across the States of India: The Global Burden of Disease Study 2017." *The Lancet Planetary Health* 3 (1): e26–e39. doi:10.1016/S2542-5196(18)30261-4.

Balazs, C. L., and I. Ray. 2014. "The Drinking Water Disparities Framework: On the Origins and Persistence of Inequities in Exposure." *American Journal of Public Health* 104 (4): 603–11. doi:10.2105/ajph.2013.301664.

Baldé, C. P., V. Forti, V. Gray, R. Kuehr, and P. Stegmann. 2017. *The Global E-waste Monitor 2017: Quantities, Flows, and Resources*. International Telecommunication Union.

Barber, L. B., J. E. Loyo-Rosales, C. P. Rice, T. A. Minarik, and A. K. Oskouie. 2015. "Endocrine Disrupting Alkylphenolic Chemicals and Other Contaminants in Wastewater Treatment Plant Effluents, Urban Streams, and Fish in the Great Lakes and Upper Mississippi River Regions." *Science of the Total Environment* 517: 195–206. doi:10.1016/j.scitotenv.2015.02.035.

British Geological Survey. 2001. Arsenic Contamination of Groundwater in Bangladesh. In *British Geological Survey Technical Report WC/00/19*, edited by D. G. Kinniburgh and P. L. Smedley. DFID and BGS.

British Geological Survey. 2001. Arsenic Contamination of Groundwater in Bangladesh. In *British Geological Survey Technical Report WC/00/19*, edited by D. G. Kinniburgh and P. L. Smedley.

Broughton, Edward. 2005. "The Bhopal Disaster and Its Aftermath: A Review." *Environmental Health: A Global Access Science Source* 4 (1): 6. doi:10.1186/1476-069X-4-6.

Browne, Mark Anthony, Phillip Crump, Stewart J. Niven, Emma Teuten, Andrew Tonkin, Tamara Galloway, and Richard Thompson. 2011. "Accumulation of Microplastic on Shorelines Woldwide: Sources and Sinks." *Environmental Science & Technology* 45 (21): 9175–79. doi:10.1021/es201811s.

Bullard, R. D. 2008. *Dumping in Dixie: Race, Class, and Environmental Quality*. 3rd ed. Westview Press.

Caldwell, B. K., J. C. Caldwell, S. N. Mitra, and W. Smith. 2003. "Searching for an Optimum Solution to the Bangladesh Arsenic Crisis." *Social Science and Medicine* 56 (10): 2089–96.

[CDC] Centers for Disease Control. 2018. "Bisphenol A (BPA) Factsheet." Accessed August 30, 2019. www.cdc.gov/biomonitoring/BisphenolA_FactSheet.html.

[CDC] Centers for Disease Control and Prevention. 2005. "Infectious Disease and Dermatologic Conditions in Evacuees and Rescue Workers After Hurricane Katrina—Multiple States, August–September 2005." *Morbidity and Mortality Weekly Report*, September 6, 2015.

[CDC] Centers for Disease Control and Prevention. 2008. "Formaldehyde Levels in FEMA-Supplied Trailers: Findings from the Centers for Disease Control and Prevention." Accessed October 23, 2019. www.fema.gov/pdf/media/2008/formaldehyde_resident_flyer_english.pdf.

[CDC] Centers for Disease Control and Prevention. 2015. "Tuskegee Study, 1932–1972." Accessed June 22. www.cdc.gov/tuskegee/index.html.

[CDC] Centers for Disease Control and Prevention. 2017. "Bisphenol A (BPA) Factsheet." Accessed June 22. www.cdc.gov/biomonitoring/BisphenolA_FactSheet.html.

Chatterjee, R. 2007. "E-waste Recycling Spews Dioxins into the Air." *Environmental Science and Technology* 41 (16): 5577.

Colborn, T. 2004. "Neurodevelopment and Endocrine Disruption." *Environmental Health Perspectives* 112 (9): 944–49. doi:10.1289/ehp.6601.

Cornwall, W. 2015. "Sewage Sludge Could Contain Millions of Dollars' Worth of Gold." *Science Magazine*. Accessed June 2, 2019. https://www.sciencemag.org/news/2015/01/sewage-sludge-could-contain-millions-dollars-worth-gold.

Cuppucci, M. 2018. "Air Pollution Skyrockets to Hazardous Levels in India." *The Washington Post*. Accessed June 22, 2019. www.washingtonpost.com/weather/2018/11/08/air-pollution-skyrockets-hazardous-levels-india/?noredirect=on&utm_term=.de6f1b556be4.

Cutter, Susan L. 1995. "Race, Class and Environmental Justice." *Progress in Human Geography* 19 (1): 111–22. doi:10.1177/030913259501900111.

Dockery, D. W., C. A. Pope, X. P. Xu, J. D. Spengler, et al. 1993. "An Association Between Air Quality and Mortality in 6 United States Cities." *New England Journal of Medicine* 329 (24): 1753–59. doi:10.1056/nejm199312093292401.

Doherty, R., M. Heal, P. Wilkinson, S. Pattenden, et al. 2009. "Current and Future Climate- and Air Pollution-Mediated Impacts on Human Health." *Environmental Health Perspectives* 8 (Suppl 1): S8.

Dris, R., J. Gasperi, M. Saad, C. Mirande, and B. Tassin. 2016. "Synthetic Fibers in Atmospheric Fallout: A Source of Microplastics in the Environment?" *Marine Pollution Bulletin* 104 (1–2): 290–93. doi:10.1016/j.marpolbul.2016.01.006.

Edwards, Marc, Simoni Triantafyllidou, and Dana Best. 2009. "Elevated Blood Lead in Young Children Due to Lead-Contaminated Drinking Water: Washington, DC, 2001 – 2004." *Environmental Science & Technology* 43 (5): 1618–23. doi:10.1021/es802789w.

[EPA] Environmental Protection Agency. 2018a. "Facts About Formaldehyde." Accessed June 22. www.epa.gov/formaldehyde/facts-about-formaldehyde#whatare.

[EPA] Environmental Protection Agency. 2018b. "Health Effects of Ozone Pollution." Accessed August 28, 2019. https://www.epa.gov/ground-level-ozone-pollution/health-effects-ozone-pollution.

[EPA] Environmental Protection Agency. 2019. "Criteria Air Pollutants." Accessed August 29, 2019. www.epa.gov/criteria-air-pollutants.

European Automobile Manufacturers Association. 2017. "Drop in Diesel Sales Offset by Increase in Petrol Vehicles." Accessed June 22. www.acea.be/statistics/article/drop-in-diesel-sales-offset-by-increase-in-petrol-vehicles.

European Automobile Manufacturers Association. 2018. "Share of Diesel in New Passenger Cars." Accessed June 22. www.acea.be/statistics/tag/category/share-of-diesel-in-new-passenger-cars.

European Public Health Alliance. 2007. "World Facing Increasing Threat of Arsenic Poisoning." Accessed June 22. https://epha.org/world-facing-increasing-threat-of-arsenic-poisoning/.

Flanagan, S. V., R. B. Johnston, and Y. Zheng. 2012. "Arsenic in Tube Well Water in Bangladesh: Health and Economic Impacts and Implications for Arsenic Mitigation." *Bulletin of the World Health Organization* 90 (11): 839–46. doi:10.2471/blt.11.101253.

Friedrich, M. J. 2011. "Air Pollution in China." *Journal of the American Medical Association* 305 (11): 1085. doi:10.1001/jama.2011.294.

Fuentes-Leonarte, Virginia, Ferran Ballester, and José Maria Tenías. 2009. "Sources of Indoor Air Pollution and Respiratory Health in Preschool Children." *Journal of Environmental and Public Health*. doi:10.1155/2009/727516.

Gallagher, J. 2012. "British Broadcasting Company." Accessed June 22. www.bbc.com/news/health-18415532.

Gioiosa, L., P. Palanza, S. Parmigiani, and F. S. Vom Saal. 2015. "Risk Evaluation of Endocrine-Disrupting Chemicals: Effects of Developmental Exposure to Low Doses of Bisphenol A on Behavior and Physiology in Mice (*Mus musculus*)." *Dose Response* 13 (4). doi:10.1177/1559325815610760.

Global Climate and Health Alliance. 2019. "Skin Cancer in Australia." Accessed August 30, 2019. http://climateandhealthalliance.org/resources/impacts/skin-cancer-in-australia/.

Greenberg, M. R. 2016. "Delivering Fresh Water: Critical Infrastructure, Environmental Justice, and Flint, Michigan." *American Journal of Public Health* 106 (8): 1358–60. doi:10.2105/ajph.2016.303235.

Guillette, E. A., M. M. Meza, M. G. Aquilar, A. D. Soto, and I. E. Garcia. 1998. "An Anthropological Approach to the Evaluation of Preschool Children Exposed to Pesticides in Mexico." *Environmental Health Perspectives* 106 (6): 347–53. doi:10.1289/ehp.98106347.

Guiloski, I. C., J. L. C. Ribas, L. D. S. Piancini, A. C. Dagostim, et al. 2017. "Paracetamol Causes Endocrine Disruption and Hepatotoxicity in Male Fish *Rhamdia quelen* After Subchronic Exposure." *Environmental Toxicology Pharmacology* 53: 111–20. doi:10.1016/j.etap.2017.05.005.

Hanna-Attisha, Mona, Jenny LaChance, Richard Casey Sadler, and Allison Champney Schnepp. 2016. "Elevated Blood Lead Levels in Children Associated with the Flint Drinking Water Crisis: A Spatial Analysis of Risk and Public Health Response." *American Journal of Public Health* 106 (2): 283–90. doi:10.2105/ajph.2015.303003.

Harlan, Sharon L., Anthony J. Brazel, Lela Prashad, William L. Stefanov, and Larissa Larsen. 2006. "Neighborhood Microclimates and Vulnerability to Heat Stress." *Social Science & Medicine* 63 (11): 2847–63. doi:10.1016/j.socscimed.2006.07.030.

Hilson, Gavin, and James Haselip. 2004. "The Environmental and Socioeconomic Performance of Multinational Mining Companies in the Developing World Economy." *Minerals & Energy—Raw Materials Report* 19 (3): 25–47. doi:10.1080/14041040410027318.

Hsu, S. 2010. "FEMA's Sale of Katrina Trailers Sparks Criticism." *The Washington Post*, March 13. Accessed June 21, 2019. www.washingtonpost.com/wp-dyn/content/article/2010/03/12/AR2010031202213.html.

Human Rights Watch. 2016. "Nepotism and Neglect the Failing Response to Arsenic in the Drinking Water of Bangladesh's Rural Poor." Accessed June 22. www.hrw.org/report/2016/04/06/nepotism-and-neglect/failing-response-arsenic-drinking-water-bangladeshs-rural.

Hunt, A., J. L. Abraham, B. Judson, and C. L. Berry. 2003. "Toxicologic and Epidemiologic Clues from the Characterization of the 1952 London Smog Fine Particulate Matter in Archival Autopsy Lung Tissues." *Environmental Health Perspectives* 111 (9): 1209–14. doi:10.1289/ehp.6114.

Huo, X., L. Peng, X. Xu, et al. 2007. "Elevated Blood Lead Levels of Children in Guiyu, an Electronic Waste Recycling Town in China." *Environmental Health Perspectives* 115(7): 1113–1117.

Jambeck, Jenna R., Roland Geyer, Chris Wilcox, Theodore R. Siegler, et al. 2015. "Plastic Waste Inputs from Land into the Ocean." *Science* 347 (6223): 768. doi:10.1126/science.1260352.

Jia, Y. L., D. Stone, W. T. Wang, J. Schrlau, et al. 2011. "Estimated Reduction in Cancer Risk due to PAH

Exposures If Source Control Measures During the 2008 Beijing Olympics Were Sustained." *Environmental Health Perspectives* 119 (6): 815–20. doi:10.1289/ehp.1003100.

Karbalaei, Samaneh, Parichehr Hanachi, Tony R. Walker, and Matthew Cole. 2018. "Occurrence, Sources, Human Health Impacts and Mitigation of Microplastic Pollution." *Environmental Science and Pollution Research* 25 (36): 36046–63. doi:10.1007/s11356-018-3508-7.

Kidd, K. A., P. J. Blanchfield, K. H. Mills, V. P. Palace, et al. 2007. "Collapse of a Fish Population After Exposure to a Synthetic Estrogen." *Proceedings of the National Academy of Science USA* 104 (21): 8897–901. doi:10.1073/pnas.0609568104.

Kolpin, Dana W., Edward T. Furlong, Michael T. Meyer, E. Michael Thurman, et al. 2002. "Pharmaceuticals, Hormones, and Other Organic Wastewater Contaminants in U.S. Streams, 1999–2000: A National Reconnaissance." *Environmental Science & Technology* 36 (6): 1202–11. doi:10.1021/es011055j.

Kosuth, M., E. Wattenberg, S. Mason, and C. Tyree. 2017. "Polymer Contamination in Global Drinking Water." Accessed June 23. https://orbmedia.org/stories/invisibles_final_report.

Kozul-Horvath, C. D., F. Zandbergen, B. P. Jackson, R. I. Enelow, and J. W. Hamilton. 2012. "Effects of Low-Dose Drinking Water Arsenic on Mouse Fetal and Postnatal Growth and Development." *PLoS One* 7 (5): e38249. doi:10.1371/journal.pone.0038249.

LaDou, J., and S. Lovegrove. 2008. "Export of Electronics Equipment Waste." *International Journal of Occupational and Environmental Health* 14 (1): 1–10. doi:10.1179/oeh.2008.14.1.1.

Lei, K., F. Qiao, Q. Liu, Z. Wei, et al. 2017. "Microplastics Releasing from Personal Care and Cosmetic Products in China." *Marine Pollution Bulletin* 123 (1–2): 122–26. doi:10.1016/j.marpolbul.2017.09.016.

Li, W. C., H. F. Tse, and L. Fok. 2016. "Plastic Waste in the Marine Environment: A Review of Sources, Occurrence and Effects." *Science of The Total Environment* 566–567: 333–49. doi:10.1016/j.scitotenv.2016.05.084.

Liu, Q., J. Cao, K. Q. Li, X. H. Miao, et al. 2009. "Chromosomal Aberrations and DNA Damage in Human Populations Exposed to the Processing of Electronics Waste." *Environmental Science and Pollution Research* 16 (3): 329–38. doi:10.1007/s11356-008-0087-z.

Loewenberg, S. 2017. "The Poisoning of Bangladesh: How Arsenic Is Ravaging a Nation." Accessed June 22. https://undark.org/article/bangladesh-arsenic-poisoning-drinking-water/.

Lohr, K. 2007. "FEMA Trailers May Be Making Residents Sick." Accessed August 31, 2019. www.npr.org/templates/story/story.php?storyId=15811496.

Maantay, J. 2007. "Asthma and Air Pollution in the Bronx: Methodological and Data Considerations in Using GIS for Environmental Justice and Health Research." *Health & Place* 13 (1): 32–56. doi:10.1016/j.healthplace.2005.09.009.

MacKenzie, D. 2002. "Fresh Evidence on Bhopal Disaster—Documents Suggest US Company Was Responsible for Plant's Design and Cut Investment to Maintain Control." *New Scientist* 176 (2372): 6–7.

Mandavilli, A. 2018. "The World's Worst Industrial Disaster Is Still Unfolding." Accessed June 23. www.theatlantic.com/science/archive/2018/07/the-worlds-worst-industrial-disaster-is-still-unfolding/560726/.

Matta, Murali K., Robbert Zusterzeel, Nageswara R. Pilli, Vikram Patel, et al. 2019. "Effect of Sunscreen Application Under Maximal Use Conditions on Plasma Concentration of Sunscreen Active Ingredients: A Randomized Clinical Trial." *JAMA* 321 (21): 2082–91. doi:10.1001/jama.2019.5586.

Mawela, Ailwei, and Geesje van den Berg. 2018. "Management of School Nutrition Programmes to Improve Environmental Justice in Schools: A South African Case Study." *South African Journal of Clinical Nutrition* 1–6. doi:10.1080/16070658.2018.1507208.

Maximenko, Nikolai, Jan Hafner, and Peter Niiler. 2012. "Pathways of Marine Debris Derived from Trajectories of Lagrangian Drifters." *Marine Pollution Bulletin* 65 (1): 51–62. doi:10.1016/j.marpolbul.2011.04.016.

McDonald, Y. J., and N. E. Jones. 2018. "Drinking Water Violations and Environmental Justice in the United States, 2011–2015." *American Journal of Public Health* 108 (10): 1401–07. doi:10.2105/ajph.2018.304621.

Millman, A., D. L. Tang, and F. P. Perera. 2008. "Air Pollution Threatens the Health of Children in China." *Pediatrics* 122 (3): 620–28. doi:10.1542/peds.2007-3143.

Mills, I. C., R. W. Atkinson, S. Kang, H. Walton, and H. R. Anderson. 2015. "Quantitative Systematic Review of the Associations Between Short-Term Exposure to Nitrogen Dioxide and Mortality and Hospital Admissions." *BMJ Open* 5 (5): e006946. doi:10.1136/bmjopen-2014-006946.

Munro, A. 2019. "Doctors Against Diesel—Mission Statement." Accessed June 22. www.medact.org/2016/actions/sign-ons/doctors-against-diesel/.

Nahar, N., F. Hossain, and M. D. Hossain. 2008. "Health and Socioeconomic Effects of Groundwater Arsenic Contamination in Rural Bangladesh: New Evidence from Field Surveys." *Journal of Environmental Health* 70 (9): 42–47.

National Atlas.gov (2010). Map Layers. Accessed November 7, 2010. www.nationalatlas.gov/maplayers.html.

National Institute of Environmental Health Sciences. 2015. "Endocrine Disruptors." Accessed June 22, 2019. www.niehs.nih.gov/health/topics/agents/endocrine/index.cfm.

National Research Council, Division on Earth and Life Studies, Commission on Life Sciences, Subcommittee on Arsenic in Drinking Water. 1999. *Arsenic in Drinking Water.* National Academies Press.

[NHS] National Health Service. 2012. "WHO: 'Diesel Exhaust Fumes Cancerous'." Accessed June 22. www.nhs.uk/news/cancer/who-diesel-exhaust-fumes-cancerous/.

Oldenkamp, R., R. van Zelm, and M. A. J. Huijbregts. 2016. "Valuing the Human Health Damage Caused by the Fraud of Volkswagen." *Environmental Pollution* 212: 121–27. doi:10.1016/j.envpol.2016.01.053.

O'Rourke, Dara, and Sarah Connolly. 2003. "Just Oil? The Distribution of Environmental and Social Impacts of Oil Production and Consumption." *Annual Review of Environment and Resources* 28 (1): 587–617. Doi:10.1146/annurev.energy.28.050302.105617.

Palmeira Wanderley, Vivianni, Fernando Luiz Affonso Fonseca, André Vala Quiaios, José Nuno Domingues, et al. 2017. "Socio-Environmental and Hematological Profile of Landfill Residents (São Jorge Landfill—Sao Paulo, Brazil)." *International Journal of Environmental Research and Public Health* 14 (1): 64.

Panko, B. 2017. "Scientists Now Know Exactly How Lead Got into Flint's Water." Accessed June 23, 2019. www.smithsonianmag.com/science-nature/chemical-study-ground-zero-house-flint-water-crisis-180962030/.

Pasquali, R., E. Stener-Victorin, B. O. Yildiz, A. J. Duleba, et al. 2011. "PCOS Forum: Research in Polycystic Ovary Syndrome Today and Tomorrow." *Clin Endocrinology* 74 (4): 424–33. Doi:10.1111/j.1365-2265.2010.03956.x.

Pearce, Jamie, and Simon Kingham. 2008. "Environmental Inequalities in New Zealand: A National Study of Air Pollution and Environmental Justice." *Geoforum* 39 (2): 980–93. Doi:10.1016/j.geoforum.2007.10.007.

Petzinger, J. 2018. "Europe's Intoxicating Love Affair with Diesel Is Dying Out." Accessed June 21. https://qz.com/1183779/europes-intoxicating-love-affair-with-diesel-is-dying-out/.

Pimentel, D., S. Cooperstein, H. Randell, D. Filiberto, et al. 2007. "Ecology of Increasing Diseases: Population Growth and Environmental Degradation." *Human Ecology* 35 (6): 653–68. Doi:10.1007/s10745-007-9128-3.

Puckett, J., and T. Smith (eds.) 2002. "Exporting Harm: The High-Tech Trashing of Asia." Basel Action Network, Silicon Valley Toxics Coalition. Accessed October 23, 2019. http://www.greenpeace.org/eastasia/Global/eastasia/publications/reports/toxics/2006/exporting-harm-the-high-tech-trashing-asia.pdf.

Revel, Messika, Amélie Châtel, and Catherine Mouneyrac. 2018. "Micro(nano)plastics: A Threat to Human Health?" *Current Opinion in Environmental Science & Health* 1: 17–23. Doi:10.1016/j.coesh.2017.10.003.

Roberts, D. E. 1997. *Killing the Black Body: Race, Reproduction, and the Meaning of Liberty*. Pantheon Books.

Rochman, Chelsea M. 2018. "Microplastics Research—From Sink to Source." *Science* 360 (6384): 28. Doi:10.1126/science.aar7734.

Rosario-Ortiz, Fernando, Joan Rose, Vanessa Speight, Urs von Gunten, and Jerald Schnoor. 2016. "How Do You Like Your Tap Water?" *Science* 351 (6276): 912. Doi:10.1126/science.aaf0953.

Rose, G. 1985. "Sick Individuals and Sick Populations." *International Journal of Epidemiology* 14 (1): 32–38. Doi:10.1093/ije/14.1.32.

Royal Chemistry Society. 2019. "Elements in Danger." Accessed August 30, 2019. www.rsc.org/campaigning-outreach/campaigning/saving-precious-elements/elements-in-danger/.

Schwarzer, S., A. De Bono, G. Giuliani, S. Kluser, and P. Peduzzi. 2005. "E-waste, the Hidden Side of IT Equipment's Manufacturing and Use." Environment Alert Bulletin 5. UNEP and GRID Europe.

Sharma, D. 2005. "Bhopal: 20 Years on." *Lancet* 36 (9454): 111–12. Doi:10.1016/s0140-6736(05)17722-8.

Shim, Won Joon, Sang Hee Hong, and Soeun Eo. 2018. "Chapter 1—Marine Microplastics: Abundance, Distribution, and Composition." In *Microplastic Contamination in Aquatic Environments*, edited by Eddy Y. Zeng, 1–26. Elsevier.

Silver, B., C. L. Reddington, S. R. Arnold, and D. V. Spracklen. 2018. "Substantial Changes in Air Pollution Across China During 2015–2017." *Environmental Research Letters* 13 (11): 8. Doi:10.1088/1748-9326/aae718.

Slack, K. 2009. "Mining Conflicts in Peru: Condition Critical." Oxfam America. Accessed June 22. 2019. www.oxfamamerica.org/explore/research-publications/mining-conflicts-in-peru-condition-critical/.

Slack, K. 2016. "New Report, Same Problems at Peruvian Mine." Accessed June 23, 2019. https://politicsofpoverty.oxfamamerica.org/2016/09/new-report-same-problems-at-peruvian-mine/.

Smith, A. H., E. O. Lingas, and M. Rahman. 2000. "Contamination of Drinking-Water by Arsenic in Bangladesh: A Public Health Emergency." *Bulletin of the World Health Organization* 78 (9): 1093–103.

Song, Young Kyoung, Sang Hee Hong, Mi Jang, Gi Myung Han, et al. 2017. "Combined Effects of UV Exposure Duration and Mechanical Abrasion on Microplastic Fragmentation by Polymer Type." *Environmental Science & Technology* 51 (8): 4368–76. Doi:10.1021/acs.est.6b06155.

Srivastava, R. K., and Sushila Godara. 2017. "Use of Polycarbonate Plastic Products and Human Health." *International Journal of Basic Clinical Pharmacology* 2 (1).

The Associated Press. 2010. "Trailers-to-Haiti Proposal Stirs Backlash." Accessed August 31, 2019. www.cbsnews.com/news/trailers-to-haiti-proposal-stirs-backlash/.

Thomas, F., and Regional Office for Europe, World Health Organization. 2017. "Pharmaceutical Waste in the Environment: A Cultural Perspective." *Public Health Panorama* 3 (1): 127–32.

Thompson, Richard C., Ylva Olsen, Richard P. Mitchell, Anthony Davis, et al. 2004. "Lost at Sea: Where Is All the Plastic?" *Science* 304 (5672): 838. Doi:10.1126/science.1094559.

Topham, G. 2018. "Hybrid Cars to Be Exempt from 2040 Petrol and Diesel Ban." *The Guardian*. www.theguardian.com/business/2018/jul/09/hybrid-cars-to-be-exempt-from-2040-petrol-and-diesel-ban.

Topham, G., S. Clark, C. Levett, P. Scruton, and M. Fidler. 2015. "The Volkswagen Emissions Scandal Explained." *The Guardian*, September 23. Accessed June 22, 2019. www.theguardian.com/business/ng-interactive/2015/sep/23/volkswagen-emissions-scandal-explained-diesel-cars.

Trotter, R. Clayton, Susan G. Day, and Amy E. Love. 1989. "Bhopal, India and Union Carbide: The Second Tragedy." *Journal of Business Ethics* 8 (6): 439–54.

Union of Concerned Scientists. 2018. "Cars, Trucks, Buses and Air Pollution: Transportation Is a Major Source of Air Pollution in the United States." Accessed

June 22. www.ucsusa.org/clean-vehicles/vehicles-air-pollution-and-human-health/cars-trucks-air-pollution.

Union of Concerned Scientists. 2019. "Diesel Engines and Public Health." Accessed June 22, 2019. www.ucsusa.org/clean-vehicles/vehicles-air-pollution-and-human-health/diesel-engines.

Union of Concerned Scientists. no date. "Vehicles, Air Pollution, and Human Health." Accessed June 22, 2019. www.ucsusa.org/clean-vehicles/vehicles-air-pollution-and-human-health.

United States General Accounting Office. 1983. *Siting of Hazardous Waste Landfills and Their Correlation with Racial and Economic Status of Surrounding Communities.* Washington, DC: United States General Accounting Office.

United States National Library of Medicine. 2009. "TOX-MAP: Environmental Health E-Maps." Accessed May 21, 2010. http://toxmap.nlm.nih.gov/toxmap/main/index.jsp.

US Census Bureau. 2009. "The 2000 United States Census." Accessed January 30, 2010. www.census.gov/main/www/access.html.

National Toxicology Program. 2016. Report on Carcinogens, Fourteenth Edition. Research Triangle Park, NC: U.S. Department of Health and Human Services, Public Health Service. https://ntp.niehs.nih.gov/go/roc14.

Vandenberg, L. N., R. Hauser, M. Marcus, N. Olea, and W. V. Welshons. 2007. "Human Exposure to Bisphenol A (BPA)." *Reproductive Toxicology* 24 (2): 139–77. Doi:10.1016/j.reprotox.2007.07.010.

Vanderheiden, S. 2008. *Atmospheric Justice: A Political Theory of Climate Change.* New York: Oxford University Press.

Wang, Jundong, Zhi Tan, Jinping Peng, Qiongxuan Qiu, and Meimin Li. 2016. "The Behaviors of Microplastics in the Marine Environment." *Marine Environmental Research* 113: 7–17. Doi:10.1016/j.marenvres.2015.10.014.

[WHO] World Health Organization. 2001. "Arsenic in Drinking Water." Accessed January 11, 2019. www.who.int/mediacentre/factsheets/fs210.en/print.html.

[WHO] World Health Organization. 2018a. "Air Pollution and Child Health: Prescribing Clean Air." Accessed June 21, 2019. www.who.int/ceh/publications/air-pollution-child-health/en/.

[WHO] World Health Organization. 2018b. "Air Pollution." Accessed June 22, 2019. https://www.who.int/en/news-room/fact-sheets/detail/ambient-(outdoor)-air-quality-and-health.

[WHO] World Health Organization. 2018c. "Household Air Pollution and Health." Accessed June 22. www.who.int/en/news-room/fact-sheets/detail/household-air-pollution-and-health.

[WHO] World Health Organization. no date. "BreatheLife." Accessed June 12, 2019. www.who.int/sustainable-development/BreatheLife-Technical-Flyer.pdf?ua=1.

[WHO] World Health Organization. Department of Public Health, Environmental, and Social Determinants of Health. 2016. *Ambient Air Pollution: A Global Assessment of Exposure and Burden of Disease.* Geneva: World Health Organization.

[WHO and UNEP] World Health Organization, and United Nations Environment Programme. 2012. *Endocrine Disrupting Chemicals*, edited by Å. Bergman, J. Heindel, S. Jobling, K. Kidd, and R. T. Zoeller. IOMC: WHO and UNEP.

WHO, and WHO Regional Office for Europe. 2000. *Quantification of Health Effects of Exposure to Air Pollution: Report of a WHO Working Group.* Bilthoven: WHO Regional Office for Europe.

Widmer, Rolf, Heidi Oswald-Krapf, Deepali Sinha-Khetriwal, Max Schnellmann, and Heinz Böni. 2005. "Global Perspectives on E-waste." *Environmental Impact Assessment Review* 25 (5): 436–58. doi:10.1016/j.eiar.2005.04.001.

Wong, M. H., S. C. Wu, W. J. Deng, X. Z. Yu, et al. 2007. "Export of Toxic Chemicals—A Review of the Case of Uncontrolled Electronic-Waste Recycling."

Environ Pollution 149 (2): 131–40. doi:10.1016/j.envpol.2007.01.044.

[WRI] World Resources Institute, United Nations Environment Programme, United Nations Development Programme, and The World Bank. 1998. *World Resources 1998–99: A Guide to the Global Environment—Environmental Change and Human Health (English).* Oxford: Oxford University Press.

Yip, M., and P. Madl. 2002. *Air Pollution in Mexico City* [Online]. Accessed January 9, 2010. www.sbg.ac.at/ipk/avstudio/pierofun/mexico/air.htm.

Yu, K. 2018. "The Good News (and Not so Good News) About China's Smoggy Air." Accessed June 23, 2019. www.npr.org/sections/goatsandsoda/2018/12/18/669757478/the-good-news-and-not-so-good-news-about-chinas-smoggy-air.

Zhang, J. L., J. S. Reid, R. Alfaro-Contreras, and P. Xian. 2017. "Has China Been Exporting Less Particulate Air Pollution Over the Past Decade?" *Geophysical Research Letters* 44 (6): 2941–48. doi:10.1002/2017gl072617.

Zhao, Bin, Haotian Zheng, Shuxiao Wang, Kirk R. Smith, et al. 2018. "Change in Household Fuels Dominates the Decrease in $PM_{2.5}$ Exposure and Premature Mortality in China in 2005–2015." *Proceedings of the National Academy of Sciences* 115 (49): 12401–6. doi:10.1073/pnas.1812955115.

only as good as the data that were used to create it and so it is important to consider carefully the source of the data presented (Box 6.1). Decisions around how the data were displayed can also influence the way the map is read. Cartographers must make gross simplifications in order to map information in a way that is relevant, interesting, and understandable to the audience, and these decisions can influence the mapreader. Think again about the map of mortality in Color plate 1. To make a readable map, the mapmaker has excluded huge amounts of information from this map, including highways, bodies of water, the location of cities, and

---

Box 6.1

SOURCES OF HEALTH DATA

Inaccuracies in health data are a significant barrier to drawing robust conclusions in health studies. Whenever **secondary data** are used in health research, even in a cursory exploration of spatial patterns, it is important to maintain a critical awareness of how the data were collected and reported so that potential biases or inaccuracies might be identified. It is, therefore, worth briefly considering some of the main sources of health data and some of their potential limitations.

Professionally collected data, such as disease registries and death certificates, tend to be of higher quality than self-reported data. In many countries, deaths must be reported by a physician, providing a relatively consistent and reliable data set. However, even death certificates are not always an accurate reflection of the true cause of death. One survey of death certificates in suburban towns in the US suggested that some form of error exists on up to 48 percent of death certificates (Cambridge and Cina 2010). Similar work in the Netherlands and the UK shows that the accuracy of death certification varies by cause of death (Harteloh, de Bruin, and Kardaun 2010; Turner et al. 2016). In low-income regions around the world, many deaths may not be recorded at all. *How* we choose to report data can also make an important difference. For instance, in 1949, the US changed its reporting of cause of death from "lethal importance" to "underlying cause." Suddenly, emphysema was reported on a dramatically larger number of death certificates—emphysema had always been a common

underlying cause of death but was rarely designated as the final cause.

Morbidity data are often even less reliable, given that many illnesses may go unreported or may be mistaken for other conditions. Again, professionally reported morbidity data are likely to be most accurate. This might include physicians' reports and hospital data. **Disease registries** are systematic efforts to document particular diseases by requiring that physicians report certain diseases to public health authorities. In the US, for example, the CDC annually publishes a list of "notifiable diseases" that must be officially reported so that the CDC can monitor disease patterns at the national level. This includes unusual diseases such as Zika virus, where an uptick in cases may indicate an impending outbreak, as well as highly communicable diseases such as measles. A disease registry can provide a wealth of recent and historical data about a disease, its incidence, and its spatial distribution, offering the potential to guide research, prevention, and control efforts. Despite the greater accuracy and reliability we would expect from officially reported data, these data should also be viewed with some caution. For example, the reporting rate could be affected by the diagnostic facilities available, the control measures that are being implemented at a particular time and place, or even physicians' familiarity with the disease.

There are also several international registries for diseases, ranging from emerging infectious diseases to chronic conditions such as cancer and heart conditions. For instance, the WHO maintains a website that includes information on underlying cause of death, collected from national disease registries (WHO 2019). Here, again, however, we must be wary of data quality issues, especially inconsistencies in how the data are reported since the database relies on national reporting systems with different reporting criteria.

A further source of health data is **morbidity** surveys in which a sample of the population is given clinical exams and interviews. The key issue here is of missing, biased, or inaccurate data, given that the information is self-reported. Individuals may not remember the details of past health events or may have a motive for deliberately misreporting information such as avoiding revealing stigmatized conditions. Gaining reliable data on sensitive topics, such as the relationship between HIV infection and intravenous drug use, can therefore be a considerable challenge. Even basic information such as height and weight is likely to be distorted when self-reported. One analysis reports that North Americans tend to take pounds off when they report their weight but exaggerate their height (Lipsky et al. 2019).

A final source of data is historical health records, including parish records of births and deaths, and disease records such as bills of mortality. This may be our only source of data if we want to extend a longitudinal analysis back before modern records, but the error introduced is likely to be substantial. Early records were likely only a snapshot of an entire population, and records were often far from systematic. In addition, many of the conditions recorded in early records may be hard to match to current conditions. One famous source of historical data, a survey of bills of mortality from London in 1632 completed by John Graunt, includes the diseases livergrown, grief, and rising of the lights, which are hard to equate to accepted modern conditions (Wynder 1975).

population density. While some of this information could be useful, the cartographers have omitted all but the most critical information to make the map easier to read and understand. The cartographer also had to make countless other decisions, such as how to **aggregate** the data, how to shade categories, whether to include the borders of neighboring countries, and so forth. In this way, map-making is a highly subjective process, involving hundreds of cartographic decisions. While most mapmakers do not deliberately mislead their readers, these decisions inevitably have consequences on how a map is perceived and understood.

Figure 6.3 provides an example of how cartographic decisions can deliberately change the message behind a map. This map uses the same data as the map in Color plate 1, but the mapmaker has purposefully tried to influence how these data are interpreted by the reader through changing the categorization scheme, the way the data are symbolized, and the title. The map highlights only areas with the lowest mortality rates, focusing attention on just one part of the data set. Additionally, the mapmaker has chosen a rather misleading title for the map: "Healthy USA," which encourages the reader to jump to the conclusion that the variable being presented (mortality rate) is synonymous with health. As we discuss in Chapter 7, mortality rates can actually be a very poor predictor of the overall health of a population.

To provide another example, by highlighting a cluster of cancer cases around a toxic waste dump we could imply that there is a causative connection between the toxic site and the cancer, even if the number of cases recorded is far too small to be statistically significant. Some websites that identify cancer clusters for lobbying purposes use maps in exactly this way. It is good practice to think about the subjective decisions behind any map. The checklist in Table 6.1 provides some questions to think about in interpreting any map—we explore these ideas in greater detail as our discussion proceeds.

Types of thematic maps

Thematic maps can be categorized in terms of the primary class of symbol used to represent the data. **Map symbols** are the points, lines, and areas that make up a graduated circles to illustrate the number of reported cases by country.

Dot density maps are similar in appearance to point distribution maps but are distinct in that points are not used to represent the *actual location* of individual cases such as clinics. Instead, the points on dot density maps are placed randomly within the boundaries of the region to which they pertain, thereby visually depicting the *density* of the phenomenon in the specified area. Figure 6.5 shows a dot density map of Lyme disease in the United States. As noted on the map itself, each dot represents one confirmed case and has been placed randomly within the county of residence of that case. The major advantage of dot density maps is that they provide a visual impression of the density of a phenomenon that can be effectively compared among regions, even if the regions are of different sizes. In this case, it is immediately clear that the Northeast and Upper Midwest are hotspots of Lyme disease activity in the US. Users must not assume that the points on the maps represent the actual locations of individuals with Lyme disease, however.

Reported Cases of Lyme Disease – United States, 2017

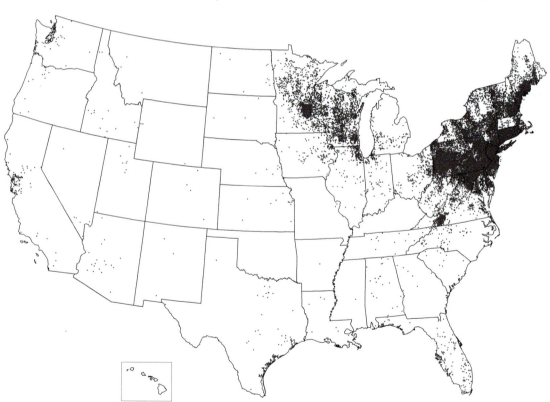

1 dot placed randomly within county of residence for each confirmed case

Figure 6.5 A dot density map of Lyme disease in the United States, 2017
Source: CDC (2018)

Line maps

Line symbols can be put to several purposes. First, they are ideal for representing direction and rate of flow. Health researchers often use **flow maps** to symbolize the path and rate of transmission of a disease, although they could also illustrate the diffusion of medical technologies or health workers across space. Figures 6.6 and 6.7 appeared in a research report on the links between civil war and the spread of HIV/AIDS in the early 1990s in Uganda. The authors demonstrated that patterns in the distribution of HIV/AIDS infection were related to ethnic patterns of recruitment into the Ugandan National Liberation Army (Smallman-Raynor and Cliff 1991). By mapping the volumes of migrant flows into different regions of Uganda, they were able to illustrate how the pathogen might be diffusing with

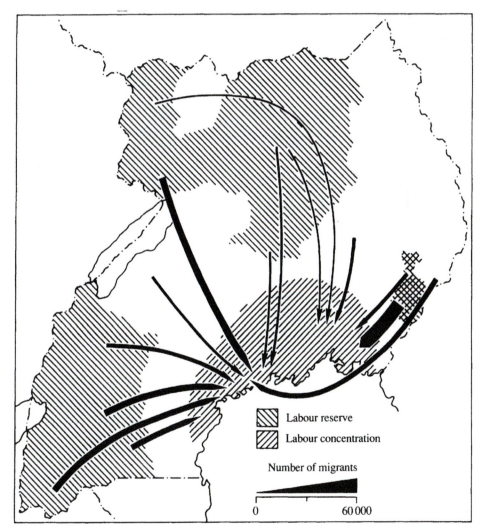

Figure 6.6 Migrant flows in Uganda

Source: Smallman-Raynor and Cliff 2009, © Cambridge Journals, reproduced with permission.

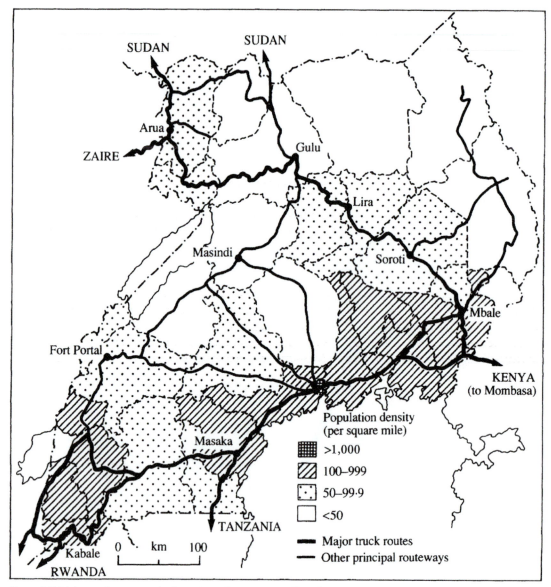

Figure 6.7 Major truck routes and other principal routeways in Uganda

Source: Smallman-Raynor and Cliff 2009, © Cambridge Journals, reproduced with permission.

These maps were used in a study published in 1991 to explore the role of Uganda's civil war in the spread of HIV/AIDS. The authors explored historical transportation and migration flows in Africa as a means of explaining the distribution of HIV/AIDS.

the migrants (Figure 6.6). They used a graduated flow-line to provide information on not only the direction of migrant movements but also the volume of these flows. In a second map, they depicted major truck routes to consider an alternative hypothesis for the spread of AIDS—the "truck town hypothesis," which suggests that major truck routes formed the primary paths along which AIDS spread. In this case, lines show connections between major population centers but do not provide information about volumes or directions of flow (Figure 6.7).

Line symbols are also the basis of **contour maps**. In this case, **isolines** connect all points of equal value on the map. We are probably most familiar with seeing contour lines on maps of elevation, but they can be used to represent any type of **continuous data**. Continuous data have a value at all points on the earth's surface, but researchers must work from a limited number of observations and then estimate the values for locations in between. For instance, in health, we could map exposure to sulfur dioxide in the air as a continuous variable. Researchers would use measurements from pollution monitoring stations and then estimate exposure values for places between the stations. This process of "filling in" data for places where there are no measurements is called **interpolation**. Techniques for interpolating data are often complicated, but modern software can do much of the work. Color plate 3 shows a map from a study that estimated exposure to benzopyrene in France (Ioannidou et al. 2018). The authors used a variety of data, including air, water, and soil measurements, to produce the estimate.

Area maps

Choropleth maps display information about the intensity of a disease or phenomenon, mapped by statistical or administrative units (e.g., see Color plate 1). This type of map is common in public health work because many health data are **aggregated** by statistical or administrative units, such as counties or provinces. Many governments and other organizations regularly collect data by administrative units (such as in national censuses),

and so aggregated data are often easier and cheaper to obtain than individual-level data. There are also fewer privacy concerns with aggregated maps, as data are not displayed for individual patients.

Choropleth maps can represent **absolute counts** of a phenomenon (e.g., number of Lyme disease cases) or data that have been converted into **rates** (e.g., number of Lyme disease cases per 100,000 population). Data are divided into categories and then each category shaded to illustrate the relative intensity of the phenomenon in different regions. It is important to be careful when mapping **absolute counts** in a choropleth map, because administrative units can differ enormously in size or population. Most of the time, counts should be converted into **rates** so that patterns do not merely reflect the distribution of the underlying population. For example, a map that showed that 246 people contracted chicken pox in a population over the course of a year does not mean much without taking overall population size into account. If 246 people contracted the disease in a village of 400, we would want to explore the cause of this major outbreak! By contrast, 246 cases in India's population of more than one billion would be little cause for concern.

Calculating rates may still be insufficient to allow useful comparisons across space and time, however. Data that have been adjusted only for population size, as described earlier, are known as **crude rates**. Because of the critical relation between age and health, most health data are better represented with data that have also been adjusted for age structure. **Age adjustment** enables comparison of mortality and morbidity statistics across populations with different age structures. The commonest approach, direct **age adjustment**, applies the age structure of a reference population to the mortality or morbidity rates of all study populations under comparison. Nearly any population can serve as the reference population: the critical idea is that the reference population's age structure is applied to all study populations before the indicator rates are compared so that we can compare them as if all populations had the same age structure. This technique for age adjustment is described in more detail in Box 6.2.

Box 6.2

AGE ADJUSTMENT

Because demographic factors can critically influence health outcomes, it is important to remove the effect of such factors from population health statistics. If this is not done, the demographic composition of the population can influence mortality and morbidity figures at the expense of revealing important disease patterns. Adjustment can be performed on any population factor, such as gender or ethnicity, but age is by far the most frequently adjusted factor. The procedure described here—the direct method—is the most commonly used.

In direct age adjustment, mortality figures are standardized to reflect the population structure of a reference population. The same reference population must be used for all populations to be compared in a particular study. Although almost any population can serve as the reference population, many health organizations consistently use the same reference population so that rates they report can be directly compared. The WHO has published a standard world reference population that it recommends for comparing rates across countries (Ahmad et al. 2001).

Direct age adjustment uses the mortality rates for different age cohorts within each study population and recalculates them as if the study populations had the same population structure as the reference population. It is therefore necessary to know: 1) the number of deaths in each age cohort of each study population, 2) the population structure of each study population, and 3) the population structure of the reference population.

In the following example, we calculate the direct age-adjusted mortality rate for England and Wales in 2006 (the *crude* mortality rate for this population is 945 per 100,000). We first calculate **age-specific mortality rates** by dividing the number of deaths in each age group in 2006 by the total population in those age groups during the same year. The results are usually expressed in terms of

England and Wales, 2006					
	(A) Number of deaths	(B) Population	(C) Rate/100,000	(D) Reference population	(E) Weighted rate
0-4	3,930	2,488,142	157.9	0.0886	14.0
5-14	760	6,572,222	11.6	0.1729	2.0
15-24	2,870	7,051,638	40.7	0.1669	6.8
25-34	4,510	7,055,315	63.9	0.1554	9.9
35-44	10,120	8,207,854	123.3	0.1374	16.9
45-54	20,400	6,912,149	295.1	0.1141	33.7
55-64	45,500	6,332,792	718.5	0.0827	59.4
65-74	83,400	4,438,840	1878.9	0.0517	97.1
75-84	163,100	3,042,648	5360.5	0.0243	130.3
85+	168,100	1,122,105	14980.8	0.0064	95.1
					465.3

the rate per 100,000 population. For example, among 0- to 4-year-olds, there were 3,930 deaths out of a total population of 2,488,142 in that age group. We therefore divide the number of deaths (A) by the population (B) and multiply the result by 100,000, yielding an age-specific mortality rate of 157.9 per 100,000 for 0- to 4-year-olds (C).

Once we have calculated age-specific mortality rates for all age groups, we must assign each rate a weight according to the proportion of people in that age group in the reference population. In this example, we use the reference population structure issued by the WHO. The numbers in column D indicate the proportion of the reference population represented by each age cohort. Our task is to give each cohort from the British figures the appropriate weight of this reference population. For example, since the 5- to 14-year group represents 0.1729 of the weighting of the total reference population, the British age-specific mortality rate for this age group should also constitute that fraction of the total age-adjusted figure for the British population. To achieve this, we multiply each age-specific rate (C) by the fraction of the reference population represented by that age group (D) to obtain the weighted rate (E). Finally, we add all the weighted rates from column E to determine the age-adjusted rate. In this case, this results in an overall age-adjusted death rate for the whole UK population of 465 per 100,000. This age-adjusted rate can be legitimately compared with the age-adjusted rates of other populations with different demographic structures, as long as the age-adjusted rates were calculated using the same reference population.

The importance of using age-adjusted rates is clear from comparing a map of crude rates of cancer mortality with one of age-adjusted rates (Figures 6.8 and 6.9). At first glance, the map of crude rates suggests that cancer risk is highest in the eastern US (Figure 6.8). After age-adjustment, however, it appears that the West has the greater risk of cancer (Figure 6.9). What explains this difference in patterns? Because cancer is most likely to occur in old age, parts of the country with a large elderly population have relatively high *crude* cancer mortality rates simply because a large part of their population is in the age group where cancer is most likely. This situation is exemplified by the far northeastern state of Maine, which has a very high crude cancer rate. Once mortality rates have been **age-adjusted**, however, Maine compares favorably with the rest of the US. The lower age-adjusted rate shows that it is the large elderly population of the state and not any intrinsic cancer risk associated with living in Maine that explains its high crude death rate. We can support this hypothesis by looking at the median age in Maine, which is 44, six years older than the US average of 38. If we consider age-adjusted rates, the southwest US has the highest cancer rates, allowing us to legitimately ask what factors in the southwest US may

contribute to high cancer incidence, given that age has been removed as an explanatory variable.

Another important set of considerations in a choropleth map is around how to symbolize and classify the data. Color plate 4 is an example of a map of **qualitative** or **categorical data**: data that can easily be categorized according to particular characteristics. In this case, we are looking at child car restraint laws in different countries. Different color hues (e.g., green, yellow, red) are effective at symbolizing these different categories, which bear no numerical relationship to one another, although color has been used to indicate the relative desirability of the categories, with red warning us of those countries not meeting best practice.

When maps communicate **quantitative data** (data that are counted or measured), the cartographer should display the data in a way that emphasizes the clear numerical order inherent in the data set. Color gradients (e.g., light blue, medium blue, dark blue) are ideal for showing the relative intensity of the different categories of the data set. For instance, Color plate 5 shows the prevalence of heavy drinking by country. Even before we have studied the legend, the map conveys the idea that heavy drinking is a problem in Russia and parts of

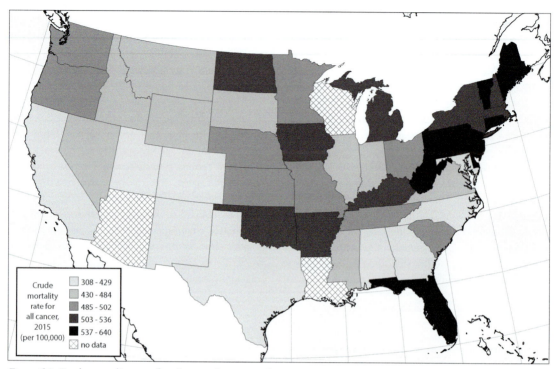

Figure 6.8 Crude mortality rates for all cancer by state in the US, 2015
Source: Data from the CDC (2018)

central and southern Africa, as the darker shading indicates higher intensity of the phenomenon.

Other analytical and cartographic techniques can transform a complex and messy map into a more legible map that clearly reveals patterns. **Data smoothing** removes local variations by averaging values across a larger area of the map. Mapping smoothed rates is particularly useful when there is a low incidence rate because random aberrations in the data can otherwise hide patterns. For example, Middleton et al. (2008) produced maps of suicide in the UK to explore regional variations (Color plate 6). To highlight patterns, the authors showed **smoothed** rates for suicide in addition to reporting the mortality rate by administrative unit. They concluded that suicide rates rise as one moves away from major cities and towards the coasts, generating hypotheses for further investigation.

The *Atlas of United States Mortality* (1996) illustrates many aspects of good cartographic practice with respect to choropleth mapping in health (see Color plate 1). The atlas was an effort to produce a comprehensive depiction of disease mortality across the US, including maps of the 18 leading causes of death, which collectively accounted for 83 percent of all deaths. Data in the atlas were mapped by Health Service Area (HSA)—administrative units that the CDC considered to be "relatively self-contained with respect to hospital care" (Pickle and National Center for Health Statistics 1996, 5). The designers of the atlas drew from cognitive research, statistical methodology, and cartographic practice to produce maps that communicate health information as effectively as possible.

Data for the atlas were taken from death certificate information collected by the National Center for Health Statistics (NCHS). Because age is a key influence on many causes of mortality, the final product shows age-adjusted mortality rates. Gender and ethnicity were also considered to be potentially significant in interpreting mortality patterns and so the atlas includes

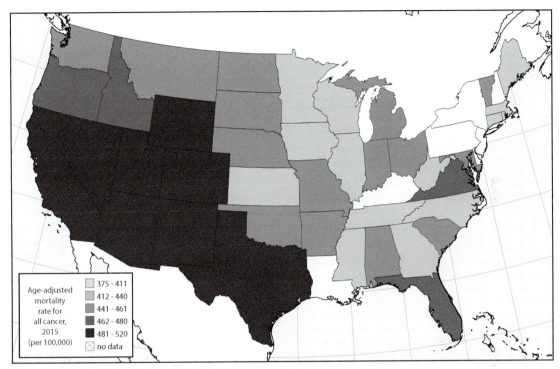

Figure 6.9 Age-adjusted mortality rates for all cancer by state in the US, 2015
Source: Data from the US Centers for Disease Control (2018)

separate maps for four different age and gender combinations (white males, white females, black males, black females). For comparison, the atlas also includes maps of **covariates**—factors that may be related to mortality rates, such as income, urbanization rates, and percentage of the population with college degrees.

The mapmakers carefully considered ways to symbolize the data to maximize users' abilities to recognize patterns as well as to enable readers to effectively compare mortality rates for different diseases and populations. Ultimately the authors decided on a double-ended color gradient scheme in which two hues are used—a brown gradient for high rates and a green gradient for low rates. The color gradient allows the reader to easily perceive the range of incidence rates, with darker shades representing the highest and lowest rates, while the two different color hues (brown and green) illustrate an important qualitative distinction: rates above the national average and those below.

The mapmakers also attempted to convey the certainty behind the information presented on their maps. In areas with low populations, small numbers of cases can dramatically inflate rates. To address this concern, the authors applied hatching over HSAs where rates were unstable to warn users about the lower reliability of the data.

Finally, the authors wished to produce maps whose focus was patterns rather than specific rates, in order to highlight regions with particularly high or low mortality rates. To this end, they produced additional maps that highlighted how death rates of each HSA compared with the US average, as well as maps showing smoothed rates (Color plate 7).

Animated cartography and interactive maps

Computer technology has vastly expanded the kinds of data that can be mapped and has paved the way for new forms of cartography. For instance, **animated maps** can display changes in data over both time and space, offering a powerful tool for visualizing phenomena such as the spread of disease. An early example of animated mapping in health shows the spread of HIV across the US (Banks et al. 2000). The animation showed weekly AIDS mortality by county for an eleven-year period from 1981 to 1992 (Color plate 8). A large amount of data was compiled and hundreds of images incorporated to produce the animation. More recent examples of online animated maps showing the spread of AIDS have been prepared by the UNESCO Culture GIS Unit (UNESCO 2017).

Computer technology has also facilitated the development of **interactive maps** that enable the user to request specific types of data or to change the way that data are displayed. For example, the WHO's *Global Health Atlas* website enables users to explore a variety of health-related data, ranging from cholera infection rates to health service availability. Interactive maps can enable users to produce maps from a repository of data that is constantly updated, turning it into a useful tool for disease monitoring.

Interactive and web mapping can also enable community groups to participate in environmental justice and other public health research (Maclachlan et al. 2007; Cromley and McLafferty 2012). Geographic information systems available on the Internet for public use can make critical spatial information accessible to communities, enabling people to produce maps from their own data. An excellent example of leveraging the power of communities with geographic information was the relief effort following the 2010 earthquake in Haiti. Volunteers from around the world successfully compiled maps and other geospatial resources to assist workers on the ground (Zook et al. 2010) (Box 6.3).

Box 6.3

LEVERAGING CROWD-SOURCED DATA IN THE 2010 HAITI EARTHQUAKE

On January 12, 2010, an extremely powerful earthquake struck Haiti, its epicenter around 25 km from Haiti's capital, Port-au-Prince. As one of the poorest countries in the world, Haiti was extremely vulnerable to the effects of the earthquake and ill-equipped to mount an effective response. The International Foundation of the Red Cross estimated that around three million Haitian people were affected by the quake (Central Broadcasting Service 2010), and the (disputed) estimate of the ultimate death toll was more than 316,000, making it the second-deadliest earthquake in human history (Ritchie 2018).

The international humanitarian response to the disaster was swift and immediate, but it faced some critical challenges. Relief workers needed to know where people needed help and had to navigate an infrastructure that was poor and neglected and had largely been destroyed by the earthquake. There was little geospatial data on the country, with little information available on transportation networks, infrastructure, or medical assets that normally form the foundation of any significant relief effort (Zook et al. 2010). In response, major geoinformatics corporations (such as Google, DigitalGlobe, and GeoEye) collaborated to release and make available high-quality satellite data, but the lack of digital geospatial infrastructure such as GIS databases continued to hinder progress.

Thankfully, the NGO CrisisCommons (crisiscommons.org) had been formed the previous year with the mission to coordinate stakeholders in the case of an international crisis precisely like the Haiti earthquake. The organization was based in the United States, but efforts quickly spread to cities around the world after the earthquake. Facilitated by a variety of social media platforms, a well-coordinated international collection of volunteers was able to provide essential mapping services to relief workers in Haiti. Their efforts included developing a map browser that could be used off line by workers on the ground (where there was little or no cellular or Internet access), providing mapped data on critical infrastructure, medical assets, and demographic data, and working to facilitate communication among workers on the ground (Zook et al. 2010). Some volunteers traveled to Haiti to provide critical up-to-date information on the condition of roads, hospitals, medical and drug supplies, and so forth (Keegan 2010). OpenStreetMap, an open-sourced geographic information portal (at openstreetmap.org), served as the key repository for much of the data. Ultimately, thanks to the collaboration afforded by modern forms of mapping, international access to data, and the goodwill and effort of thousands of digitally connected volunteers, a geospatial information infrastructure that would normally have taken years to build was achieved in a matter of days by people in physically distant locations around the world (Zook et al. 2010).

Mapping as an analytical tool

While mapping is an excellent tool for *communicating* and *exploring* patterns in health and disease, it can also facilitate *analysis* of spatial patterns. Although even a cursory examination of a map may reveal interesting patterns, statistical methods enable much more rigorous and meaningful analysis. As spatial data become more accessible and as more health researchers discover the power of spatial approaches, the study of **spatial epidemiology** (or "geographic epidemiology") has gained traction. Although the study of spatial epidemiology involves methods well beyond the purview of this book, here we briefly visit some fundamental spatial-analytical approaches, concepts, and problems.

Mapping is an especially useful tool in **ecological studies**—observational studies in which sets of aggregated data are compared in order to test for a relationship. If we suspect that heart disease is related to poverty, for example, we could investigate whether counties with high rates of poverty are more likely to have a high incidence of heart disease. While we must be careful to avoid the **ecological fallacy**—and should never assume that an observed association between **aggregated data** implies that one factor *causes* another—ecological studies can serve as a useful way to explore these kinds of relationships when it is impractical to obtain individual-level data.

Studies that use aggregated data must avoid several other pitfalls, however. A common spatial problem is the **modifiable areal unit problem**, which is a "geographic manifestation of the ecological fallacy in which conclusions based on data aggregated to a particular set of districts may change if one aggregates the same underlying data to a different set of districts" (Waller and Gotway 2004, 104). In other words, the results of a mapping study that uses data from statistical or administrative districts might change if the units for which the data are aggregated are changed (Openshaw and Taylor 1979). This problem is particularly troublesome in health studies because the statistical and administrative units that yield a lot of health data are often arbitrarily drawn. For instance, Monmonier used John Snow's well-known cholera maps to demonstrate that the relationship between cholera cases and the Broad Street pump can completely disappear if the data are aggregated into districts in one way, but that the association remains if the districts are redrawn in another way (Monmonier 2018). Because of the ubiquity of common statistical administrative boundaries, it is likely that the modifiable areal unit problem has led to confusion in geographic health research, and addressing the problem could form an important means of improving future work.

An additional problem derives from the application of traditional statistical methods to spatial data. A key assumption in most parametric statistics is that events are independent from one another. However, when an event occurs in space, other events are often more likely to occur near to the initial event than in more distant locations, and so the events cannot be considered independent. **Spatial autocorrelation** refers to this phenomenon, in which "pairs of observations taken nearby are more alike than those taken far apart" (Waller and

climate conditions are unusual, such as in cases of flooding or warmer-than-usual temperatures, there is often a concurrent rise in the incidence of specific diseases. After these events have been carefully studied, remotely sensed data can provide warning of future situations when risk levels are raised (Palaniyandi 2012).

GIS applications in health studies

In this final section, we focus on the specific ways that mapping and GIS have contributed to the study of linkages between health and place, with a focus on four key areas: environmental exposures, disease surveillance, cluster analysis, and studies of healthcare distribution and access.

Exposure assessment

Assessing exposures to environmental toxins is a basic challenge in environmental health. Suppose that a factory in the middle of a city is suspected of affecting the health of a nearby community because it emits nitrous oxides. Accurately estimating exposure to nitrous oxide is important if we wish to link individual health outcomes (such as cancer or asthma) to exposure to pollution from that factory or if we want to construct policies to address the health risk.

One of the challenges of exposure assessment comes from the fact that airborne environmental pollution is a continuous phenomenon. If a pollutant is frequently monitored in multiple locations, as is often done for major pollutants such as sulfur dioxide or ozone, assessment can utilize data from monitoring stations. GIS then enables data from stations to be **interpolated** using sophisticated statistical methods, resulting in a continuous data surface that provides exposure estimates for all places.

Although empirical work has refined the analysis of exposure data and improved our understanding of variability in exposures across time and space, high-quality modeling remains tremendously difficult. For unusual pollutants that are not monitored on a widespread or continual basis, the challenge is even greater. Researchers must often rely on models that estimate exposure from a specific source, incorporating information about other factors

such as emission volumes and wind flow. GIS is frequently used in **fate and transport studies**, which use data on pollution sources in combination with meteorological and other environmental data to estimate the transport of pollutants and their ultimate destination through the soil, water, and atmosphere (e.g., Pistocchi 2008).

We could assume that the closer someone lives to a source of pollution, the more likely that person is to be exposed to it. Although proximity clearly has an influence, the real world is generally far too complex for such a simple model to suffice. If the prevailing winds come from the west, for instance, people living to the east of the pollution source may have greater exposure. In addition, many people spend time outside their home neighborhood, working or attending school, for example. Living near a source of pollution will therefore lead to greater exposures for some individuals than others. Some people who live in the area may work near another source of pollution. Perhaps the factory only operates during the day, meaning that people who work outside the neighborhood have relatively little exposure. For many pollutants, there are also multiple pathways of exposure—different ways that people could be exposed to a toxin—potentially including airborne, waterborne, and foodborne routes. While GIS analysis is seldom able to incorporate all these factors, it has facilitated more accurate exposure estimates.

In a study typical of early endeavors of spatial epidemiology to model contaminant diffusion, Brown et al. (1984) conducted a case-control study to determine whether proximity to a zinc smelter and steel manufacturing company was associated with increased risk for lung cancer. They defined their study area as three contiguous counties in Pennsylvania where these two plants, containing more than one hundred emission stacks, were located (Figure 6.11). They identified 360 deaths in the three counties attributable to lung cancer over a two-year period and matched these cases with controls from other places by age, gender, and residence. They collected information on smoking history, occupational history, and other relevant covariates with questionnaires.

The authors estimated exposure from location of residence by dividing the study area into a grid of 1-km cells. Each grid cell was assigned an exposure value based

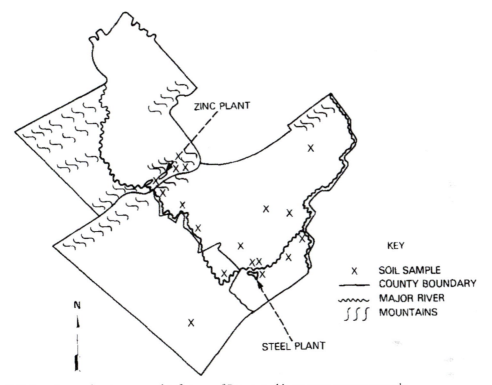

Figure 6.11 Locations and major geographic features of Brown et al.'s exposure assessment study
Source: Reprinted from *Environmental Research*, vol. 34, Brown, L, Pottern, L., and W. Blot, "Lung cancer in relation to environmental pollutants emitted from industrial sources," page 251, © 1984, with permission from Elsevier.

on its proximity to and direction from a polluting stack, as well as measurements of heavy metal contamination from soil samples (Figure 6.12). The authors also incorporated basic information on wind direction into the estimates; for example, the wind tended to flow to the northwest in the study area, so residents who lived to the northwest of an industrial source were given a higher exposure estimate (Figure 6.13). Because data were not available on emission rates from the stacks, the authors incorporated the tenuous assumption that each stack emitted the same amount of pollutant.

After adjusting for smoking and occupational status, the authors discovered a mild increase in cancer risk associated with proximity to zinc smelters. A significant advantage of the authors' use of this geographic model is that it allowed them to estimate exposures in areas in which there was no consistent monitoring or measurement during the period of interest. In other words, this model enabled them to reconstruct exposure from the data that were available.

An example of a more developed exposure analysis is a study of radiation exposure from aboveground nuclear test sites in Utah (Simon et al. 1995). The intent of this study was to explicitly address shortcomings of previous studies on the topic, which had simply used county of residence to estimate radiation exposure in an ecological model. After the authors identified qualifying cases and controls for their study, they recorded the residence history of each subject and modeled location-specific fallout estimates. Total dose for each case and control subject was then estimated by summing the likely exposures for each time and place of residence, considering

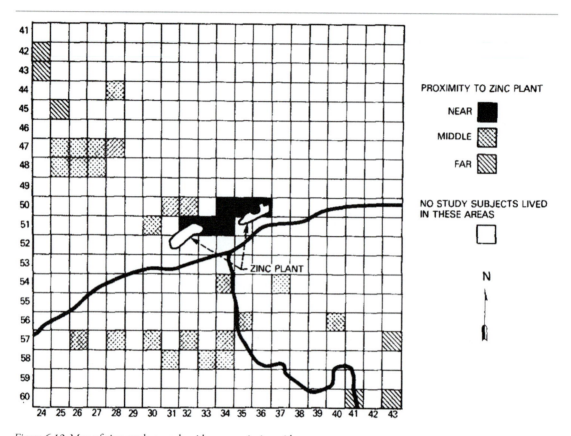

Figure 6.12 Map of zinc smelters and residence proximity grid

Source: Reprinted from *Environmental Research*, vol. 34, Brown, L, Pottern, L., and W. Blot, "Lung cancer in relation to environmental pollutants emitted from industrial sources," page 256, © 1984, with permission from Elsevier.

factors such as the residence building type and location of employment. Given the many factors used to assess exposure in the study, the need to quantify uncertainty became important to the study's integrity (Simon et al. 1995, 464). For example, uncertainty in residential history data resulted from missing or imprecise data and unrecorded time spent away from home, such as business trips or vacations. The authors quantified uncertainty by identifying a sample of their study group for more intensive and accurate data collection. They used these results to estimate the degree to which their data contained error. The results of this analysis have since been used to test whether radiation exposure levels were related to a variety of other health outcomes, including thyroid disease (Till et al. 1995) and childhood cancers (Lloyd 1998).

Disease surveillance and risk mapping

Pre-GIS initiatives to map diseases using traditional maps yielded invaluable experience in monitoring infectious disease. This work faced significant challenges, however, since traditional maps cannot be easily updated, and it is difficult to use them for systematic comparisons between diseases and related spatial patterns. A **disease surveillance** system uses systematic data collection to monitor

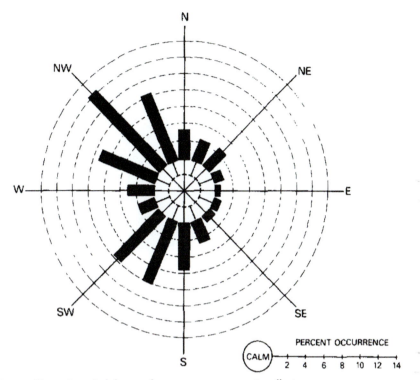

Figure 6.13 Diagram illustrating wind data used to assess exposure to air pollution

Source: Reprinted from *Environmental Research*, vol. 34, Brown, L, Pottern, L., and W. Blot, "Lung cancer in relation to environmental pollutants emitted from industrial sources," page 252, © 1984, with permission from Elsevier.

for signals that could indicate new outbreaks of disease or increasing incidence of disease. Surveillance systems often incorporate knowledge about the natural history of a disease, geographic information about the population at risk, and information about environmental conditions. Disease surveillance can serve as a tool to direct disease prevention efforts in ways that have the greatest impact. Close surveillance of disease distributions and incidence rates is especially important in places experiencing rapid environmental change.

Disease surveillance played a critical role in guiding efforts to address the 2015–2016 Zika virus epidemic in Latin America. National and regional governments in the countries hardest hit (notably Brazil and Colombia) actively collected clinical data to monitor the status and progression of the outbreak and developed longer-term systems to monitor the transmission of the disease

into the future. This was critical for planning vector control and other efforts, including travel advisories (Rodriguez-Morales et al. 2017). Figure 6.14 shows an example from a mapping project developed to monitor the status of the disease.

Research in disease ecology can be used to produce **risk maps** that show where the factors normally associated with the occurrence of a disease are present. Zika virus is transmitted by mosquitos of the *Aedes* genus, which also act as vectors for dengue and yellow fever. Fortunately, there is a significant body of research from around the world on the spatial risk factors for dengue (e.g., Dhewantara et al. 2015; Eisen and Lozano-Fuentes 2009; Leta et al. 2018), which include high population density, proximity to suitable mosquito breeding habitats (particularly warm temperatures and the availability of standing water), and poverty. Investigators were able

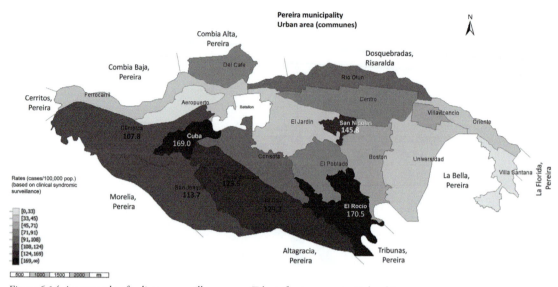

Figure 6.14 An example of a disease surveillance map: Zika infection rates in Colombia
Source: Rodriguez-Morales et al. (2017)

to take these factors into consideration to develop risk maps for Zika. In Colombia, geographic surveillance systems that integrated entomological data on mosquitos, data from national public health institutes, and social survey data into a single system enabled weekly reports on Zika risk (Ocampo et al. 2019).

With a good understanding of the factors that drive patterns in disease vectors, it is possible to make rapid and effective predictions about their distribution from carefully analyzed satellite data (Bergquist and Rinaldi 2010). Remote sensing and GIS are continually used to model a variety of other vector-borne diseases, including schistosomiasis (Ajakaye, Adedeji, and Ajayi 2017), malaria (Jeanne et al. 2018), leishmaniasis (Kavur and Artun 2017), as well as other neglected tropical diseases (e.g., Pleydell et al. 2008).

In addition to monitoring *existing* cases, disease surveillance can work in a proactive way by identifying environments that are likely to serve as *future* habitat for disease vectors and hosts. In China, for example, direct surveillance of schistosomiasis is not an option because of the low density of infected snails and insufficient funding (Zhang et al. 2008). Using remotely sensed environmental data and GIS analysis, however, it is possible to identify likely habitats for the snails that act as intermediate hosts for the disease. While these methods do not reliably predict where high rates of schistosomiasis will occur, they can assist policymakers with decisions about where to allocate time and money.

Cluster analysis

A **disease cluster** is a group of disease cases or high incidence rates beyond what we would normally expect to observe in a particular time and place (CDC 1990). The term **hotspot** is a synonymous term often used in spatial statistical work. The terms are generally reserved for unusual groupings of cases of non-infectious diseases—particularly cancers. Because disease clusters are tied to places, an environmental exposure is often assumed to be the culprit. However, it is important to remember that a spatial cluster or hotspot is not inevitably caused by a physical-environmental factor. For instance, a cluster could equally indicate that a disease is caused by an infectious agent.

While it is tempting to identify disease clusters by casually examining maps, care must be taken to determine the likelihood that they could have occurred randomly. Clusters often become rallying cries for patients and advocate groups for whom the perception of a cluster causes alarm. As one group of researchers writes:

> Every day, citizens call local and state health departments because they have been, or their spouse or child has been, diagnosed with cancer. Most callers know of one or more similar cases in their neighborhood, school, or place of work. These people feel strongly that a cancer cluster exists, and they want it investigated. They fear additional family members, neighborhoods, and friends are at risk, and they want these fears addressed immediately.
> (Schneider et al. 1993, 753)

Objective epidemiologic analysis, with the help of spatial-analytical software, can test the legitimacy of perceived clusters. While the doubling of the incidence rate of a disease in a small town may seem alarming, if four cases are observed in a particular year rather than an expected number of two, the additional two cases could easily have occurred by chance. Indeed, observed disease clusters that occur on a local scale very seldom lead to any real insight into the true causes of the disease (Elliott et al. 2001), leading some researchers to suggest that individual cluster reports should not be investigated at all, or only when the risk levels in the cluster are extremely high (Rothman 1990).

Cancer provides a particularly challenging disease for cluster analysis. The term "cancer" refers to a complex set of health conditions, each with a unique set of etiologies (causes) and risk factors. When an analysis of cancer clusters is conducted, teasing apart these multiple causes makes identifying a cause extremely complex, and the limitations in this kind of work should be carefully considered. The challenges around identifying a disease cluster are exemplified well in the case of Toms River, New Jersey (Fagin 2013). After a century of hosting industrial worksites for dye production and other industries, members of the community became concerned about what they perceived as unusually high rates of cancer in the community, particularly among children. The community advocated for the termination of a variety of industrial practices that they believed were responsible, such as the direct release of chemicals into the rivers and the illegal burial of hundreds of drums of chemicals in the town, which had leached into private wells and drinking water. The problem ultimately centered on determining the likelihood that cancer rates among the residents of the town could be as high as they were observed to be, which culminated in a complex series of investigations that involved the community, the corporations responsible for the pollution, government organizations, journalists, and scholars. Ultimately, the investigators were unable to conclude that there was a definitive cancer cluster, and the families of the cancer victims settled with the companies for an undisclosed amount in exchange for agreeing not to claim that the companies were responsible for the deaths.

Spatial cluster studies can nonetheless yield important insights on a variety of health topics, including both infectious and non-infectious diseases. Researchers have used GIS to identify hotspots in lung cancer (Zhang and Tripathi 2018), dengue (Stewart-Ibarra et al. 2014), leishmaniasis (Mollalo et al. 2015), fall injuries (Yiannakoulias et al. 2003), and dementia risk (Bagheri et al. 2018), among other health topics. A study by Chi et al. (2008) exemplifies the utility of GIS in cluster detection. The authors identified 135 cases of birth abnormalities between 1998 and 2001 in rural Heshun County, China. Trained healthcare workers administered questionnaires to mothers to collect data on the circumstances of the cases. Cases were **geocoded** and GIS was used to apply spatial statistical methods to identify and test the statistical significance of birth anomaly clusters (Color plate 10). Results identified one potentially significant cluster in the study region, and the authors proceeded to generate several hypotheses to explain it, including exposure to chemical fertilizers and poor water quality.

The same cluster analysis techniques can be used to identify groupings of cases that, while not particularly mysterious or unusual, may indicate the need for more resources. For example, in sub-Saharan Africa, where two of the leading causes of death are malaria and HIV/AIDS, understanding where high rates of both diseases coincide can offer critical guidance for intervention.

Gwitira et al. (2018) performed cluster detection techniques on both diseases in Zimbabwe. Although they concluded that the distribution of primary clusters of HIV/AIDS and malaria were not spatially coincident, they identified five districts in the country where secondary clusters of the diseases overlapped, with potential policy implications (Color plate 11).

Healthcare provision and access

GIS can contribute to the study of healthcare by mapping healthcare services, evaluating access to services, and analyzing utilization patterns (Cromley and McLafferty 2012). Each of these areas of work has distinct methodologies. Here we discuss four methods of modeling healthcare access as examples: density mapping, network analysis, least cost path analysis, and allocation modeling.

Ratio or **density mapping** has been performed for a long time but has been significantly improved by GIS. Density mapping in healthcare study uses population and physician data collected for administrative units roughly equivalent to the distance a person could reasonably be expected to travel to visit a healthcare service. Data are then combined to produce a choropleth map showing the number of people per physician for each area. The method has many important limitations: it does not consider variations in population density or terrain within the areas of analysis, it is susceptible to the modifiable area unit problem, and does not take into account the transportation network (Talen and Anselin 1998). However, density mapping provides a straightforward and accessible analytical tool.

GIS can help to overcome some of these limitations by calculating distance between health facilities and healthcare users. Euclidean (straight-line) distance is often inadequate for measuring the physical aspects of healthcare access, and so a variety of other factors must be taken into consideration. Do healthcare users mainly walk to clinics or take a car? Do public transportation routes serve clinics well? Can people choose which healthcare provider to use? Are financial or cultural barriers a problem, meaning that some people are unable to use services nearby? Many of these factors can be modeled spatially. **Network analysis** can evaluate the spatial accessibility of facilities by calculating distance via established road networks or along public transit routes. In a study in South Africa, for example, Kapwata and Manda (2018) used a GIS to calculate the distance of the most direct route to a health facility among cardiovascular patients. Their work showed that the western parts of the country generally had better spatial access to healthcare facilities and that the poorest people in the country had to travel the longest distances, which could translate into increased mortality from cardiovascular disease.

A more sophisticated methodology uses a form of network analysis called **least cost path analysis** (LCPA). This method uses information about the transportation network to calculate the shortest distance and travel time required to go from any point to the nearest general practice physician. Sophisticated LCPAs may include information about barriers in the network (such as road works), directionality (such as one-way streets), and travel time (which could take into account speed limits, traffic signals, and even traffic volume). For instance, in one study in New Zealand, LCPA was used to assess the impacts of healthcare privatization on access to emergency services (Brabyn and Beere 2006). The authors produced maps showing populations who lived sixty or more minutes of travel time from the nearest hospital emergency department (Figure 6.15). They concluded that significant parts of the rural population have been excluded from emergency services as a consequence of the closure of hospitals associated with healthcare restructuring, although they note that the results may also be influenced by changes in population distribution.

Location-allocation modeling is a variation of least cost path analysis. In addition to calculating distance and travel time via a transportation network, this model considers the capacity of the nearest service. Patients are allocated to the nearest clinic until the capacity of that clinic is reached. Once that threshold is met, patients are allocated to the next-nearest clinic where there is space. Location-allocation models are often performed to examine access to specialized health services such as maternal care, in order to identify gaps in a healthcare system (e.g., Schuurman 2009). Other examples include evaluating healthcare in critically stressed regions, such as in Syria following civil war (Mic, Koyuncu, and Hallak 2019).

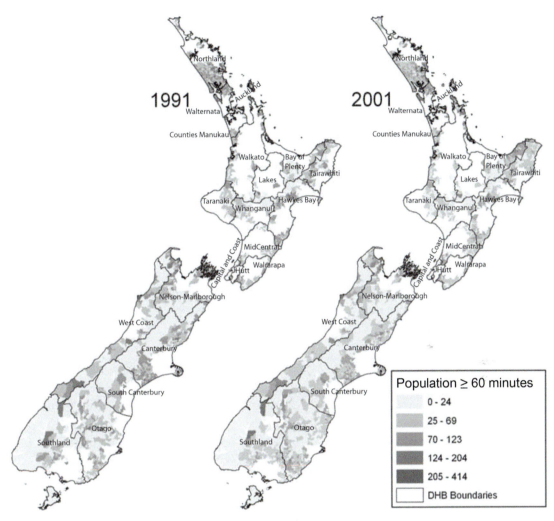

Figure 6.15 Travel times to emergency health services in New Zealand, 1991 and 2001

Source: Reprinted from *Health Informatics Journal*, vol. 12, Brabyn, L., and P. Beere, "Population Access to Hospital Emergency Departments and the Impacts of Health Reform in New Zealand," page 234, © 2008 by SAGE. Reprinted by Permission of SAGE.

CONCLUSION

Health geographers use maps to communicate, explore, and analyze relationships between health and space. Maps enable us to identify patterns that would otherwise remain invisible, allowing us to generate hypotheses about spatial relationships. From map analysis, we can begin to consider risk factors for disease, seek connections between health and environment, and investigate equity of healthcare provision, among other things.

GIS has greatly increased the scope of our ability to grapple with questions of health and place, by making it possible to store, display, and analyze huge quantities of spatial data. Health geographers must always bear in mind that GIS is only useful to the study of health as a *tool*, however. Without an understanding of the biomedical, ecological, demographic, social, political, and cultural facets of health, GIS may provide answers but can offer little insight. One critique that has emerged about the use of GIS is its reliance on empirical data, which can ultimately limit the topics studied and arguably constrains approaches to the study of health. In the next section of the book, we turn to the social approaches that deepen and extend our understanding of the relationship between place and health.

DISCUSSION QUESTIONS

1   Can you find a map that demonstrates how health indicators in your area compare with other places? What information can you glean from this map? How might you critique this map in terms of factors such as the reliability of its data and how effectively the data were presented?

2   Discuss which type of map you would recommend for exploring the following health issues. What kinds of symbols do you think would best communicate these topics?

   a   The location of hospitals in a particular country

   b   The diffusion of plague across Europe in the Middle Ages

   c   Infant mortality rates in Central America

3   Find an example of a map that is being used to intentionally make a political point about the connection between an exposure and a specific health problem. Discuss how this map makes this political point explicit. Do you consider this map to be a reliable source of information? Why or why not?

4   In the text, we discussed examples of health data that are subject to bias if self-reported. What other kinds of data quality issues do you think are important in health mapping? How might these issues lead to biases in research findings?

SUGGESTED READING

Dartmouth Institute for Health Policy and Clinical Practice. 2019. *The Dartmouth Atlas of Health Care* [Online]. www.dartmouthatlas.org/.

Jerrett, M., S. Gale, and C. Kontgis. 2010. "Spatial Modeling in Environmental and Public Health Research." *International Journal of Environmental Research and Public Health* 7 (4): 1302–29.

Koch, T. 2016. *Cartographies of Disease: Maps, Mapping, and Medicine*. Esri Press.

Pickle, L. W., and National Center for Health Statistics. 1996. *Atlas of United States Mortality*. Hyattsville, MD: National Center for Health Statistics, Centers for Disease Control and Prevention, U.S. Department of Health and Human Services. www.cdc.gov/nchs/products/other/atlas/atlas.htm.

Shaw, Nicola T. 2012. "Geographical Information Systems and Health: Current State and Future Directions." *Health Informatics Research* 18 (2): 88–96.

WHO. 2019. *Global Health Observatory (GHO) Data: Map Gallery* [Online]. www.who.int/gho/map_gallery/en/.

REFERENCES

Ahmad, O., C. Boschi-Pinto, A. D. Lopez, C. J. L. Murray, R. Lozano, and M. Inoue. 2001. *Age Standardization of Rates: A New WHO Standard*. Geneva: World Health Organization.

Ajakaye, O. G., O. I. Adedeji, and P. O. Ajayi. 2017. "Modeling the Risk of Transmission of Schistosomiasis in Akure North Local Government Area of Ondo State, Nigeria Using Satellite Derived Environmental Data." *PLoS Neglected Tropical Diseases* 11 (7): 20. doi:10.1371/journal.pntd.0005733.

Bagheri, N., K. Wangdi, N. Cherbuin, and K. J. Anstey. 2018. "Combining Geospatial Analysis with Dementia Risk Utilising General Practice Data: A Systematic Review." *Journal of Prevention of Alzheimer's Disease* 5 (1): 71–77. doi:10.14283/jpad.2017.33.

Banks, B., T. Cote, M. Golden, R. Lake, H. Meij, R. Rogers, and P. Rosenberg. 2000. AIDS Mortality in U.S. Counties: Small Count Counties Aggregate. AIDS Data Animation Project, Animation for Weekly AIDS Mortality in the United States Jan 1981-Dec 1992. Center for International Earth Science Information Network (CIESIN), Columbia University. Accessed November 14, 2010 www.ciesin.columbia.edu/datasets/cdc-ncl/continental.html.

Bergquist, R., and L. Rinaldi. 2010. "Health Research Based on Geospatial Tools: A Timely Approach in a Changing Environment." *Journal of Helminthology* 84 (1): 1–11. doi:10.1017/s0022149x09990484.

Boulos, M. N. K., G. C. Peng, and T. VoPham. 2019. "An Overview of GeoAI Applications in Health and Healthcare." *International Journal of Health Geographics* 18: 9. doi:10.1186/s12942-019-0171-2.

Brabyn, Lars, and Paul Beere. 2006. "Population Access to Hospital Emergency Departments and the Impacts of Health Reform in New Zealand." *Health Informatics Journal* 12 (3): 227–37.

Brown, L. M., L. M. Pottern, and W. J. Blot. 1984. "Lung-Cancer in Relation to Environmental-Pollutants Emitted from Industrial Sources." *Environmental Research* 34 (2): 250–61. doi:10.1016/0013-9351(84)90093-8.

Cambridge, B., and S. J. Cina. 2010. "The Accuracy of Death Certificate Completion in a Suburban Community." *American Journal of Forensic Medicine and Pathology* 31 (3): 232–35. doi:10.1097/PAF.0b013e3181e5e0e2.

[CDC] Centers for Disease Control. 1990. "Guidelines for Investigating Clusters of Health Events." *Morbidity and Mortality Weekly Report* 39 (RR-11): 1–16.

[CDC] Centers for Disease Control and Prevention. 2018. United States Cancer Statistics (USCS).

Ceccato, P., S. J. Connor, I. Jeanne, and M. C. Thomson. 2005. "Application of Geographical Information Systems and Remote Sensing Technologies for Assessing and Monitoring Malaria Risk." *Parassitologia* 47 (1): 81–96.

Central Broadcasting Service. 2010. "Red Cross: 3M Haitians Affected by Quake." Accessed August 28, 2019. www.cbsnews.com/news/red-cross-3m-haitians-affected-by-quake/.

Chi, W., J. Wang, X. Li, X. Zheng, and Y. Liao. 2008. "Analysis of Geographical Clustering of Birth Defects in Heshun County, Shanxi Province." *International Journal of Environmental Health Research* 18 (4): 243–52. doi:10.1080/09603120701824524.

Craglia, M., and R. Maheswaran. 2016. *GIS in Public Health Practice*. CRC Press.

Cromley, E. K., and S. McLafferty. 2012. *GIS and Public Health*. Guilford Publications.

Dartmouth Institute for Health Policy and Clinical Practice. 2019. "Dartmouth Atlas Project." Accessed May 31, 2019. www.dartmouthatlas.org/interactive-apps/.

Dhewantara, P. W., A. Ruliansyah, M. E. A. Fuadiyah, E. P. Astuti, and M. Widawati. 2015. "Space-time Scan Statistics of 2007–2013 Dengue Incidence in Cimahi City, Indonesia." *Geospatial Health* 10 (2): 255–60. doi:10.4081/gh.2015.373.

Eisen, L., and S. Lozano-Fuentes. 2009. "Use of Mapping and Spatial and Space-Time Modeling Approaches in Operational Control of *Aedes aegypti* and Dengue." *PLoS Neglected Tropical Diseases* 3 (4): 7. doi:10.1371/journal.pntd.0000411.

Elliott, P., J. C. Wakefield, N. G. Best, and D. Briggs. 2001. *Spatial Epidemiology: Methods and Applications*. Oxford University Press.

Fagin, D. 2013. *Toms River: A Story of Science and Salvation*. Random House Publishing Group.

Gwitira, I., A. Murwira, J. Mberikunashe, and M. Masocha. 2018. "Spatial Overlaps in the Distribution of HIV/AIDS and Malaria in Zimbabwe." *BMC Infectious Diseases* 18: 10. doi:10.1186/s12879-018-3513-y.

Harteloh, P., K. de Bruin, and J. Kardaun. 2010. "The Reliability of Cause-of-Death Coding in the Netherlands." *European Journal of Epidemiology* 25 (8): 531–38. doi:10.1007/s10654-010-9445-5.

Ioannidou, D., L. Malherbe, M. Beauchamp, N. P. A. Saby, et al. 2018. "Characterization of Environmental Health Inequalities Due to Polyaromatic Hydrocarbon Exposure in France." *International Journal of Environmental Research and Public Health* 15 (12): 20. doi:10.3390/ijerph15122680.

Jeanne, I., L. E. Chambers, A. Kazazic, T. L. Russell, et al. 2018. "Mapping a Plasmodium Transmission Spatial Suitability Index in Solomon Islands: A Malaria Monitoring and Control Tool." *Malaria Journal* 17. doi:10.1186/s12936-018-2521-0.

Kapwata, T., and S. Manda. 2018. "Geographic Assessment of Access to Health Care in Patients with Cardiovascular Disease in South Africa." *BMC Health Services Research* 18: 10. doi:10.1186/s12913-018-3006-0.

Kavur, H., and O. Artun. 2017. "Geographical Information Systems in Determination of Cutaneous Leishmaniasis Spatial Risk Level Based on Distribution of Vector Species in Imamoglu Province, Adana." *Journal of Medical Entomology* 54 (5): 1175–82. doi:10.1093/jme/tjx102.

Keegan, V. 2010. "Meet the Wikipedia of the Mapping World." Accessed August 28. www.theguardian.com/technology/2010/feb/04/mapping-open-source-victor-keegan.

Koch, T. 2005. *Cartographies of Disease: Maps, Mapping, and Medicine*. ESRI Press.

Koch, T., and K. Denike. 2009. "Crediting His Critics' Concerns: Remaking John Snow's Map of Broad Street Cholera, 1854." *Social Science & Medicine* 69 (8): 1246–51. doi:10.1016/j.socscimed.2009.07.046.

Leta, S., T. J. Beyene, E. M. De Clercq, K. Amenu, et al. 2018. "Global Risk Mapping for Major Diseases Transmitted by *Aedes aegypti* and *Aedes albopictus*." *International Journal of Infectious Diseases* 67: 25–35. doi:10.1016/j.ijid.2017.11.026.

Lipsky, L. M., D. L. Haynie, C. Hill, T. R. Nansel, et al. 2019. "Accuracy of Self-Reported Height, Weight, and BMI Over Time in Emerging Adults." *American Journal of Preventive Medicine* 56 (6): 860–68. doi:10.1016/j.amepre.2019.01.004.

Lloyd, K. 1998. "Ethnicity, Social Inequality, and Mental Illness—In a Community Setting the Picture Is Complex." *British Medical Journal* 316 (7147): 1763. doi:10.1136/bmj.316.7147.1763.

Lorant, V., I. Thomas, D. Deliege, and R. Tonglet. 2001. "Deprivation and Mortality: The Implications of Spatial Autocorrelation for Health Resources Allocation." *Social Science & Medicine* 53 (12): 1711–19. doi:10.1016/s0277-9536(00)00456-1.

Maclachlan, J. C., M. Jerrett, T. Abernathy, M. Sears, and M. J. Bunch. 2007. "Mapping Health on the Internet: A New Tool for Environmental Justice and Public Health Research." *Health & Place* 13 (1): 72–86. doi:10.1016/j.healthplace.2005.09.012.

Mic, P., M. Koyuncu, and J. Hallak. 2019. "Primary Health Care Center (PHCC) Location-Allocation with Multi-Objective Modelling: A Case Study in Idleb, Syria." *International Journal of Environmental Research and Public Health* 16 (5): 23. doi:10.3390/ijerph16050811.

Middleton, N., J. A. C. Sterne, and D. J. Gunnell. 2008. "An Atlas of Suicide Mortality: England and Wales, 1988–1994." *Health & Place* 14 (3): 492–506. doi:10.1016/j.healthplace.2007.09.007.

Mollalo, A., A. Alimohammadi, M. R. Shirzadi, and M. R. Malek. 2015. "Geographic Information System-Based Analysis of the Spatial and Spatio-Temporal Distribution of Zoonotic Cutaneous Leishmaniasis in Golestan Province, North-East of Iran." *Zoonoses and Public Health* 62 (1): 18–28. doi:10.1111/zph.12109.

Monmonier, M. 2018. *How to Lie with Maps*. 3rd ed. University of Chicago Press.

Ocampo, C. B., N. J. Mina, M. I. Echavarria, M. Acuna, et al. 2019. "VECTOS: An Integrated System for Monitoring Risk Factors Associated with Urban Arbovirus Transmission." *Global Health-Science and Practice* 7 (1): 128–37. doi:10.9745/ghsp-d-18-00300.

Openshaw, S., and P. Taylor. 1979. "A Million or so Correlation Coefficients: Three Experiments on the Modifiable Areal Unit Problem." In *Statistical Applications in the Spatial Sciences*, edited by N. Wrigley. Pion.

Palaniyandi, M. 2012. "The Role of Remote Sensing and GIS for Spatial Prediction of Vector-Borne Diseases Transmission: A Systematic Review." *Journal of Vector Borne Diseases* 49 (4): 197–204.

Pickle, Linda Williams, and National Center for Health Statistics. 1996. *Atlas of United States Mortality.* Hyattsville, MD: National Center for Health Statistics, Centers for Disease Control and Prevention, U.S. Department of Health and Human Services.

Pistocchi, A. 2008. "A GIS-Based Approach for Modeling the Fate and Transport of Pollutants in Europe." *Environmental Science & Technology* 42 (10): 3640–47. doi:10.1021/es071548.

Pleydell, D. R. J., Y. R. Yang, F. M. Danson, F. Raoul, et al. 2008. "Landscape Composition and Spatial Prediction of Alveolar Echinococcosis in Southern Ningxia, China." *PLoS Neglected Tropical Diseases* 2 (9): 10. doi:10.1371/journal.pntd.0000287.

QGIS. 2019. "QGIS: A Free and Open Source Geographic Information System." Accessed June 8. www.qgis.org/en/site/.

Ritchie, H. 2018. "What Were the World's Deadliest Earthquakes?" Accessed August 28. https://ourworldindata.org/the-worlds-deadliest-earthquakes.

Robinson, A. H., and B. B. Petchenik. 1976. *The Nature of Maps: Essays Toward Understanding Maps and Mapping.* Chicago: University of Chicago Press.

Rodenwaldt, E., and H. J. Jusatz. 1952. *Welt-Seuchen-Atlas: Weltatlas der Seuchenverbreitung und Seuchenbewegung.* Falk-Verlag.

Rodriguez-Morales, A. J., P. Ruiz, J. Tabares, C. A. Ossa, et al. 2017. "Mapping the Ecoepidemiology of Zika Virus Infection in Urban and Rural Areas of Pereira, Risaralda, Colombia, 2015–2016: Implications for Public Health and Travel Medicine." *Travel Medicine and Infectious Disease* 18: 57–66. doi:10.1016/j.tmaid.2017.05.004.

Rothman, K. J. 1990. "A Sobering Start for the Cluster Busters' Conference." *American Journal of Epidemiology* 132 (Suppl 1): S6–13. doi:10.1093/oxfordjournals.aje.a115790.

Sadler, R. C., C. Hippensteel, V. Nelson, E. Greene-Moton, and C. D. Furr-Holden. 2019. "Community-Engaged Development of a GIS-Based Healthfulness Index to Shape Health Equity Solutions." *Social Science & Medicine* 227: 63–75. doi:10.1016/j.socscimed.2018.07.030.

Schneider, D., M. R. Greenberg, M. H. Donaldson, and D. Choi. 1993. "Cancer Clusters: The Importance of Monitoring Multiple Geographic Scales." *Social Science & Medicine* 37 (6): 753–59. doi:10.1016/0277-9536(93)90369-f.

Schuurman, N. 2009. "The Effects of Population Density, Physical Distance, and Socio-Economic Vulnerability on Access to Primary Health Care in Rural and Remote British Colombia, Canada." In *Primary Health Care: People, Practice, Place*, edited by V. A. Crooks and G. J. Andrews. Farnham, Surrey: Ashgate.

Simon, S. L., J. E. Till, R. D. Lloyd, R. L. Kerber, et al. 1995. "The Utah Leukemia Case-Control Study—Dosimetry Methodology and Results." *Health Physics* 68 (4): 460–71. doi:10.1097/00004032-199504000-00003.

Smallman-Raynor, M. R., and A. D. Cliff. 1991. "Civil War and the Spread of AIDS in Central Africa." *Epidemiology and Infection* 107 (1): 69–80. doi:10.1017/s095026880004869x.

Snow, J. 1855. *On the Mode of Communication of Cholera.* John Churchill.

Stewart-Ibarra, A. M., A. G. Munoz, S. J. Ryan, E. B. Ayala, et al. 2014. "Spatiotemporal Clustering, Climate Periodicity, and Social-Ecological Risk Factors for Dengue During an Outbreak in Machala, Ecuador, in 2010." *BMC Infectious Diseases* 14: 16. doi:10.1186/s12879-014-0610-4.

Talen, E., and L. Anselin. 1998. "Assessing Spatial Equity: An Evaluation of Measures of Accessibility to Public Playgrounds." *Environment and Planning A* 30 (4): 595–613. doi:10.1068/a300595.

Till, J. E., S. L. Simon, R. Kerber, R. D. Lloyd, et al. 1995. "The Utah Thyroid Cohort Study—Analysis of the Dosimetry Results." *Health Physics* 68 (4): 472–83. doi:10.1097/00004032-199504000-00004.

Turner, E. L., C. Metcalfe, J. L. Donovan, S. Noble, et al. 2016. "Contemporary Accuracy of Death Certificates for Coding Prostate Cancer as a Cause of Death: Is Reliance on Death Certification Good Enough? A Comparison with Blinded Review by an Independent Cause of Death Evaluation Committee." *British Journal of Cancer* 115 (1): 90–94. doi:10.1038/bjc.2016.162.

[UNESCO] United Nations Educational, Scientific and Cultural Organization. 2017. "Trafficking and HIV/AIDS Project: HIV/AIDS Animated Maps." Accessed October 24, 2019. http://www.unescobkk.org/culture/diversity/trafficking-hiv/projects/gis-linked-social-sentinel-surveillance-project/hivaids-animated-map-new/.

[USGS] United States Geological Survey. 2019. "What Is a Geographic Information System (GIS)?" Accessed May 31, 2019. www.usgs.gov/faqs/what-a-geographic-information-system-gis?qt-news_science_products=1#qt-news_science_products.

Waller, Lance A., and Carol A. Gotway. 2004. *Applied Spatial Statistics for Public Health Data, Wiley Series in Probability and Statistics*. Hoboken, NJ: John Wiley & Sons.

[WHO] World Health Organization. 2018a. *Global Status Report on Alcohol and Health 2018*. Geneva: World Health Organization.

[WHO] World Health Organization. 2018c. *Global Tuberculosis Report 2018*. Geneva: WHO.

[WHO] World Health Organization. 2018b. *Global Status Report on Road Safety 2018*. Geneva: World Health Organization.

[WHO] World Health Organization. 2019. "WHO Mortality Database." Accessed June 16, 2019. www.who.int/healthinfo/mortality_data/en/.

Wynder, E. 1975. "John Graunt 1620–1674: The Father of Demography." *Preventive Medicine* 4 (1): 85–88.

Yiannakoulias, N., B. H. Rowe, L. W. Svenson, D. P. Schopflocher, K. Kelly, and D. C. Voaklander. 2003. "Zones of Prevention: The Geography of Fall Injuries in the Elderly." *Social Science & Medicine* 57 (11): 2065–73. doi:10.1016/s0277-9536(03)00081-9.

Zhang, H. R., and N. K. Tripathi. 2018. "Geospatial Hot Spot Analysis of Lung Cancer Patients Correlated to Fine Particulate Matter (PM2.5) and Industrial Wind in Eastern Thailand." *Journal of Cleaner Production* 170: 407–24. doi:10.1016/j.jclepro.2017.09.185.

Zhang, Z. J., T. E. Carpenter, Y. Chen, A. B. Clark, et al. 2008. "Identifying High-Risk Regions for Schistosomiasis in Guichi, China: A Spatial Analysis." *Acta Tropica* 107 (3): 217–23. doi:10.1016/j.actatropica.2008.04.027.

Zook, Matthew, Mark Graham, Taylor Shelton, and Sean Gorman. 2010. "Volunteered Geographic Information and Crowdsourcing Disaster Relief: A Case Study of the Haitian Earthquake." *World Medical & Health Policy* 2 (2): 7–33. doi:10.2202/1948-4682.1069.

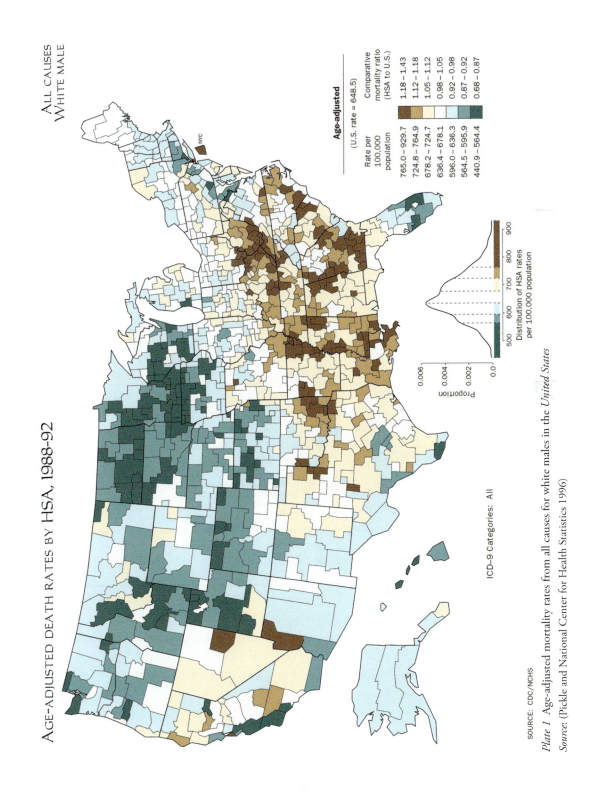

AGE-ADJUSTED DEATH RATES BY HSA, 1988-92

ALL CAUSES
WHITE MALE

Age-adjusted

(U.S. rate = 648.5)

Rate per 100,000 population	Comparative mortality ratio (HSA to U.S.)
765.0 – 929.7	1.18 – 1.43
724.8 – 764.9	1.12 – 1.18
678.2 – 724.7	1.05 – 1.12
636.4 – 678.1	0.98 – 1.05
596.0 – 636.3	0.92 – 0.98
564.5 – 595.9	0.87 – 0.92
440.9 – 564.4	0.68 – 0.87

Distribution of HSA rates per 100,000 population

ICD-9 Categories: All

SOURCE: CDC/NCHS

Plate 1 Age-adjusted mortality rates from all causes for white males in the *United States*

Source: (Pickle and National Center for Health Statistics 1996)

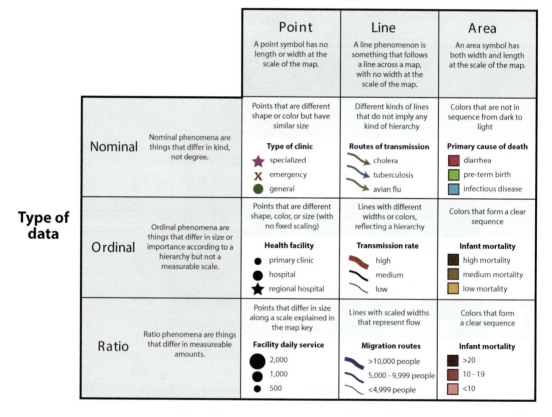

Plate 2 Types of map symbols

A variety of map symbols can be used to represent different geographic phenomena. This figure lists some examples of map symbols that are generally appropriate for different kinds of data.

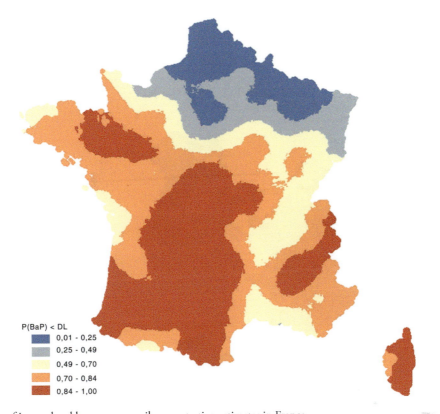

Plate 3 Map of interpolated benzopyrene soil concentration estimates in France

Source: Ioannidou, D., L. Malherbe, M. Beauchamp, N. P. A. Saby, et al. (2018). "Characterization of Environmental Health Inequalities Due to Polyaromatic Hydrocarbon Exposure in France." *International Journal of Environmental Research and Public Health* 15 (12): 20.

The map above shows an example of an isopleth map that uses shading to indicate values. This map was part of a study that estimated benzopyrene in soils in France.

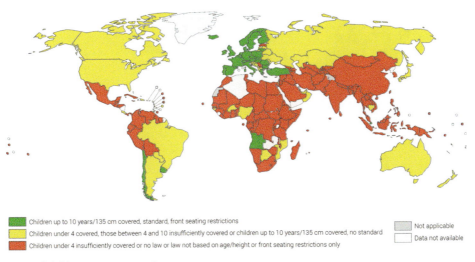

Plate 4 A map of child restraint practices by country

Source: World Health Organization. 2018b. Global Status Report on Road Safety 2018. Geneva: World Health Organization.

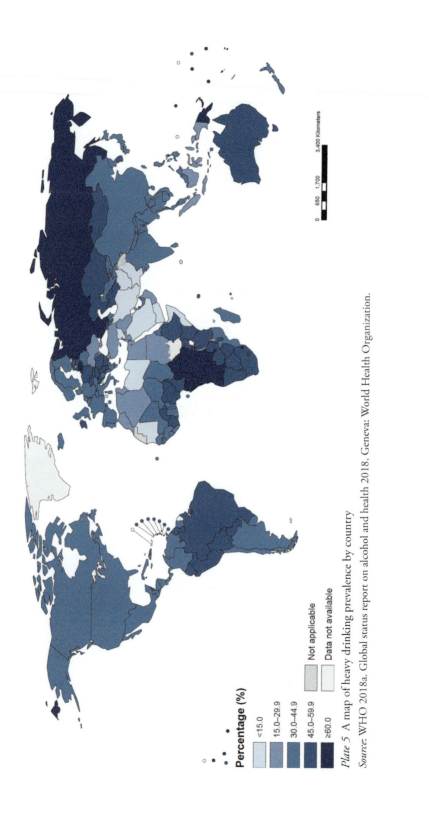

Plate 5 A map of heavy drinking prevalence by country

Source: WHO 2018a. Global status report on alcohol and health 2018. Geneva: World Health Organization.

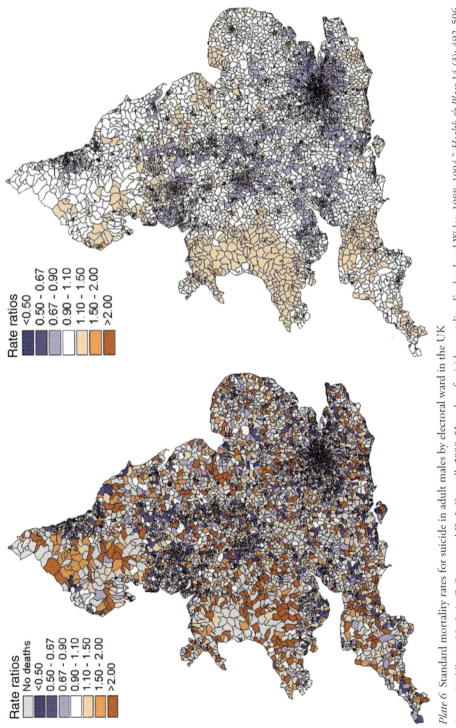

Plate 6 Standard mortality rates for suicide in adult males by electoral ward in the UK

Source: Middleton, N., J. A. C. Sterne, and D. J. Gunnell. 2008. "An atlas of suicide mortality: England and Wales, 1988–1994." *Health & Place* 14 (3): 492–506. doi:10.1016/j.healthplace.2007.09.007.

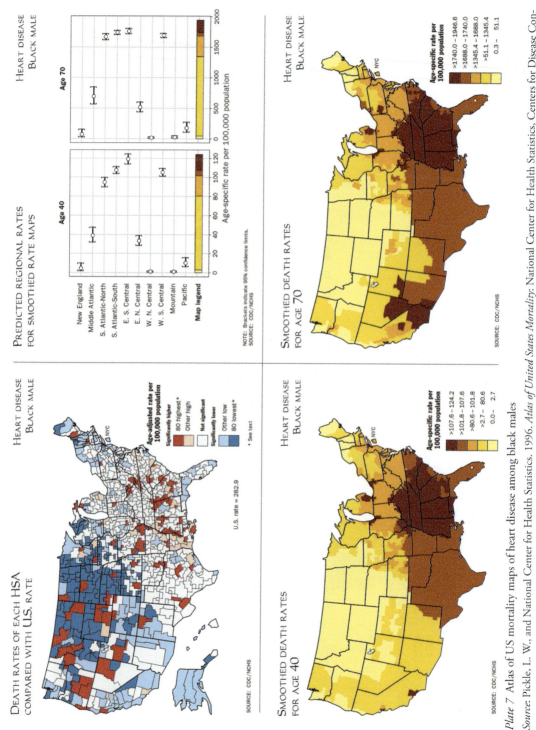

Source: Pickle, L. W., and National Center for Health Statistics. 1996. *Atlas of United States Mortality*: National Center for Health Statistics, Centers for Disease Control and Prevention, U.S. Dept. of Health and Human Services.

Plate 7 Atlas of US mortality maps of heart disease among black males

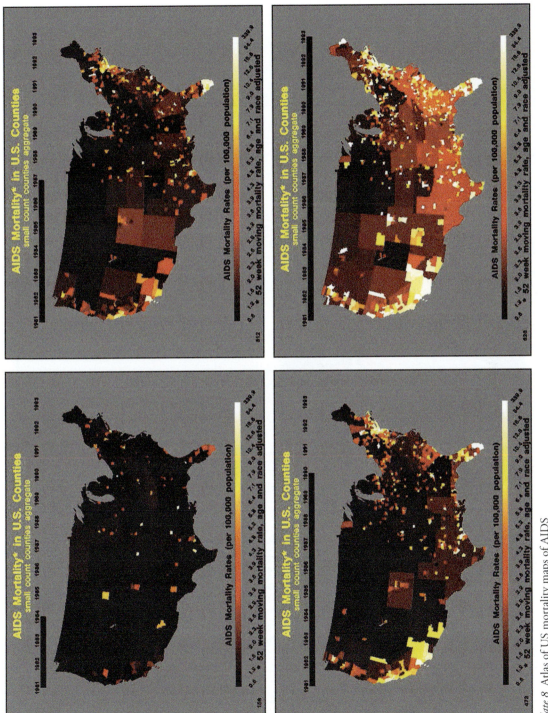

Plate 8 Atlas of US mortality maps of AIDS

Source: Pickle, L. W., and National Center for Health Statistics. 1996. *Atlas of United States Mortality*: National Center for Health Statistics, Centers for Disease Control and Prevention, U.S. Dept. of Health and Human Services.

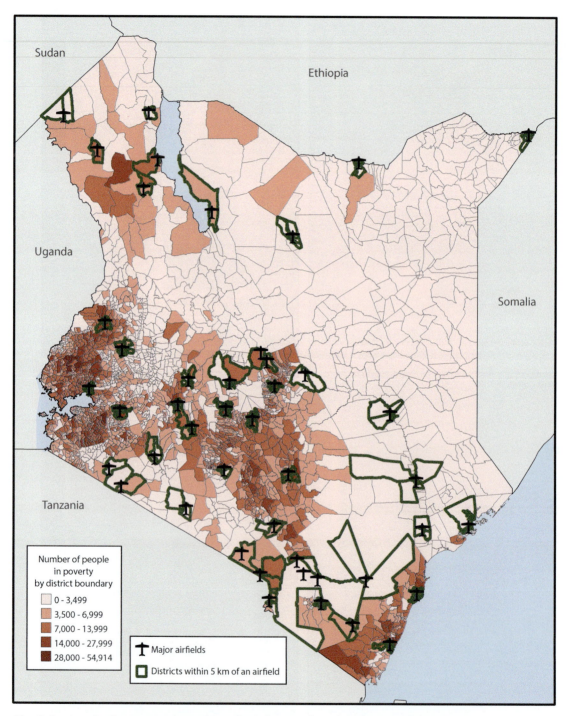

Plate 9 An example of a query to estimate the total population within 5 km of major airfields in Kenya

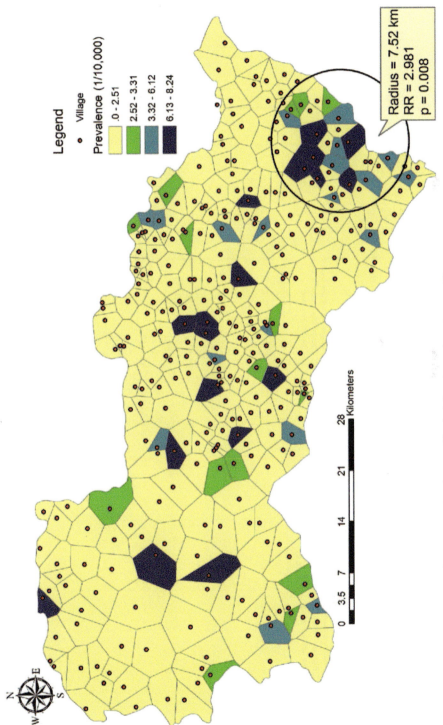

Plate 10 Map of statistically significant geographic clusters of birth defects in Heshun County, China

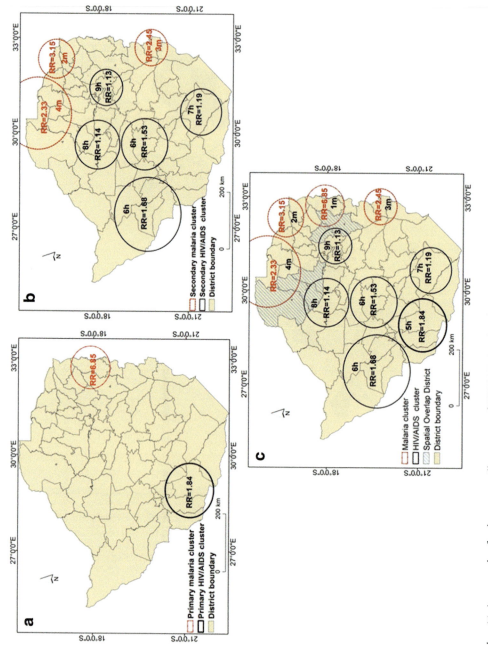

Plate 11 An example of a cluster surveillance map of HIV/AIDS and malaria in Zimbabwe

Source. Gwitira, I., A. Murwira, J. Mberikunashe, and M. Masocha. 2018. "Spatial overlaps in the distribution of HIV/AIDS and malaria in Zimbabwe." *BMN Infectious Diseases* 18:10. doi:10.1186/s12879-018-3513-y.

"*Distribution of HIV/AIDS and malaria primary and secondary clusters and their relative risk in Zimbabwe. Black circles indicate HIV/AIDS clusters, while the red circles indicate malaria clusters. Districts with hatching pattern indicate areas in which HIV/AIDS and malaria overlap in space. The letters 'h' and 'm' on the cluster numbers indicate HIV/AIDS and malaria clusters, respectively. 'RR' denotes relative risk.*" (Gwitera et al. 2018, 4)

SECTION II

SOCIAL APPROACHES TO HEALTH AND HEALTHCARE

In the previous section, we introduced the importance of ecological approaches to human health, noting the significance of *natural, built,* and *social* environments. In this section, we expand on the role of the social environment.

In the West, focus on social influences on disease was weakened with the rise of germ theory, which led to the assumption that health could be explained as a simple relationship between a causative agent and host. This **reductionist** approach has since been criticized as failing to incorporate other influences on disease, ranging from the impact of broader structures of society such as political and economic institutions, to the influence of culture and differing worldviews on how we perceive and treat disease and those who are ill. In response, research within the social sciences has begun to emphasize the importance of factors such as culture, identity, and political and economic environments, resulting in a renewed focus on the social context of health and healthcare.

Social approaches to health and healthcare incorporate a variety of different theoretical and philosophical approaches to answering health questions. The **positivist**, **quantitative** methods used by epidemiology and clinical **biomedicine**—the dominant paradigm of Western medicine—are useful for answering some social questions. For instance, a researcher could do a quantitative analysis using census data to find out whether populations with a high income have better provision of health facilities than poorer districts. Many social analysts argue that these quantitative studies miss a lot of critical information, however, such as why people make the decisions they do and how people's actions are situated in particular social contexts. This information can often only be uncovered with **qualitative** research, which tends to use smaller sample sizes and techniques such as interviews, focus groups, and observation of people's behavior to achieve an in-depth understanding of the experiences, attitudes, and decisions of particular individuals. Researchers often record these experiences to illustrate interactions within society rather than to make claims that these experiences are representative of a broader population.

Much of the work undertaken in this area focuses on issues of equality and difference, emphasizing how people's health experiences are influenced by factors such as their position in society, socioeconomic status, gender, or race. Social approaches therefore move beyond considering "society" as an undifferentiated whole to focus on the experiences of particular social groups such as women, immigrants, or people with disabilities. Some scholars have argued that we need to move even further in understanding the importance of geographies of difference to considering the lived experiences of *individuals*, rather than groups of people. This perspective suggests that individual **"life-worlds"**—the sum of one's lived experiences—are important in understanding the uniqueness of health outcomes and responses to them.

We refer to these many approaches that have emerged since (and often in reaction to) positivism as "post-positivist approaches." In trying to make sense of the diversity of approaches that fall outside the positivist mold, we identify two main scales of analysis. First, a focus on social organization can uncover how cultural, political, and economic structures of society influence access to resources and the structural constraints within which people operate. This is sometimes referred to as a structuralist or **political economy** approach. Applied to health, we can argue that causes of ill health are rooted within structures of society and, significantly, the inequalities inherent in those structures. For instance, a political economy approach might argue that the HIV/AIDS epidemic was caused in part by the marginalized economic and political situation of many groups affected by the disease and a lack of interest in overturning the structures of society that disempowered them.

A second movement, the **humanistic** approach, promotes analysis at the scale of the individual, emphasizing "the *feelings*, purposes, and goals of individuals," in search of "*understanding* rather than *explanation*" (Gesler 1991, 183, emphasis original). A humanistic approach focuses on individual **agency** (the power of the individual) and experience, emphasizing the beliefs, values, perceptions, experiences, and social networks of individuals, and arguing that meaning is constructed out of everyday lived experiences.

Consistent with this emphasis on the individual and subjective nature of experience, even the very meaning of "knowledge" has been challenged. Geography in the 1950s and '60s was characterized by a heavy emphasis on quantitative techniques, associated with **modernist** ideas that value rationality and objectivity above all else. In reaction to this, a new form of geography emerged that emphasized the limitations and assumptions inherent in this form of science. This **postmodern** movement draws attention to the inherent biases in scientific understandings of the world, emphasizing how the ways that we construct knowledge are framed by cultural and historic settings, and the power relations that come with them. Arguing that the modern world has been characterized by an overestimation of what we can really know or prove, postmodern theorists argue that knowledge is instead socially constructed and value laden. Proponents of postmodernism argue that our understandings of the world are situated in socio-cultural settings and are influenced by the beliefs and value systems of the communities that construct them. In this framework, we can talk about the existence of multiple **"knowledges**," each created within its own social context.

Feminist approaches provide an example of a distinct thread of analysis that questions the very structure of society and how it promotes certain understandings of the world. Feminist approaches all share an interest in understanding how society is structured in ways that lead women and men to have different experiences but may otherwise fall across a wide spectrum of philosophical approaches. For instance, some feminist approaches emphasize the problematic position of women in society, viewing **patriarchy** as a societal structure that disadvantages women. Other feminists argue that it is important to understand the lived experiences of women and the uniqueness of their social interactions, a more humanistic perspective. Today, feminist scholars often also consider issues of **intersectionality**, whereby different aspects of identity interact in complex ways, often reinforcing and deepening inequalities. In this way, feminists are often interested in other aspects of identity, such as ethnicity and sexuality.

Philosophical approaches to social study are not mutually exclusive—one can use both a feminist and humanistic framework at the same time, for instance. Furthermore, all of these approaches have advantages and drawbacks and may be more or less appropriate depending on the particular research question in mind. We must therefore consider both the benefits and shortcomings of any approach before applying it. For instance, one of the main criticisms leveled against post-positivist approaches is that they lack objectivity. While many people might agree that there is more room for interpretation of qualitative data than statistical output, a postmodern approach suggests that the interpretation of quantitative data is itself deeply value laden and subjective. Many have argued, for instance, that the scientific method of hypothesis testing is situated within a particular system of constructing knowledge that carries its own biases. Many postmodernists therefore argue that quantitative analyses can be even more problematic than qualitative approaches since the biases of the researcher are often hidden behind the guise of objectivity. Another common criticism of qualitative approaches is that a focus on individuals and their experiences does not produce generalizable knowledge. Advocates of these approaches might respond, however, that their goal is not to make broad generalizations but to detail the experiences of a small group of individuals to see what we can learn from them.

Overall, approaches to generating knowledge are hotly contested. We encourage you to consider the approaches detailed here with an open mind and to think about the situations in which different approaches might be more or less appropriate. In the long run, we argue that a judicious combination of approaches is the most effective way to tackle the huge diversity of important questions in the realm of health and healthcare—a topic we revisit in the concluding chapter.

Consistent with this goal of emphasizing the value of integrating theoretical approaches, we have divided this social approaches section thematically rather than by philosophy although, as you will see, certain topics lend themselves more easily to some philosophical approaches than others. We begin in Chapter 7 by thinking about the social and economic environments within which individuals live, and that constrain or enable different health-related behaviors. We then turn to the importance of culture and identity, in both constructing knowledge and in influencing health experiences and responses. In Chapter 9, we focus on the political environment, viewing health and healthcare as the product of particular configurations of power. Finally, we turn to consider healthcare in detail, considering what drives the approaches different communities take to healing, and what enables or precludes access to these facilities.

DISCUSSION QUESTIONS

1   How would you define a social approach to health? Does this definition match that of your neighbor?

2   Which of the theoretical approaches described earlier do you think would be most valuable for:

   a   deciding the location of a new hospital?

   b   assessing the importance of childhood experiences in causing illness in adulthood?

   c   uncovering issues related to inequalities in access to drugs?

3   Provide an example of a study that you could do that would use each of the following approaches:

   a   humanistic

   b   political economy

   c   feminist

   d   positivist

REFERENCE

Gesler, W. 1991. *The Cultural Geography of Health Care.* Pittsburgh, PA: University of Pittsburgh Press.

7

SOCIOECONOMIC ENVIRONMENTS

In this chapter, we discuss the complex dynamics of socio-economic environments and how they relate to human health. Why are the poor likely to experience worse health than the rich? How do economic factors interact with the social environment in ways that influence health? How might inequality and discrimination be important above and beyond the impact of deprivation?

The relationship between lack of material resources—such as food, housing, and medical facilities—and poor health may seem straightforward. Up to a point, providing people with the resources required for a healthy life is likely to improve health outcomes directly. Even where most families have access to the basic resources needed to stay healthy, there often remains a stark gradient in health related to socioeconomic status, however. Non-material factors, such as behavior, structural barriers, and psycho-social factors must therefore play a role in mediating the relationship between material affluence and health. In this chapter, we begin by exploring direct connections between health and wealth, as well as some of the measures we use to quantify health status. We then turn to less tangible aspects of socioeconomic environments such as inequality and discrimination, asking how they might influence health outcomes.

HEALTH AND WEALTH

Poverty and poor health are linked: poor countries usually have poorer health outcomes than rich ones, and socially and economically disadvantaged groups within countries tend to be less healthy than richer communities (Wagstaff 2000; Pearce and Dorling 2009). This health gap forms a clear *gradient*, with health status increasing incrementally if we stratify individuals or areas by measures of poverty or socioeconomic status (Pearce and Dorling 2009). For example, in a large survey of health differences across different income quintiles in the Global South, Gwatkin et al. (2007) note that under-age-5 mortality, malnutrition, and fertility all consistently decline as income increases, while antenatal care and immunization rates consistently increase with income.

This relation between health and wealth is apparent through a variety of health indicators and can easily be explored graphically. Gapminder.org provides a free online tool that enables users to graph health data in ways that make these sorts of connections evident. We provide some examples from Gapminder here and encourage you to explore these graphs for yourself online. If we

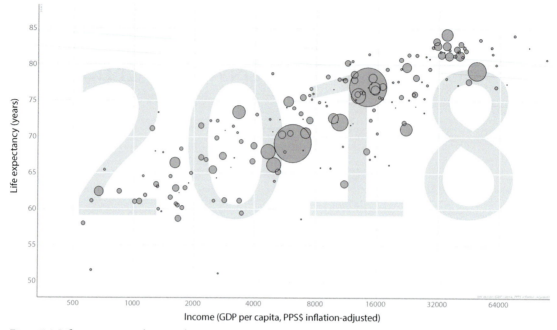

Figure 7.1 Life expectancy and income by country, 2018

Source: Free material from www.gapminder.org. Data source: www.gapminder.org/data/

This scatterplot graph shows the relation between income and life expectancy. The points on the graph represent countries, sized proportionally by population size.

compare life expectancy and income, for instance, we see a clear health gradient—the world's poorest countries generally have life expectancies in the 50s and 60s, in middle-income countries life expectancies are typically in the upper 60s and 70s, while in many affluent countries life expectancy exceeds 80 (Figure 7.1). This represents a strong positive **correlation**—as one variable (wealth) rises, so does the other (life expectancy)—suggesting that the two variables are linked.

Several factors may explain this association between health and wealth. For individuals, material affluence can provide the means to obtain sufficient food, sanitary and safe living conditions, access to healthcare, and the option to avoid dangerous working conditions. At a societal scale, material wealth enables the construction of well-functioning infrastructure, the funding of health care, and the development and enforcement of health-promoting legislation. Beyond these obvious

connections, however, the nature of the relationship between health and wealth is complex.

Gapminder also allows us to consider changes in health over time. Graphing child mortality against income, for instance, we can compare data from 1960 and 2018 (Figures 7.2 and 7.3). There is a clear correlation between child mortality and income in both time periods. In this case, the two variables show a *negative* correlation (as income increases, child mortality decreases). Although this relation persists across both years, mortality rates have dropped significantly over this time period, indicating improvements in child health at the global scale. In 1960, child mortality rates by country ranged from 20 to 448 deaths per 1,000 births; by 2018, rates ranged from just two to 126. It is important to note that *inequalities* among countries have increased over this time period, however. In 1960, the country with the highest child mortality rate (Mali)

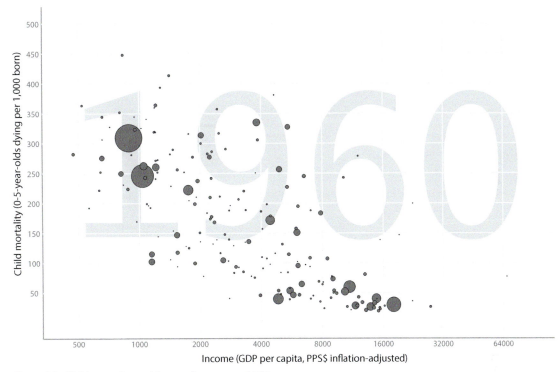

Figure 7.2 Child mortality and income by country, 1960

Source: Free material from www.gapminder.org. Data source: www.gapminder.org/data/

had a rate about twenty times greater than the country with the lowest rate (Sweden). By 2018, Somalia and Chad's child mortality rates were about sixty times greater than rates reported for Finland and Iceland. In other words, improvements in child health occurred across almost all countries during this time period, but some countries made significantly greater progress than others. Income inequalities also increased over this period. We see a sixty-fold difference in income between the richest and poorest countries in 1960, which balloons to a 150-fold difference by 2018 (after removing the influence of **outliers**).

There are three fundamental explanatory frameworks for why the poor bear a high health burden (Marmot et al. 1997). So far, we have been focusing on the idea of "social causation," which suggests that poor health is a consequence of low income and social position. This explanation focuses attention on the **social determinants of health**—the social, political, and economic conditions that influence health (Figure 7.4). The idea of "health selection" posits the inverse: that poverty is actually a *consequence* of poor health. For example, poor health may limit job opportunities or force people to spend much of their income on healthcare, leading to poverty. Finally, "indirect selection" suggests that factors present early in life affect both wealth and health. For example, lacking a good education could both impair an individual's chance to secure a high income and be linked to unhealthy behaviors. In reality, the relationship between health and wealth is probably a combination of these contributory factors. In this chapter, we concentrate on the social determinants of health, which is the primary focus of geographic research as it focuses on the idea that social and physical environments influence health outcomes.

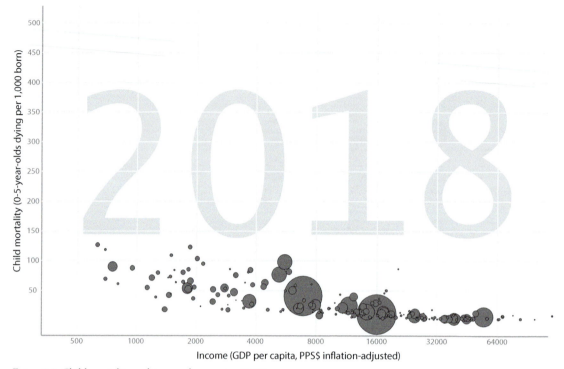

Figure 7.3 Child mortality and income by country, 2018
Source: Free material from www.gapminder.org. Data source: www.gapminder.org/data/

To fully explore the relationship between health and wealth, we must consider processes at different geographic scales because factors significant at one scale may have a different effect, or indeed no effect, at another scale. At the global scale, international trade networks, patterns of mobility, and the policies of non-governmental organizations (NGOs) and international governmental organization (IGOs) are some of the key drivers of health. Explicitly health-focused organizations such as the World Health Organization have clearly had a positive influence on global health, and yet we tend to overestimate their impacts, given the tight budgets they must work within. By contrast, we often *under*estimate the role of indirect effects of global processes on health, such as the impacts of **globalization** (the integration of the world's economic system). On one hand, the dramatic growth of the global economy generated by globalization has been associated with rising incomes for many populations, providing improved access to basic needs and thereby improving

health. This may be one of the key explanations for the generally improving health standards since the 1960s that we identified earlier. On the other hand, the globalized economy raises health concerns as affluent consumers exploit cheap labor in the Global South, deepening inequalities and effectively "outsourcing" the burdens of industrial production (e.g., pollution and work-related injuries) onto poorer communities. Greater international mobility associated with global modes of production also contributes to the spread of disease. Globalization is also spreading cultural practices with health implications such as eating processed and fast foods, which is contributing to the global increase in obesity and diabetes.

Other forces drive the relationship between health and wealth at other scales. At the national or regional scale, factors such as the structure of the economy, healthcare policies, and environmental and occupational health legislation have crucial impacts on health. At the household scale, differences in health may best be

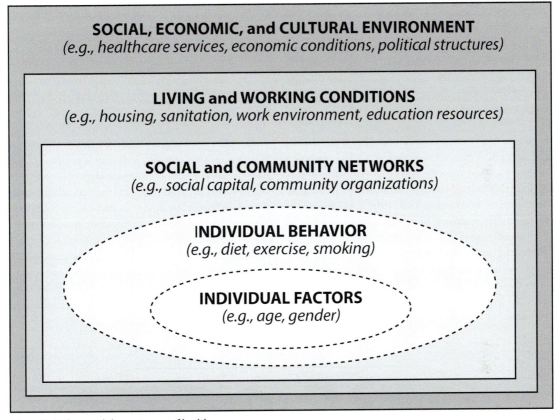

Figure 7.4 The social determinants of health

Source: Adapted from Dahlgren and Whitehead (1991)

The social determinants of health are cultural, political, and economic conditions that affect health. Individual factors, such as behavior and genetics, exist in this social context, mediating the relationship between the social environment and health.

explained by employment and education patterns or the relative empowerment of women.

There are also differences in *how* we assess wealth and health at different scales. Global-scale analyses often require the use of large aggregated data sets to summarize the health of entire populations. Consequently, researchers often estimate health at these scales using **health indicators**: specific measurable components of a population's health such as life expectancy and child mortality. As the scale of analysis narrows to communities or individuals, researchers often turn to **qualitative** methods such as focus groups or personal narratives to consider factors such as the influence of relative position in society and inequality. We discuss these kinds of

qualitative methods in Chapter 8 and focus here on the use of **quantitative** health indicators.

Health indicators

Health indicators provide estimates of different facets of population health and can be applied in a standard way to enable comparisons among populations. Indicators are typically given as a **rate**—the number of cases per unit of population (often per 1,000 or per 100,000 people)—to remove the influence of population size. Small population sizes can present a challenge when working with aggregated data because a single event in a small population can appear very significant. For instance, we would expect there to be no deaths from road traffic accidents in any given year in a small village of 400 people. However, if there were, by chance, a traffic accident that killed five people in our village, the death rate for that year would be incredibly high—5 per 400 inhabitants, or 1,250 deaths per 100,000! Clearly, this figure could be misleading if it were used to suggest that this village was a traffic accident hotspot based on one isolated incident. For this reason, researchers often **aggregate** data by combining data from larger populations or longer time scales. Although this reduces the **resolution** of the data, it also limits the potentially misleading impact of isolated or unusual events. If the goal is to identify temporal patterns in a **longitudinal** study, it makes sense to aggregate data by space. If, by contrast, our primary interest is in spatial patterns, we could aggregate data over time to preserve the **spatial resolution** of our data.

Many health indicators estimate either **morbidity** or **mortality**. Morbidity rates report the proportion of a population with a disease or infirmity, often given as a prevalence or incidence rate. The **prevalence** rate refers to the total number of cases within a specified population at a point in time. The **incidence** rate reports the number of *new* cases within a population over a specified period. Mortality rates (or death rates) indicate the likelihood of death in a population—either from all causes or from a specific cause. In 2017, for instance, the annual mortality rate from diarrhea for children under age 5 in Africa was roughly six per thousand (WHO 2018).

Mortality or morbidity rates calculated without any statistical manipulation or adjustment beyond accounting for population size are called **crude rates**. As noted in the previous chapter, crude rates can be misleading because they are influenced by the age structure of the population. We must therefore always consider two possible explanations for high death rates: 1) the population is genuinely unhealthy, leading to premature death for many people, or 2) the population has a large proportion of people in older age **cohorts** where death is most likely. If we examine a map of crude death rates worldwide (Figure 7.5), we can see two regions that stand out with particularly

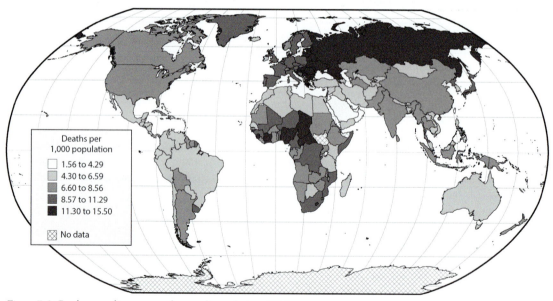

Figure 7.5 Crude mortality rate per thousand population, 2016
Source: World Bank (2019)

This map shows the crude mortality rate per thousand population in 2016. This is calculated by dividing the number of deaths in a year by the total population.

high rates: central Africa and the former Soviet bloc. In sub-Saharan Africa, war, economic crisis, debt servicing, government corruption, and poor commodity prices have exacerbated conditions of poverty, leading to some of the world's highest death rates (Kalipeni and Oppong 1998). In this case, high death rates reflect a genuinely problematic situation with respect to health. In the countries of the former Soviet Union, economic instability and the undermining of public health and social institutions have played an important role in declining health outcomes, but high death rates in ex-Soviet states are also due in part to age structure. A dramatic decline in birth rates since the fall of the Soviet Union has resulted in rapidly aging populations in many of the ex-Soviet states, increasing the death rate. The regions with the lowest death rates are generally middle- and low-income countries that have made strides towards improving living conditions and healthcare provision, but which also have very young populations. Malaysia and Morocco, for instance, both had a death rate of only around five per thousand in 2018 (World Bank 2019). High-income regions such as Europe and North America tend to have intermediate mortality rates. Generally good health circumstances in

these regions have resulted in long life expectancies, but the large proportions of elderly people in these countries inflates otherwise low raw death rates. Owing to this influence of age structure on the population, crude death rates are *not* considered to be a good indicator of the relative healthiness of a place.

If we wish to compare the impact of living conditions on mortality between regions effectively, it is important to use **age-adjusted data**, as we introduced in Chapter 6. Deaths from specific diseases are also often more likely to occur in certain age groups and so we must again address the issue of age structure when comparing *cause-specific* mortality rates between communities. Heart disease, for instance, is more common in older cohorts, while diarrheal infections and some types of leukemia are more likely in children.

Another method for removing the effects of age is to examine **age-specific mortality rates**. In this case, we avoid the problem of different health outcomes occurring at different ages by considering only one age group. For example, the WHO publishes data on the lifetime mortality rate for adults between the ages of 15 and 60 (Figure 7.6). Because this map considers only adults in

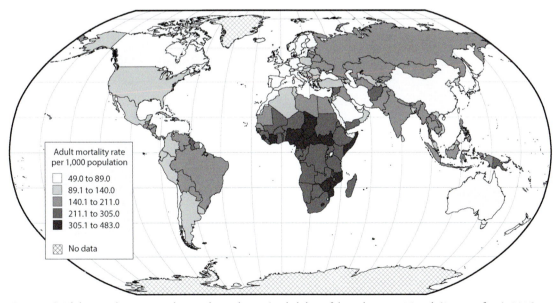

Figure 7.6 Adult mortality rate per thousand population (probability of dying between 15 and 60 years of age), 2016
Source: World Bank (2019)

an age group where we do not expect people to die, it provides a fair basis upon which to compare the relative "healthiness" of different countries. Now that we have excluded the elderly from our data set, we see that we have the lowest rates of mortality in the most affluent countries of the Global North, with intermediate rates in middle-income countries, and the highest rates in low-income countries, reflecting a genuine gradient in the healthiness of conditions in these countries. If we compare Figures 7.5 and 7.6, we see that this adult mortality map confirms that high *crude* death rates in Eastern Europe result in part from the age structure of the population because the rates are lower once the elderly have been excluded in the second map.

Another commonly used age-specific mortality rate is the **infant mortality rate**. The infant mortality rate is the rate at which children die before the age of 1, reported as infant deaths per thousand live births. Infant mortality rates are often considered to be one of the best indicators of the health status of an entire population because so many factors bear upon an infant's likelihood of survival, including sanitation, access to maternal and perinatal care, and the education and empowerment of women. As such, high infant mortality rates indicate a society struggling to meet basic needs and have even been used as an indicator of state failure (King and Zeng 2011). At the global scale, infant mortality rates vary widely, reflecting stark inequalities among countries (Figure 7.7), although these disparities have decreased over the past fifty years. In 2018, the Central African Republic (CAR) had the highest infant mortality rate in the world, with a rate of 85 deaths per thousand live births (PRB 2018). With political instability and conflict over mineral resources threatening the provision of basic needs, the CAR struggles with fundamental social and health problems. The problematic situation in the CAR is clear when we compare it with other countries; currently, the average infant mortality rate for "less developed" countries as a group is about forty per thousand, and for "developed" countries just five per thousand (Population Reference Bureau 2018). It should be noted, however, that as recently as ten years ago several countries still had infant mortality rates well over one hundred, with the highest rates in Afghanistan and

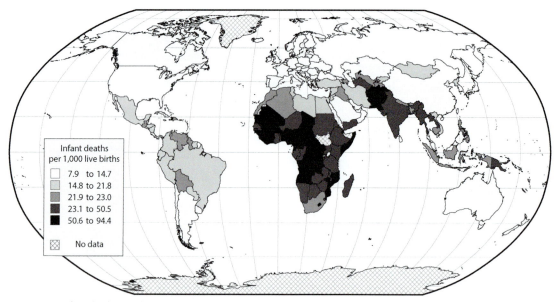

Figure 7.7 Infant deaths per thousand live births by country, 2017
Source: World Bank (2019)

Sierra Leone topping 150. The significant declines in infant mortality in many countries over the past decade demonstrate that improvements in health are being made, even in some of the world's poorest countries.

Another common health metric is **life expectancy**, the "average number of years that a newborn is expected to live if current mortality rates continue to apply" (WHO 2009). Worldwide, average life expectancy at birth was 70 years for men and 74 for women in 2018, although life expectancy is almost ten years longer in "developed" than in "less developed" countries (82 and 73 years, respectively) (PRB 2018). Average life expectancy ranges from more than 80 in much of the Global North to less than 60 in some African countries such as Somalia and Nigeria (*ibid.*) (Figure 7.8). Life expectancy is strongly influenced by infant mortality because infant deaths reduce a country's average lifespan. This does not mean that all individuals in societies with a short life expectancy will die young, however. Those who survive the perilous years of childhood may have a good likelihood of surviving into relatively old age. In such cases, mortality may show a **bimodal distribution**, with many

deaths occurring among children and then most of the remainder not occurring until late adulthood.

Other health indicators target specific groups in a population such as women or mothers. The **maternal mortality rate** refers to the risk of death associated with pregnancy and childbirth and is often used as an indicator of women's health and status in society more broadly. Maternal mortality is calculated by counting the number of women who die within 42 days of childbirth (regardless of cause), among women who bore a child in that particular year. Given good health circumstances, women are expected to have a longer biological lifespan than men, reflected in the six-year difference in life expectancy between men and women in high-income countries where women and men are treated relatively equitably. In the group of countries with the lowest income, by contrast, the difference between male and female life expectancy is only three years (PRB 2018). Here, high rates of maternal mortality combine with the toll of health inequalities produced by patriarchal societies to reduce the additional years women are expected to live.

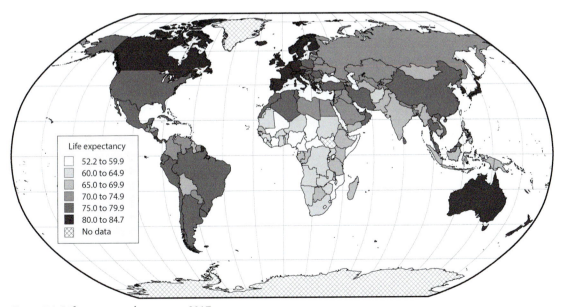

Figure 7.8 Life expectancy by country, 2017
Source: World Bank (2019)

Another way to estimate a population's health status is with the **Disability Adjusted Life Years (DALY)** statistic. This measure was born out of the WHO's Global Burden of Disease Study (see Ezzati et al. 2003), whose purpose was to quantify the impacts of different burdens to health. The DALY is reported as the number of years of healthy, working life lost to disease, disability, or premature death. The WHO explains this as "a measurement of the gap between current health status and an ideal health situation where the entire population lives to an advanced age, free of disease and disability" (WHO 2019a). Because the intent of this measure is to quantify the amount of time lost because of poor health, debilitating disease elevates the DALY statistic, and the death of a child carries more weight than the death of an adult.

Complexities in the health-wealth relationship

We have seen that there are significant global disparities in health and that poorer countries generally suffer a higher overall disease burden, higher infant mortality, and shorter life expectancies. We generally expect to see improvements in health with increasing wealth, but events can disrupt this progress. For instance, the HIV/AIDS epidemic prevented anticipated declines in infant mortality rates and increases in life expectancies in many countries of sub-Saharan Africa in the 1990s and early 2000s (Kalipeni 2000). In parts of Southern Africa, life expectancies dropped into the forties and even thirties in the early part of the twenty-first century, before the development of anti-retroviral drugs slowly reversed the trend. Progress towards better health can also be disrupted by economic and political crises, as was illustrated after the fall of the Soviet Union and by Latin America's "lost decade" of the 1980s, when a debt crisis led to the withdrawal of numerous social programs. Today, growing social inequality is increasingly being held responsible for stagnant or worsening health outcomes, even where national affluence is increasing.

The health benefits of affluence therefore do not always accrue in a unidirectional or even linear fashion. Indeed, "it is well known that, among rich countries, there is little correlation between gross national product (GNP) per person and life expectancy" (Marmot et al. 1997, 1102), with additional health benefits generally diminishing as income increases (Rodgers 1979). Instead, among the affluent countries of the Global North, health-related and social policies are often more significant indicators of health than income. This helps to explain why the US falls behind other Western countries in many health indicators, despite its high income. Stark inequalities in society combined with the market-based approach taken by the US to health service provision and access take a significant toll on the health of its citizens, compared with the more egalitarian societies of countries such as Finland and Sweden.

A scatterplot graph can make this relation clear. Figure 7.9 illustrates gross national income plotted against infant mortality. Here, infant mortality is used as a **proxy** for the overall healthiness of a society. The data form a curve: the steep part of the line on the left shows the dramatic improvements in health outcomes that we observe with even small increases in wealth among poor countries. The flatter portion of the graph on moving to the right shows the weak relationship between health and wealth among richer countries, with countries of differing affluence showing very similar infant mortality rates in many cases (e.g., Ukraine, the UK, Australia, and the US). The graph also allows us to consider why certain countries have higher or lower infant mortality rates than we would expect, given their income. For instance, all countries with a per capita income of at least US$20,000 per year have an infant mortality rate under twenty, except for Equatorial Guinea, which is an **outlier**, with a dramatically higher infant mortality rate than would be expected for its level of affluence. In countries such as this, circumstances such as war or political instability may be the culprit. Corruption or mismanagement of funds may also be to blame, resulting in underfunding of health programs. Equatorial Guinea's wealth has been generated rapidly from oil since the 1990s, making it the wealthiest country in sub-Saharan Africa. However, this wealth is controlled by an elite group against whom claims of corruption are common, and the wealth has not been invested in social programs that would benefit the majority of the population. By contrast, national policies that ensure access to **primary healthcare** for the entire population can explain

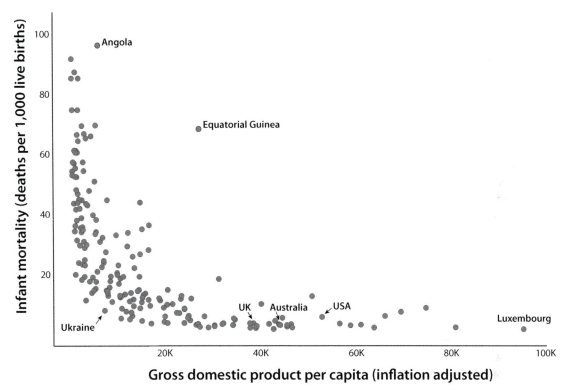

Figure 7.9 Gross national income per capita and infant mortality rates, 2015

Source: Institute for Metrics and Evaluation (2019)

indicators that are better than expected, given income (Starfield, Shi, and Macinko 2005).

For some health issues, the relationship between health and wealth is even more complex. In the case of road traffic accidents, for example, we know that the very poorest populations tend to have relatively low rates of traffic-related deaths as there are few cars on the roads.

As affluence increases and car ownership burgeons, road traffic accidents become increasingly common until a certain point where society is willing to invest in road safety measures, after which deaths on the roads go down again. We therefore see a non-linear relationship between deaths from road traffic accidents and income (Box 7.1).

Box 7.1

INEQUALITY AND ROAD TRAFFIC ACCIDENTS

Traffic-related deaths provide an example of the influence of economic inequalities on health. With increasing affluence and rising rates of car ownership, road traffic deaths have been increasing around the world. Globally, 1.35 million deaths from traffic accidents occurred in 2016, and road traffic injury is now the leading cause of death among children and young adults, 5 to 29 years

continued

Box 7.1 *continued*

(WHO 2019b). The WHO recognizes road traffic deaths as a key target for future interventions. Indeed, one of the WHO's sustainable development goals is to reduce deaths from road traffic accidents by half by 2020—a target we are not on track to meet (WHO 2019b).

We might expect traffic-related deaths to be concentrated in richer countries where car ownership per capita is highest, but this is not the case. Per capita vehicle-related fatality rates are concentrated in low-income countries, particularly in Africa, although some affluent countries in the Middle East also have very high traffic-related death rates (Figure 7.10). High-income countries account for just 15 percent of global population but 40 percent of car ownership and yet only 7 percent of road traffic deaths; the remaining 93 percent of deaths on the roads occur in low- and middle-income countries with just 60 percent of vehicles (WHO 2019b).

While growing affluence in a country initially leads to more road accidents as more cars are put on the roads, safety legislation can dramatically reduce vehicle-related deaths. The WHO (2019b) identifies legislation to regulate vehicle safety features, speeding, drunken driving, and the use of motorcycle helmets, seatbelts, and child restraints as particularly significant. South Korea illustrates the dramatic influence that legislation can have on road traffic fatalities; a road safety initiative focused on schools is credited with contributing to a 95 percent decrease in road traffic deaths in children under age 14 between 1998 and 2012 (WHO 2019b). As of 2018, however, only one billion of the world's nearly eight billion people lived in countries where road traffic laws met best practice (WHO 2019b), with the countries of the Global North most likely to have adequate legislation. Color plate 4 shows global variations in child vehicle restraint laws, for instance, although child restraints are just one aspect of best practice for road safety.

Most vehicle-related legislation relates to the safety of drivers and passengers in cars, and yet

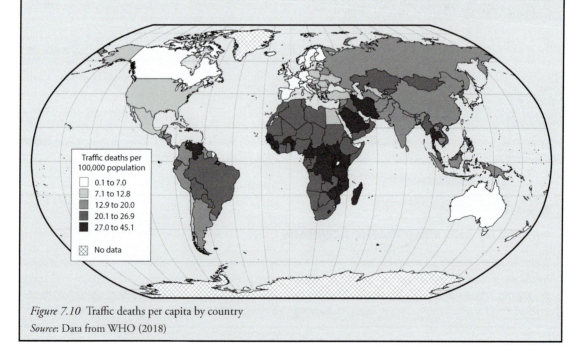

Figure 7.10 Traffic deaths per capita by country

Source: Data from WHO (2018)

more than half of all road traffic deaths occur among pedestrians, cyclists, and motorcyclists (WHO 2019b), revealing a significant inequality—the health burden from road traffic accidents often falls disproportionately on populations who may not benefit from car ownership. This reasoning could also be extended to include morbidity and mortality associated with air pollution from cars, as it is often poorer communities who live along major highways or in congested urban areas that are at the greatest risk from vehicle pollution, many of whom contribute little to the problem. In summary, although affluent populations benefit most from car ownership, it is often the poor who bear many of the health burdens of automobile use.

Health disparities exist at multiple scales, complicating the picture even further. At the sub-national scale, wealthy populations in low-income countries often have health indicators similar to wealthy populations in industrialized countries, enjoying long life expectancies but also suffering from **diseases of affluence** such as diabetes and cancer. Meanwhile, poorer segments of the population in the Global South struggle to meet basic needs, resulting in problems such as malnutrition and infectious disease, leaving many poorer countries suffering from a "**double burden of disease**." Different patterns in health at different scales remind us that we must avoid committing the **ecological fallacy**—the assumption that a relationship observed at an aggregated scale exists at other scales. For example, if we observe that richer countries report higher rates of obesity, we should not assume that wealthier individuals within those countries are more likely to be obese. Patterns in rates of mental health conditions are similarly complex (Box 7.2).

Box 7.2

PATTERNS IN MENTAL HEALTH

At the global scale, mental health and substance use disorders are often reported at higher rates in the Global North, although the association varies among conditions (Ritchie and Roser 2018). Indeed, a wide variety of different conditions fall under the umbrella of mental health (Table 7.1), calling into question the degree to which patterns in mental health as a whole are meaningful. Nonetheless, the fact that rich countries report some relatively poor mental health indicators remains a puzzling observation. Part of the issue is probably **reporting bias**, with many cases in the Global South going undiagnosed or unreported because of limited funding for mental healthcare. Furthermore, chronic mental health conditions may go unheeded by individuals in poor communities, given more pressing, acute health problems. A wide range of hypotheses have also been put forward related to risk factors of living in Western societies, including weak community and family structures, competitive work and education environments, and high levels of social inequality, but the many influences on mental health make this a hugely challenging area to study. The situation becomes even more complex as we note that some *poorer* countries report very high rates of certain conditions. For instance, Iran and Algeria report high rates of anxiety disorders; China reports high rates of schizophrenia (Ritchie and Roser 2018). Challenging political situations or conflict may help explain some of these situations, but the complexity of the patterns we see at the global scale requires carefully nuanced analyses. We encourage you to explore some different mental health indicators at the global scale available at https://ourworldindata.org/mental-health to get a feel for these complex patterns.

continued

Box 7.2 *continued*

Table 7.1 Global prevalence of mental health and substance use disorders, 2017

Disorder	Proportion of population affected (%)
Anxiety disorders	3.76
Depression	3.44
Alcohol use disorders	1.40
Drug use disorders	0.94
Bipolar disorder	0.60
Schizophrenia	0.25
Eating disorders	0.21

"Share of the total population with a given mental health or substance use disorder. Figures attempt to provide a true estimate (going beyond reported diagnosis) of disorder prevalence based on medical, epidemiologic data, surveys and meta-regression modelling" (Richie and Roser 2019).

This complexity is increased further if we look to sub-national scales, where many mental health and substance use disorders are often more common among *poorer* segments of the population. This may be related to the stress of coping with financial uncertainty or feelings of inadequacy related to marginalization in a society in which others appear to succeed. Certain self-destructive mental health behaviors such as suicide and substance abuse are rising in affluent communities experiencing growing economic inequalities, suggesting that inequality may be important. The causes of mental health disorders are so complex and often so deeply embedded in social context that explanations remain elusive and broad generalizations can be misleading, however. Even the direction of the relationship between mental health and affluence is sometimes unclear. For instance, in the US, schizophrenia is recorded at higher rates among those with low socioeconomic status, and yet schizophrenia itself acts as a risk factor for unemployment and homelessness, sometimes leading to a vicious cycle of poverty and mental health problems. Mental health disorders remain a significant and under-researched area of global health.

Despite significant improvements in health across many world regions, economic disparities have been generally widening over the past twenty years (Pearce and Dorling 2009; Pampalon, Hamel, and Gamache 2008), although a very recent **meta-analysis** suggests that inequalities *within* many countries of the Global South may finally be decreasing as efforts to reach poor and rural populations show some success (Gwatkin 2017). Economic inequalities nonetheless remain significant and challenge the idea that health simply improves with affluence. In the US, for example, the inverse relationship between socioeconomic status and mortality from cardiovascular disease (i.e., people with low incomes are more likely to suffer from heart disease) has increased substantially since the early 1960s (Singh et al. 2008). Even in countries with **universal healthcare**, the poor may remain disadvantaged with respect to health outcomes. In a Dutch study, for instance, researchers found that low socioeconomic status was significant in explaining increased cardiovascular mortality risk among patients undergoing surgery in a Dutch hospital, despite the country's excellent record of healthcare access (Ultee et al. 2018). The researchers identify a list of risk factors associated with low socioeconomic status that could have influenced outcomes, including stress, poor compliance to medical or lifestyle restrictions, low physical activity, poor diet, and air pollution in deprived neighborhoods. Providers may also bear some responsibility by stereotyping patients and struggling with communication. This wide-ranging list of potential explanatory factors illustrates the difficulty of sorting through multiple interacting factors that can confound relationships between socioeconomic status and health. A growing literature suggests that economic *inequality* may affect health in its own right, however, in ways that go beyond access to material resources.

HEALTH AND INEQUALITY

Once an individual has the resources to sustain a healthy lifestyle, inequalities in income and social status may influence health outcomes as much as material wealth. It is easy to imagine how *poor* individuals in unequal societies may suffer ill health from inequality through factors such as stress, resentment, marginalization, and exposure to negative stereotypes. However, research suggests that even *affluent* members of non-egalitarian societies might suffer poorer health than their counterparts in more equal societies (Pearce and Dorling 2009). This pattern may be related to the fear and anxiety generated in unequal societies. Affluent countries with high rates of income inequality such as the US, UK, and Australia also tend to have high rates of mental illness (Wilkinson and Pickett 2007), reinforcing the idea that living in unequal societies may cause or exacerbate psychological stresses. Although research has attempted to tease apart how inequality and health are related, the nature of the relationship has proven difficult to illuminate.

The *Whitehall Studies* from the UK represent a major effort to investigate the relationship between social inequalities and health. *Whitehall I*, initiated in 1967, examined the health of more than 17,000 male civil servants over the course of ten years and identified a social gradient in health outcomes (Marmot, Shipley, and Rose 1984). A second study in the 1980s, *Whitehall II*, followed ten thousand civil servants to further investigate the findings of the first study (Marmot et al. 1991). The authors found a negative association between mortality and morbidity rates and employment grade, and that this relation existed at all levels of employment. In other words, people with high-status jobs were more likely to live longer and healthier lives than people in middle-grade jobs, who in turn were more likely to live longer, healthier lives than people in low-grade jobs. Men in the lowest occupational grade were three times more likely to suffer coronary disease than their high-grade counterparts, for example. This work contradicted a widely held belief at the time that top executives were most likely to suffer stress-related conditions such as heart disease because of the seemingly more stressful nature of their responsibilities.

Several hypotheses have sought to explain these and similar findings. One line of reasoning suggests that relative position in society may influence an individual's likelihood of participating in risky behaviors. There is clear evidence that the poor engage in more risky behaviors such as smoking than the wealthy. These health behaviors probably account for as much as a third of the difference in income-based health disparities (Marmot et al. 1997). High rates of tobacco use may be a "self-medicating" response to being poor (Barnes and Smith 2009), the result of ignorance about the impacts of smoking (Robinson and Kirkcaldy 2007), or reflect poor access to smoking cessation programs (Tsourtos and O'Dwyer 2008). More broadly, the shame and distrust that result from having low socioeconomic status may lead to unhealthy behaviors such as smoking or drinking, while financial pressures may leave little emotional energy for practicing healthy behaviors.

A second hypothesis suggests that low status in a social hierarchy harms health through stress. It has been theorized that having low social status might initiate a physiological response in which the body's stress hormones are in a constant state of alert and that chronic exposure to these stress hormones has a variety of negative physical implications (Mayer and Sarin 2005). A third, but related, idea focuses on the negative emotions that accompany an individual's perception of having a low income relative to others. According to this hypothesis, it is more stressful to live in poverty where others are affluent than it is to be poor among others in poverty (Ellaway, Macintyre, and Kearns 2001). Frustrations associated with relative deprivation may explain why deprived areas such as impoverished inner cities often have particularly poor health indicators.

Regardless of the precise cause-and-effect mechanism, a recent rise in mental health and substance use disorders in the Global North remind us that health is about meeting more than just material needs. The term **"deaths of despair"** recognizes mortality from a group of causes that have been linked to individuals in psychological distress, including accidental poisonings (particularly drug overdose), suicide, and chronic liver disease (associated with alcoholism) (Case and Deaton 2015). This group of issues has had a particularly large impact in the US, where 2015 marked the first decline in life expectancy in more than twenty years, followed by further declines in 2017 and 2018 (Acciai and Firebaugh

2017). White, non-Hispanic, middle-aged Americans without a college degree experienced some of the greatest increases in mortality, in direct contrast to older and more educated groups who have continued to see improvements in survival. This has led to the suggestion that cumulative disadvantages triggered by deteriorating job opportunities for people with low levels of education may have caused a health crisis for this group (Case and Deaton 2015). Other researchers have suggested that this explanation may overemphasize despair when the real issue is largely opioid misuse (Masters, Tilstra, and Simon 2017) (Box 7.3). Regardless of the precise mechanisms behind these findings, the fact that life expectancies have begun to decline in the US is concerning.

Box 7.3

THE CRISIS OF OPIOID ADDICTION IN THE US

with Hailey Macrander

An epidemic of misuse of opioids has been identified as a key contributor to an unexpected reduction in life expectancy in the US that began in 2015 (National Center for Health 2018). Opioids caused 68 percent of the more than 70,000 drug overdose fatalities in 2017, representing a six-fold increase in opioid-related deaths since 1999 (CDC 2019a). That same year, the US government declared the opioid epidemic to be a public health emergency (Jones et al. 2018). This rapid rise in opioid-related deaths illustrates the complex ways in which health is connected to the social environment.

In response to widespread abuse of opioid-based substances throughout the 1800s and early 1900s, opiates were stigmatized in many countries by the early twentieth century. Recognizing that opiates were highly addictive, doctors began to turn away from them, and commentators noted a corresponding undertreatment of pain in Europe and North America throughout much of the twentieth century (Jones et al. 2018). In the 1980s, the WHO highlighted the undertreatment of postoperative and cancer-related pain (WHO 1986), leading to a reconsideration of the pain-relieving potential of opioids, initially for cancer but increasingly for chronic pain as well (Melzack 1990). US medical organizations began to develop new pain management standards that included opioids (Melzack 1990), and opioid prescriptions increased rapidly as physicians and hospitals began to fear accusations that they were under-treating pain. Concurrently, patient satisfaction with the effectiveness of opioids increased demand. The pharmaceutical industry, for its part, emphasized the "humane" use of opioids to address pain and developed new formulations of the lucrative drugs (*ibid.*) Thus, the cultural framework around pain management during the late twentieth century switched the debate from the safety of opioids to a renewed emphasis on the need for opioids as an effective treatment for moderate to severe pain, with lucrative consequences for many in the opioid business.

The result of this cultural shift was a dramatic increase in opioid consumption, which rose from 47,000 kg in 2000 to a peak of 165,500 kg in 2012 (Manchikanti et al. 2017). Although opioid prescriptions have decreased somewhat since then, there were still 59 opioid prescriptions written per hundred Americans in 2017, and the number of days prescriptions are taken has been increasing (CDC 2019b). The introduction of synthetic opioids such as fentanyl (which is eighty to one hundred times stronger than morphine) further fueled the use of opioids as legal access to prescription drugs began to be supplemented with illicitly manufactured opioids.

Although opioids have proved to be very successful in the treatment of pain, they have also lived up to their early reputation of being addictive, with many studies now documenting the highly addictive nature of opioids for certain individuals (e.g., Frieden and Houry 2016; Just et al. 2018). In 2014, more than twenty million Americans had substance use disorders, among whom two million involved prescription opioids and a further 586,000 the illegal drug heroin (Hedden et al. 2015). The pipeline from prescription opioids to the illegal opioid heroin is now well documented, with an estimated 80 percent of new heroin users beginning by abusing prescription opioids (Hedegaard et al. 2018). Of the 47,055 drug overdose fatalities in 2014, 18,893 were caused by prescription opioids and 10,574 by heroin (US Department of Health and Human Services et al. 2016). The switch to heroin often occurs when access to prescription opioids is cut off, as doctors are once again encouraged to limit opioid prescriptions in response to concerns over opioid abuse (Chopra and Marasa 2017). In summary, the CDC identifies three main waves of opioid use: the first, starting in the late 1990s, attributed to an increase in opioid prescriptions for pain management; the second, beginning in 2010, associated with increased heroin use; and the third, from 2013, characterized by synthetic opioid use (CDC 2019a) (Figure 7.11).

The social nature of the opioid crisis is also reflected in *who* is affected. Prescription opioid misuse initially occurred chiefly in rural areas and remains problematic in many rural areas, such as parts of Appalachia and Maine (Rigg and Monnat 2015). This pattern is often explained as a symptom of economic decline in rural areas but may also be related to the fact that other drugs are less available in rural areas and by drug-marketing strategies that initially targeted rural and suburban populations (Rigg and Monnat 2015; Santoro and Santoro 2018). Increasingly, opioid misuse is now also an urban problem, however (Rigg and Monnat 2015), with prescription opioid overdoses occurring in economically disadvantaged areas in both rural and urban areas (Pear et al. 2019). People with manual occupations are at particularly high risk (Rigg and Monnat 2015), probably owing to the intersection of low socioeconomic status and the likelihood of injury leading to chronic pain.

Thus, the burden of opioid addiction is often added to other socioeconomic disadvantages.

Race is often also highlighted as significant to the opioid story. Many commentators have noted that the opioid crisis has, to date, occurred largely among the white majority in the US, although recent evidence suggests that this gap may be closing as opioid abuse rises in urban areas, where communities of color are also likely to be affected (Santoro and Santoro 2018). This racial history highlights significant biases. Some evidence suggests that whites are prescribed opioids at higher rates, possibly associated with conscious or unconscious stereotyping that suggests that the black body is somehow physically different from the white body and better able to withstand pain or more likely to suffer from addiction (Hoffman et al. 2016; Santoro and Santoro 2018). Responses to the opioid epidemic also reveal patterns by race, with drug use among minority populations often criminalized, while white middle-class opioid users have more often been depicted as victims of substance use disorders (Santoro and Santoro 2018). Even treatment patterns reveal racial biases, as white communities generally have greater access to the healthcare resources that provide treatment. Even if individuals of color do seek treatment through the healthcare system, they are less likely to be prescribed the most up-to-date treatment options and are more likely to be treated by a general health practitioner rather than an addiction specialist (Hansen et al. 2013; Santoro and Santoro 2018).

In response to rising concerns over opioid misuse, the US Food and Drug Administration has recently redrafted its guidelines for opioid prescriptions, improved warnings on opioid labels, and begun supporting alternative pain management strategies. Many hospitals now encourage alternatives to opioids to manage pain after surgery, such as nerve blocks and non-steroidal drugs (Jones et al. 2018). The pharmaceutical industry has also been widely criticized for promoting opioids inappropriately, and some efforts have been made to call companies to account. In 2007, Purdue Pharma pleaded

continued

Box 7.3 *continued*

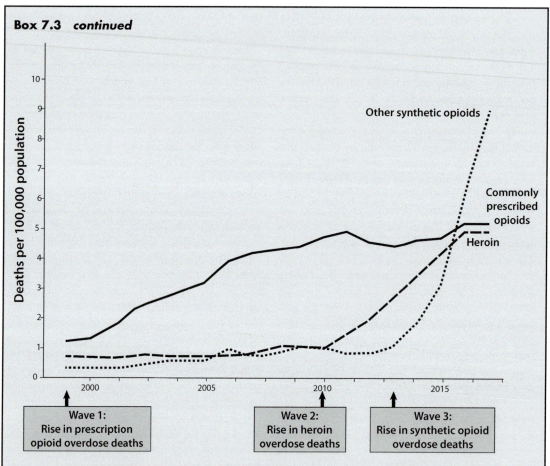

Figure 7.11 Trends in opioid overdose deaths in the US, 1999–2017

Source: Data from Scholl et al. (2019); graphic adapted from CDC (2019a)

Almost 400,000 people in the US died from an opioid overdose between 1999 and 2017 (Scholl et al. 2019). Over time, the source of opioids has changed. Initially, most overdoses were from prescription opioids, but more recently an increasing proportion of overdoses were associated with heroin and then other synthetic opioids such as fentanyl.

guilty to misleading physicians and the healthcare industry about the benefits of OxyContin (*ibid.*). More recently, Purdue and several other firms were accused of using deceptive practices to sell opioids (BBC 2019). While efforts to control future use of opioids are clearly valuable, for the thousands of Americans already addicted, recovery is going to be hard fought.

Area (or neighborhood) effects

Since the 1990s, work focusing on deprived areas has supported the idea that factors at the neighborhood scale may partially explain the relationship between health and socioeconomic inequality. **Area effects** (or **neighborhood effects**) consider the "net change in the contribution to life-chances made by living in one area rather than another" (Atkinson and Kintrea 2001, 2278). Central to the idea is the notion that neighborhood-level characteristics are significant to health *beyond* individual-level characteristics (Meijer 2013). In the context of mental health studies, for instance, we can ask whether spatial groupings of disease incidence are a result of an agglomeration of susceptible individuals, or whether characteristics of the area itself play an independent role in the onset of mental illness (Kearns and Joseph 1993). As Flowerdew et al. (2008, 1242) explain:

> There is little doubt that health statistics in an area are influenced by the composition of the local population. Age, education, employment, ethnicity, housing, social class, and other factors may all influence individuals' health. But are these compositional factors sufficient to account for geographical variations, or is there a contextual effect, some sort of local influence which makes an area's health better (or worse) than what would be expected from population composition?

It is difficult to disentangle the role of neighborhood environments from the characteristics and behavior of the individuals who live there and so investigating area effects is challenging (Meijer 2013). Many studies have nonetheless identified discernible impacts of local context on health. For instance, one national study in the US found neighborhood socioeconomic status to be more closely related to health status than individual or family socioeconomic status (Robert 1998). In Peru, Alderman et al. (2003) found that children's nutrition status was more closely associated with the average level of education of the neighborhood than with the education level of the child's own mother. However, other work has found little evidence of area effects (e.g., Roos et al. 2004; Malmstrom, Johansson, and Sundquist 2001).

One way to make sense of this high variability is to turn to meta-analyses that evaluate the findings of a group of studies. Meijer (2013) examined 24 studies exploring **area effects** and found that 16 of the 24 studies identified significant area level impacts on health. Large or longitudinal studies may also have the statistical power to reveal small effects. In the Moving to Opportunity for Fair Housing (MTO) demonstration, the US Department of Housing and Urban Development offered a randomly selected group of families living in public housing the opportunity to move to a less-disadvantaged neighborhood. From 1994 to 1998, more than four thousand families enrolled in the program in Baltimore, Boston, Chicago, Los Angeles, and New York. One third of the families were given standard vouchers to assist with the cost of housing; one third were given housing vouchers that required that they move to a census tract with low poverty rates; the remaining third—the control group—received no assistance through MTO. After about ten years, there were measurable improvements in the risk of diabetes and extreme obesity among people who had moved to a less-disadvantaged neighborhood. By contrast, there was no effect among participants who received standard housing vouchers, which suggests that it was moving to a less-disadvantaged neighborhood, and not simply receiving help with housing, that had a health impact (Ludwig et al. 2011). Moving to a less-disadvantaged neighborhood was also associated with marginally significant impacts on a broad range of mental health outcomes (*ibid.*). The socioeconomic status of the participating families did not change during the study, suggesting that the critical factor was not merely affluence. While the observed effects were small, this work provides evidence that neighborhood effects can be influential.

In summary, it appears that area effects may have an impact, but the impact is relatively small and hard to pin down definitively. Further investigation of these issues is important as the significance of environmental context has important policy implications, particularly given increasing spatial concentration of poverty and deprivation in many places. More work is also needed on this topic in the Global South.

Given current evidence on **area effects**, researchers have generated several hypotheses to explain *how* the local environment might influence health. Meijar (2013) suggests four potential pathways that could connect neighborhood to health outcomes: 1) health policy and health-related resources, which differ among areas; 2) the impact of neighborhood context on health-related behaviors such as smoking, diet, physical activity, and alcohol consumption; 3) perceptions of factors such as neighborhood safety and **social cohesion**; and 4) physical quality of the neighborhood, including factors such as traffic noise, water quality, and housing characteristics. It is helpful to divide these ideas into factors of the neighborhood that influence health *directly* such as air pollution or traffic noise, and those that operate *indirectly*, for instance by constraining residents' ability to practice healthy behaviors (Stock and Ellaway 2013).

Considerable work has examined "**landscapes of deprivation**," urban areas in which area effects may come into play with respect to underfunding of public services, limited employment opportunities, and the breakdown of physical infrastructure. Here, health is affected through mechanisms such as exposure to pollutants, poor access to healthcare, and limited physical activity. It is not just cities that suffer poor health, however. Studies suggest that urban residents and people living in remote rural areas tend to have the worst health, suggesting a "U-shaped association between urbanisation and health" (van Hooijdonk et al. 2007).

Rural-urban connections with health are complex and often place-specific, however. For instance, US studies often indicate lower birthweights in rural areas, while in the UK rural birthweights tend to be higher, probably related to the different character of urban and rural settings in the two countries. The complexity of these connections is illustrated by a Scottish study that compared birthweights of babies born on remote rural Scottish islands with birthweights from mainland Scotland. Several risk factors for low birthweight, including smoking and low socioeconomic status, were found to be less influential on islands than on the mainland. In this case, something about island life appears to provide a protective buffer for fetal health that counteracts the potential disadvantages associated with remote rural living (e.g., poorer access to specialist health facilities). The researchers hypothesize that high levels of community support in

island communities could be instrumental in supporting perinatal health. Additionally, pressures from greater social surveillance among residents in island communities may mean that women who smoked during pregnancy in island communities could have been smoking less than their mainland counterparts (Halliday, Clemens, and Dibben 2019).

Evidence such as this suggest that characteristics of a community's *social* environment are critical to health. Neighborhoods are not only sites where daily activities and routines take place, but also places where individuals interact (Ming, Browning, and Cagney 2007). Social isolation, a lack of role models, the development of a culture with deviant social norms, and high turnover of residents have all been identified as potential problems of deprived neighborhoods that can damage social cohesion (see Wilkinson and Pickett 2007). By contrast, places with high social cohesion such as island communities may buffer people from health insults, as suggested by the Scottish birthweight study. People's *perceptions* of their neighborhoods may also be significant (Stock and Ellaway 2013). In one study of four neighborhoods in Glasgow, Scotland, for example, perceptions of neighborhood cohesion were positively associated with health outcomes, while reports of neighborhood problems were negatively correlated with health (Ellaway, Macintyre, and Kearns 2001). Although the most affluent neighborhoods had the most social cohesion and fewest health problems (with the inverse for the poorest neighborhoods), middle-income neighborhoods did not follow a clear linear pattern, suggesting that other factors, perhaps housing tenure and employment patterns, may complicate these relationships.

The term **social capital** refers to the value of connections among individuals in a community. Communities can use their social capital to improve access to healthcare services, lobby for health-promoting facilities, or advocate for a clean environment. **Collective efficacy** refers to the idea that individuals in a group or community believe that they have the institutional capacity to operate towards the greater good. Social scientists have found that collective efficacy is lower in structurally disadvantaged neighborhoods characterized by poverty, single-parent households, high residential mobility, and ethnic heterogeneity. Neighborhoods that have high levels of collective efficacy, particularly where individuals are actively involved with groups that serve their community, offer benefits to residents. These include reduced crime (Sampson, Raudenbush, and Earls 1997), vibrant social networks, and organizational and economic resources that can help to provide a healthy environment (Robert 1998), all of which may ultimately lead to improved health outcomes (Cagney, Browning, and Wen 2005). When there is attachment to others in a community, individuals' health problems are also more likely to be noticed and acted upon by the broader community (Putnam 2001). In all these ways, "a psychological sense of community" may produce benefits to health (Ellaway, Macintyre, and Kearns 2001).

The social environment may also influence risk-taking behaviors above and beyond residents' socioeconomic status. A significant body of work on behavior and social context has been conducted on HIV/AIDS. Explanations of vulnerability to HIV have traditionally focused on identifying risky behaviors such as engaging in unprotected sex. More recently, the notion of the **risk environment** has been adopted to explore the influence of socioeconomic context on behavior. This structural approach emphasizes how "the riskiness of the behavior is a characteristic of the environment rather than of the individuals or the particular practices" (Barnett and Whiteside 2002, 81). For instance, places with high rates of urban poverty are often high-risk environments for HIV/AIDS because they often host a drug trade and produce opportunities for strangers to engage in unprotected sex in locations where risky behaviors are unlikely to be challenged (Fullilove 2004). This is not to suggest that all people living in impoverished neighborhoods will inevitably take part in risky behaviors, but rather that the social environment facilitates risky activities.

The mechanisms through which these factors bear upon health remain elusive. Nonetheless, living in conditions of poverty clearly has important health impacts in terms of absolute material deprivation, the psychological impacts of income inequalities, and the role of contexts of deprivation. While research across the social sciences continues to further elucidate these links, one unique approach that has emerged from the geographic literature is investigation of the significance of "sense of place."

SENSE OF PLACE AND HEALTH

Place can be viewed as merely a container for social processes or as having an active role in shaping those processes (Curtis 2004). Humanist perspectives start with the idea that poor environments might have not only physical impacts (such as exposing individuals to pollutants) but also psychological impacts (such as generating fear or anxiety), shaping people's interactions with particular places. **Sense of place** considers subjective perceptions of and relationships with particular places, including factors such as "rootedness, belonging, place identity, meaningfulness, place satisfaction and emotional attachment" (Eyles and Williams 2008, 5). In this context, there is growing recognition that the same setting may be experienced in different ways by different people (Curtis and Riva 2010).

As Eyles (1985) notes, sense of place is related not only to one's day-to-day experience in a place but can also be tied to one's "place-in-the-world." Sense of place can therefore include a feeling of "fitting in" to society. Immigrants are especially likely to experience limited connections to place if they are not feeling well integrated into their host society, for example (Agyekum and Newbold 2016). In this way, sense of place also has a critical temporal component, with individuals' feelings related to place needing time to develop. The notion of "psychological rootedness" reflects the strong sense of connection to place that develops over time (Williams 2002, 146). Population displacements can damage individuals' sense of place, with deleterious impacts on health and well-being, such as when people are forced to relocate through urban renewal programs (Fullilove 2004).

Sense of place is often especially significant for indigenous communities, whose identity is often closely tied to ancestral lands. Richmond and Ross (2009) use the term "environmental dispossession" to refer to the processes that denied First Nation and Inuit peoples in Canada access to the resources of their traditional environments. The respondents in their study conceptualized the physical environment as inextricably linked to their culture and traditional way of life, with decreased access to traditional lands and resources viewed as a "precursor for life imbalance, loss of life control, changing forms of education, lacking material resources, and strain on social life" (*ibid.*, 149).

Sense of place has also contributed to studies of landscapes of deprivation. For instance, Williams et al. (2003) investigated how neighborhood stereotypes influence the health of the people who live there. Santana and Noguiera (2009) analyzed how sense of place is related to weight gain in urban areas. Collins et al. (1998) found that mothers' perceptions of their living conditions (e.g., cleanliness and safety) were associated with their babies' birth weights, even after controlling for maternal behaviors. In some circumstances, people's perceptions of their surroundings may be even more influential than objective environmental conditions. For instance, Fuller et al. (1993) concluded that housing satisfaction was significantly related to more measures of health than were objective housing conditions.

A more recent body of work looks at the influence of environmental change, particularly climate change, on sense of place. This literature suggests that the rapid changes that many communities are experiencing associated with climate change affect individuals' attachment to place, with detrimental health consequences. For instance, Cunsolo Willox et al. (2012) examined how processes associated with climate change have changed feelings of place attachment among Inuit populations in Nunatsiavut, Canada. In this case, climate change has led to rapidly changing environmental conditions that disrupt traditional lifeways such as hunting, foraging, and traveling. People in the region report feelings of sadness, fear, anxiety, and distress associated with these changes, leading to associated impacts on physical, mental, and emotional health. The authors call for further work that explores "place-attachment as a vital indicator of health and well-being" (Cunsolo Willox et al. 2012, 538). Non-indigenous communities may also develop close attachments to the land. In a study of Australian family farmers, for instance, Ellis and Albrecht (2017) identify similarly negative health consequences, including heightened risk of depression and suicide, associated with damage to place-based attachments and identities from climate change.

VULNERABILITY AND RESILIENCY

While all individuals must adapt to changing social and physical environments, economic stability may greatly assist with weathering change. The notions of **vulnerability** and **resiliency** highlight the fact that some individuals and groups have more resources for coping with sudden changes than others. For instance, vulnerable coastal communities with limited financial resources and little political power may be unable to move to safer locations, adapt their houses to new conditions, or lobby governments to install new infrastructure to cope with rising sea levels. By contrast, financial resources give affluent communities more resiliency by giving them the ability to escape or adapt to changing conditions, as well as the political power to demand new legislation, infrastructure, and other assistance from government.

The plight of the Marshall Islands illustrates the challenges faced by communities with limited economic and political power (Harner 2018). Recognized by the United Nations as a small island developing state (SIDS), the Marshall Islands share a number of vulnerabilities with other SIDS, including limited resources, remoteness, dependence on international trade, and susceptibility to natural disasters. Although analyses suggest that a large proportion of the land area of this South Pacific island nation will likely be submerged due to climate change, the international community has remained deaf to repeated calls from activists that addressing climate change is vital to protecting the lifeways of the 55,000 Marshallese people. Emigration from the islands may soon be the only option, with potentially devastating consequences for the traditional lifeways and sense of place of the Marshallese.

Vulnerability and environmental hazards

Although climate change provides compelling examples of relative vulnerabilities, the notion of vulnerability was originally developed in scholarly literature on environmental hazards, with a group of critical scholars arguing that political and economic structures influence who is most affected by hazards such as earthquakes and hurricanes (Wisner et al. 2004). Vulnerability refers to the characteristics of an individual or community that influence how quickly and effectively they are able to recover from a hazard event. While the environmental event itself can be viewed as "natural," the way in which it affects people is seen as mediated by social factors. As such, we can only understand how an environmental event becomes a "disaster" if we consider the social context within which the hazard occurred (*ibid.*).

We often think of environmental hazards such as earthquakes as the ultimate social leveler, and yet hazard-related deaths occur disproportionately in low-income countries. Poorer households are also usually the most affected by any particular hazard. The term "class-quake" was coined in 1976 to refer to an earthquake in Guatemala that affected poor and indigenous groups far more significantly than the rich (O'Keefe, Westgate, and Wisner 1976). Most of the 22,000 people killed lived in unsafe housing in rural areas or urban slums. By contrast, upper- and middle-class areas were left virtually unscathed (Wisner et al. 2004). In numerous subsequent events, including floods, earthquakes, and hurricanes, self-built housing on marginal land was reported as a significant risk factor for morbidity and mortality associated with the event (Figure 7.12).

The impacts of relative vulnerability were apparent in the aftermath of Hurricane Katrina, which had deeply rooted class- and race-based aspects. Many low-income African American communities were at a significant disadvantage in coping with the effects of the hurricane because they were located in some of the lowest-lying land in New Orleans, and many residents lacked personal transportation, which policymakers had assumed

Figure 7.12 Shanty town on the outskirts of Lima, Peru
Credit: Heike Alberts

people would use to evacuate (Powell et al. 2006; Smith 2006). Although many have argued that class was a more important factor than race, Powell et al. (2006) contend that longstanding histories of racism have constructed the landscape of New Orleans in ways that left African Americans particularly vulnerable to the effects of the hurricane. In addition, they prompt us to consider why the African American population has low average socioeconomic status in the first place, suggesting that institutionalized racism can explain why many African Americans lacked the means to protect themselves from the worst impacts of Katrina. As Smith (2006) states, "At all phases, up to and including reconstruction, disasters don't simply flatten landscapes, washing them smooth. Rather they deepen and erode the ruts of social difference they encounter." Klinenberg (2002) provides a further class- and race-based analysis of a heat wave that hit Chicago in 1995 (Box 7.4). Case studies such as these

---

Box 7.4

THE 1995 CHICAGO HEAT WAVE AND THE URBAN POOR

Klinenberg (2002) used a vulnerability perspective to analyze the impacts of a heat wave that struck Chicago in July 1995, arguing that socioeconomic factors and landscapes of deprivation were critical factors in driving who died from the heat. As Klinenberg (2002, 11) states, "We have collectively created the conditions that made it possible for so many Chicago residents to die in the summer of 1995, as well as the conditions that make these deaths easy to overlook and forget." The final count of **excess deaths** (deaths beyond the number that we would normally expect) associated with the Chicago heat wave totaled 739 (*ibid.*). While it is often suggested that heat waves precipitate deaths among frail individuals who would have died soon regardless, analysis of Chicago's 1995 heat wave suggested that most of the people who perished had many years of life ahead of them.

Nonetheless, the majority of fatalities were among the elderly—73 percent of heat-related deaths were people over age 65—particularly poor, elderly people living alone with limited social contacts (Klinenberg 2002, 18). Most deaths occurred in impoverished inner-city neighborhoods with high-density housing and a lack of air-conditioning. Additionally, inner-city neighborhoods had fewer air-conditioned facilities such as malls where residents could seek refuge from high temperatures, and concerns about crime left many elderly residents reluctant to leave their homes.

Demographic characteristics were also relevant. More than twice as many men as women died from the heat, despite the far larger number of women in the elderly cohorts most vulnerable to heat. Klinenberg hypothesizes that this may be related in part to the social isolation that many elderly men experience, whereas women are more likely to maintain close ties to friends and family into old age. There were also notable differences among ethnic groups. African Americans were one-and-a-half times more likely to die from the heat than whites. In contrast, there were comparatively few deaths in the Latino population. Klinenberg (2002, 91) argues that "place-specific social ecology" was important here, with African American communities suffering from "the dangerous ecology of abandoned buildings, open spaces, commercial depletion, violent crime, degraded infrastructure, low population density, and family dispersion undermin[ing] the viability of public life and the strength of local support systems, rendering older residents particularly vulnerable to isolation." Although many Latinos also lived in deprived neighborhoods, Latino communities fared remarkably well, illustrating the complex intersection of physical, cultural, and psycho-social characteristics in generating vulnerability. Klinenberg suggests that high levels of social cohesion among Latino populations may have served as a critical protective factor for many vulnerable elderly residents.

illustrate how tightly interwoven socioeconomic status is with other aspects of identity such as race and class. We extend these ideas in the next chapter.

CONCLUSION

Wealth affects health *directly*, through enabling access to resources such as high-quality healthcare and nutritious foods, and *indirectly*, through the influence of aspects of the built and social landscape on health outcomes. Health researchers have begun to appreciate the subtle ways in which social factors mediate and complicate the relationship between health and wealth. For example, while absolute income has a significant influence on health outcomes, health inequalities also put some individuals at a disadvantage. In the next chapter we expand on the role of the social environment by considering how culture and identity can have health impacts at both societal and individual scales.

DISCUSSION QUESTIONS

1   Why are crude death rates *not* considered to be a good indicator of the overall "healthiness" of a particular community? What other health indicators might give a better idea of a community's health?

2   From what you have read about economic patterns in health, do you think that our traditional division between the Global North and the Global South makes sense anymore? How else might we divide the world in ways that more accurately reflect current patterns of health in an economic framework?

3   Is it important whether socioeconomic status is related to health through direct deprivation or indirectly through aspects of the social environment? What are some of the different policy implications that might result from these different perspectives?

4   Do you feel that you have emotional attachments to particular places? Has "sense of place" influenced your health or well-being? In what ways?

5   What kinds of natural hazards affect the region in which you live? What makes people vulnerable to their effects? How could we improve the resiliency of vulnerable populations?

SUGGESTED READING

Cunsolo Willox, S. L. Harper, J. D. Ford, K. Landman, et al. 2012. "'From This Place and of This Place:' Climate Change, Sense of Place, and Health in Nunatsiavut, Canada." *Social Science and Medicine* 75 (3): 538–47.

Day, P., Pearce, J., and Dorling, D. 2008. "Twelve Worlds: A Geo-Demographic Comparison of Global Inequalities in Mortality." *Journal of Epidemiology and Community Health* 62: 1002–10.

Klinenberg, E. 2002. *Heat Wave: A Social Autopsy of Disaster in Chicago*, Chicago: University of Chicago Press.

Pearce, J., and Dorling, D. 2009. "Tackling Global Health Inequalities: Closing the Health Gap in a Generation." *Environment and Planning A*, 41: 1–6.

Ritchie, H., and Roser, M. 2019. "Mental Health." *Our WorldInData.org*. https://ourworldindata.org/mental-health.

VIDEO RESOURCES

"Unnatural Causes: Is Inequality Making Us Sick?" 2008. *California Newsreel*. More information at: www.unnaturalcauses.org.

REFERENCES

Acciai, F., and G. Firebaugh. 2017. "Why Did Life Expectancy Decline in the United States in 2015? A Gender-Specific Analysis." *Social Science & Medicine* 190: 174–80. doi:10.1016/j.socscimed.2017.08.004.

Agyekum, Boadi, and K. Bruce Newbold. 2016. "Sense of Place and Mental Wellness of Visible Minority

Immigrants in Hamilton, Ontario: Revelations from Key Informants." *Canadian Ethnic Studies* 48 (1): 21. doi:10.1353/ces.2016.0001.

Alderman, H., J. Hentschel, and R. Sabates. 2003. "With the Help of One's Neighbors: Externalities in the Production of Nutrition in Peru." *Social Science & Medicine* 56 (10): 2019–31.

Atkinson, Rowland, and Keith Kintrea. 2001. "Disentangling Area Effects: Evidence from Deprived and Non-Deprived Neighbourhoods." *Urban Studies* 38 (12): 2277–98. doi:10.1080/00420980120087162.

Barnes, Michael, G., and Trenton Smith, G. 2009. Tobacco Use as Response to Economic Insecurity: Evidence from the National Longitudinal Survey of Youth. *The B.E. Journal of Economic Analysis & Policy.*

Barnett, T., and A. Whiteside. 2002. *AIDS in the Twenty-First Century: Disease and Globalization.* Basingstoke: Palgrave Macmillan.

British Broadcasting Company. 2019. "Sackler-Owned Purdue Pharma Settles Opioid Lawsuit for $270m." Accessed August 22, 2019. www.bbc.com/news/business-47710332.

Cagney, Kathleen A., Christopher R. Browning, and Ming Wen. 2005. "Racial Disparities in Self-Rated Health at Older Ages: What Difference Does the Neighborhood Make?" *The Journals of Gerontology: Series B* 60 (4): S181–90. doi:10.1093/geronb/60.4.S181.

Case, Anne, and Angus Deaton. 2015. "Rising Morbidity and Mortality in Midlife Among White Non-Hispanic Americans in the 21st Century." *Proceedings of the National Academy of Sciences* 112 (49): 15078–83. doi:10.1073/pnas.1518393112.

[CDC] Centers for Disease Control and Prevention. 2019a. "Understanding the Epidemic." Accessed August 22, 2019. www.cdc.gov/drugoverdose/epidemic/index.html.

[CDC] Centers for Disease Control and Prevention. 2019b. "Prescription Opioid Data." Accessed August 22, 2019. www.cdc.gov/drugoverdose/data/prescribing.html.

Chopra, N., and L. H. Marasa. 2017. "The Opioid Epidemic: Challenges of Sustained Remission." *International Journal of Psychiatry in Medicine* 52 (2): 196–201. doi:10.1177/0091217417720900.

Collins, J. W., Jr., R. J. David, R. Symons, A. Handler, S. Wall, and S. Andes. 1998. "African-American Mothers' Perception of Their Residential Environment, Stressful Life Events, and Very Low Birthweight." *Epidemiology* 9 (3): 286–89.

Cunsolo Willox, A., S. L. Harper, J. D. Ford, K. Landman, et al. 2012. "'From This Place and of This Place': Climate Change, Sense of Place, and Health in Nunatsiavut, Canada." *Social Science & Medicine* 75 (3): 538–47. doi:10.1016/j.socscimed.2012.03.043.

Curtis, Sarah. 2004. *Health and Inequality: Geographical Perspectives.* London and Thousand Oaks, CA: Sage Publications.

Curtis, Sarah, and Mylène Riva. 2010. "Health Geographies II: Complexity and Health Care Systems and Policy." *Progress in Human Geography* 34 (4): 513–20. doi:10.1177/0309132509336029.

Dahlgren, G., and M. Whitehead. 1991. "Policies and Strategies to Promote Social Equity in Health. Background Document to WHO—Strategy Paper for Europe." Institute for Future Studies. Accessed October 25, 2019. core.ac.uk/download/pdf/6472456.pdf.

Ellaway, Anne, Sally Macintyre, and Ade Kearns. 2001. "Perceptions of Place and Health in Socially Contrasting Neighbourhoods." *Urban Studies* 38 (12): 2299–316. doi:10.1080/00420980120087171.

Ellis, N. R., and G. A. Albrecht. 2017. "Climate Change Threats to Family Farmers' Sense of Place and Mental Wellbeing: A Case Study from the Western Australian Wheatbelt." *Social Science & Medicine* 175: 161–68. doi:10.1016/j.socscimed.2017.01.009.

Eyles, J. 1985. *Senses of Place.* Silverbook Press.

Eyles, J., and A. Williams. 2008. *Sense of Place, Health and Quality of Life.* Ashgate.

Ezzati, M., A. D. Lopez, A. Rodgers, and C. J. L. Murray. 2003. *Comparative Quantification of Health Risks:*

Global and Regional Burden of Diseases Attributable to Selected Major Risk Factors. Geneva: WHO.

Flowerdew, R., D. J. Manley, and C. E. Sabel. 2008. "Neighbourhood Effects on Health: Does It Matter Where You Draw the Boundaries?" *Social Science & Medicine* 66 (6): 1241–55. doi:10.1016/j.socscinied.2007.11.042.

Frieden, Thomas R., and Debra Houry. 2016. "Reducing the Risks of Relief—The CDC Opioid-Prescribing Guideline." *New England Journal of Medicine* 374 (16): 1501–4. doi:10.1056/NEJMp1515917.

Fuller, T. D., J. N. Edwards, S. Sermsri, and S. Vorakitphokatorn. 1993. "Housing, Stress, and Physical Well-Being: Evidence from Thailand." *Social Science & Medicine* 36 (11): 1417–28.

Fullilove, M. T. 2004. *Root Shock: How Tearing Up City Neighborhoods Hurts America, and What We Can Do About It*. One World/Ballantine Books.

Gwatkin, D. R. 2017. "Trends in Health Inequalities in Developing Countries." *The Lancet Global Health* 5 (4): e371–e372. doi:10.1016/S2214-109X(17)30080-3.

Gwatkin, D. R., S. Rutstein, K. Johnson, E. Suliman, A. Wagstaff, and A. Amouzou. 2007. "Socio-Economic Differences in Health, Nutrition, and Population Within Developing Countries: An Overview." *Nigerian Journal of Clinical Practice* 10 (4): 272–82.

Halliday, K., T. Clemens, and C. Dibben. 2019. "Are Islands Equigenic Places? Maternal Residence on Remote Scottish Islands and Its Consequences for Birthweight." *18th International Medical Geography Symposium*, Queenstown, NZ, July 4.

Hansen, H. B., C. E. Siegel, B. G. Case, D. N. Bertollo, et al. 2013. "Variation in Use of Buprenorphine and Methadone Treatment by Racial, Ethnic, and Income Characteristics of Residential Social Areas in New York City." *The Journal of Behavioral Health Services & Research* 40 (3): 367–77. doi:10.1007/s11414-013-9341-3.

Harner, J. 2018. "Rising Seas Give Island Nation a Stark Choice: Relocate or Elevate." National Geographic. Last modified November 19, 2018. Accessed July 11, 2019. www.nationalgeographic.com/environment/2018/11/rising-seas-force-marshall-islands-relocate-elevate-artificial-islands/.

Hedden, S., J. Kennet, R. Lipari, G. Medley, et al. 2015. *Behavioral Health Trends in the United States: Results from the 2014 National Survey on Drug Use and Health*, edited by Department of Health and Human Services. Rockville, MD: Substance Abuse and Mental Health Services Administration.

Hedegaard, H, A. Miniño, and M. Warner. 2018. "Drug Overdose Deaths in the United States, 1999–2017." NCHS Data Brief 329. Accessed October 25, 2019. https://www.cdc.gov/nchs/data/databriefs/db329-h.pdf.

Hoffman, Kelly M., Sophie Trawalter, Jordan R. Axt, and M. Norman Oliver. 2016. "Racial Bias in Pain Assessment and Treatment Recommendations, and False Beliefs About Biological Differences Between Blacks and Whites." *Proceedings of the National Academy of Sciences* 113 (16): 4296–301. doi:10.1073/pnas.1516047113.

Institute for Health Metrics and Evaluation. 2019. "Global Health Data Exchange." Accessed August 22, 2019. http://ghdx.healthdata.org/.

Jones, Mark R., Omar Viswanath, Jacquelin Peck, Alan D. Kaye, et al. 2018. "A Brief History of the Opioid Epidemic and Strategies for Pain Medicine." *Pain and Therapy* 7 (1): 13–21. doi:10.1007/s40122-018-0097-6.

Just, J. M., L. Bingener, M. Bleckwenn, R. Schnakenberg, and K. Weckbecker. 2018. "Risk of Opioid Misuse in Chronic Non-Cancer Pain in Primary Care Patients—A Cross Sectional Study." *BMC Family Practice* 19 (1): 92. doi:10.1186/s12875-018-0775-9.

Kalipeni, E. 2000. "Health and Disease in Southern Africa: A Comparative and Vulnerability Perspective." *Social Science & Medicine* 50 (7–8): 965–83.

Kalipeni, E., and J. Oppong. 1998. "The Refugee Crisis in Africa and Implications for Health and Disease: A Political Ecology Approach." *Social Science & Medicine* 46 (12): 1637–53.

Kearns, Robin A., and Alun E. Joseph. 1993. "Space in Its Place: Developing the Link in Medical Geography." *Social Science & Medicine* 37 (6): 711–17. doi:10.1016/0277-9536(93)90364-A.

King, Gary, and Langche Zeng. 2011. "Improving Forecasts of State Failure." *World Politics* 53 (4): 623–58. doi:10.1353/wp.2001.0018.

Klinenberg, E. 2002. *Heat Wave: A Social Autopsy of Disaster in Chicago*. Chicago: University of Chicago Press.

Ludwig, Jens, Lisa Sanbonmatsu, Lisa Gennetian, Emma Adam, et al. 2011. "Neighborhoods, Obesity, and Diabetes—A Randomized Social Experiment." *New England Journal of Medicine* 365 (16): 1509–19. doi:10.1056/NEJMsa1103216.

Malmstrom, M., S. E. Johansson, and J. Sundquist. 2001. "A Hierarchical Analysis of Long-Term Illness and Mortality in Socially Deprived Areas." *Social Science & Medicine* 53 (3): 265–75.

Manchikanti, L., A. M. Kaye, N. N. Knezevic, H. McAnally, et al. 2017. "Responsible, Safe, and Effective Prescription of Opioids for Chronic Non-Cancer Pain: American Society of Interventional Pain Physicians (ASIPP) Guidelines." *Pain Physician* 20 (Suppl 2): S3–92.

Marmot, M. G., C. D. Ryff, L. L. Bumpass, M. Shipley, and N. F. Marks. 1997. "Social Inequalities in Health: Next Questions and Converging Evidence." *Social Science & Medicine* 44 (6): 901–10.

Marmot, M. G., M. J. Shipley, and G. Rose. 1984. "Inequalities in Death—Specific Explanations of a General Pattern?" *Lancet* 1 (8384): 1003–6. doi:10.1016/s0140-6736(84)92337-7.

Marmot, M. G., G. D. Smith, S. Stansfeld, C. Patel, et al. 1991. "Health Inequalities Among British Civil Servants: The Whitehall II Study." *Lancet* 337 (8754): 1387–93. doi:10.1016/0140-6736(91)93068-k.

Masters, Ryan K., Andrea M. Tilstra, and Daniel H. Simon. 2017. "Mortality from Suicide, Chronic Liver Disease, and Drug Poisonings Among Middle-Aged U.S. White Men and Women, 1980–2013." *Biodemography and Social Biology* 63 (1): 31–37. doi:10.1080/19485565.2016.1248892.

Mayer, S. E., and A. Sarin. 2005. "Some Mechanisms Linking Economic Inequality and Infant Mortality." *Social Science & Medicine* 60 (3): 439–55. doi:10.1016/j.socscimed.2004.06.005.

Meijer, M. 2013. "Neighborhood Context and Mortality: An Overview." In *Neighbourhood Structure and Health Promotion*, edited by C. Stock and A. Ellaway. Springer.

Melzack, R. 1990. "The Tragedy of Needless Pain." *Scientific American* 262 (2): 27–33. doi:10.1038/scientificamerican0290-27.

Ming, Wen, Christopher R. Browning, and Kathleen A. Cagney. 2007. "Neighbourhood Deprivation, Social Capital and Regular Exercise During Adulthood: A Multilevel Study in Chicago." *Urban Studies* 44 (13): 2651–71. doi:10.1080/00420980701558418.

National Center for Health, Statistics. 2018. "Health, United States." In *Health, United States, 2017: With Special Feature on Mortality*. Hyattsville, MD: National Center for Health Statistics (US).

O'Keefe, Phil, Ken Westgate, and Ben Wisner. 1976. "Taking the Naturalness out of Natural Disasters." *Nature* 260 (5552): 566–67. doi:10.1038/260566a0.

Pampalon, R., D. Hamel, and P. Gamache. 2008. "Recent Changes in the Geography of Social Disparities in Premature Mortality in Quebec." *Social Science & Medicine* 67 (8): 1269–81. doi:10.1016/j.socscimed.2008.06.010.

Pear, V. A., W. R. Ponicki, A. Gaidus, K. M. Keyes, et al. 2019. "Urban-Rural Variation in the Socioeconomic Determinants of Opioid Overdose." *Drug Alcohol Depend* 195: 66–73. doi:10.1016/j.drugalcdep.2018.11.024.

Pearce, Jamie, and Danny Dorling. 2009. "Tackling Global Health Inequalities: Closing the Health Gap in a Generation." *Environment and Planning A: Economy and Space* 41 (1): 1–6. doi:10.1068/a41319.

Powell, J., H. Jeffries, D. Newhard, and E. Steins. 2006. "Towards a Transformative View of Race: The Crisis and Opportunity of Katrina." In *There Is No Such Thing as a Natural Disaster: Race, Class, and Hurricane Katrina*, edited by C. W. Hartman and G. D. Squires. Routledge.

[PRB] Population Reference Bureau. 2018. "World Population Data." Accessed July 2, 2019. www.world popdata.org/.

Putnam, R. D. 2001. *Bowling Alone: The Collapse and Revival of American Community*. Simon & Schuster.

Richmond, C. A. M., and N. A. Ross. 2009. "The Determinants of First Nation and Inuit Health: A Critical Population Health Approach." *Health Place* 15 (2): 403–11. doi:10.1016/j.healthplace.2008.07.004.

Rigg, Khary K., and Shannon M. Monnat. 2015. "Urban vs. Rural Differences in Prescription Opioid Misuse Among Adults in the United States: Informing Region Specific Drug Policies and Interventions." *The International Journal on Drug Policy* 26 (5): 484–91. doi:10.1016/j.drugpo.2014.10.001.

Ritchie, H., and M. Roser. 2018. "Mental Health." *OurWorldInData.org*. Accessed August 31, 2019. https:// ourworldindata.org/mental-health.

Robert, S. A. 1998. "Community-Level Socioeconomic Status Effects on Adult Health." *Journal of Health and Social Behavior* 39 (1): 18–37.

Robinson, J., and A. J. Kirkcaldy. 2007. "'You Think That I'm Smoking and They're Not': Why Mothers Still Smoke in the Home." *Social Science & Medicine* 65 (4): 641–52. doi:10.1016/j.socscimed.2007.03.048.

Rodgers, G. B. 1979. "Income and Inequality as Determinants of Mortality: An International Cross-Section Analysis." *Population Studies* 33 (2): 343–51. doi:10.108 0/00324728.1979.10410449.

Roos, L. L., J. Magoon, S. Gupta, D. Chateau, and P. J. Veugelers. 2004. "Socioeconomic Determinants of Mortality in Two Canadian Provinces: Multilevel Modelling and Neighborhood Context." *Social Science & Medicine* 59 (7): 1435–47. doi:10.1016/j. socscimed.2004.01.024.

Sampson, Robert J., Stephen W. Raudenbush, and Felton Earls. 1997. "Neighborhoods and Violent Crime: A Multilevel Study of Collective Efficacy." *Science* 277 (5328): 918–24. doi:10.1126/science.277.5328.918.

Santana, Paula, Rita Santos, and Helena Nogueira. 2009. "The Link Between Local Environment and Obesity: A Multilevel Analysis in the Lisbon Metropolitan Area, Portugal." *Social Science & Medicine* 68.

Santoro, Taylor N., and Jonathan D. Santoro. 2018. "Racial Bias in the US Opioid Epidemic: A Review of the History of Systemic Bias and Implications for Care." *Cureus* 10 (12): e3733. doi:10.7759/cureus.3733.

Scholl, L., P. Seth, M. Kariisa, N. Wilson, and G. Baldwin. 2019. "Drug and Opioid-Involved Overdose Deaths—United States, 2013–2017." *Morbidity and Mortality Weekly Report (MMWR)* 67: 1419–27. doi:10.15585/mmwr.mm675152e1.

Singh, G. K., M. D. Kogan, P. C. Van Dyck, and M. Siahpush. 2008. "Racial/Ethnic, Socioeconomic, and Behavioral Determinants of Childhood and Adolescent Obesity in the United States: Analyzing Independent and Joint Associations." *Annals of Epidemiology* 18 (9): 682–95. doi:10.1016/j.annepidem.2008.05.001.

Smith, N. 2006. "There's No Such Thing as a Natural Disaster." Accessed July 11, 2019. http://understanding katrina.ssrc.org/Smith/.

Starfield, B., L. Shi, and J. Macinko. 2005. "Contribution of Primary Care to Health Systems and Health." *Milbank Quarterly* 83 (3): 457–502. doi:10.1111/j. 1468-0009.2005.00409.x.

Stock, C., and A. Ellaway. 2013. *Neighbourhood Structure and Health Promotion*. Springer.

Tsourtos, G., and L. O'Dwyer. 2008. "Stress, Stress Management, Smoking Prevalence and Quit Rates in a Disadvantaged Area: Has Anything Changed?" *Health Promotion Journal of Australia* 19 (1): 40–44.

Ultee, K. H. J., E. K. M. Tjeertes, F. Bastos Goncalves, E. V. Rouwet, et al. 2018. "The Relation Between Household Income and Surgical Outcome in the Dutch Setting of Equal Access to and Provision of Healthcare." *PLoS One* 13 (1): e0191464. doi:10.1371/journal.pone.0191464.

US Department of Health and Human Services, Centers for Disease Control and Prevention, and National Center for Health Statistics. 2016. "Number and

Age-Adjusted Rates of Drug-Poisoning Deaths Involving Opioid Analgesics and Heroin: United States, 1999–2014." Accessed August 22, 2019. www.cdc.gov/nchs/data/health_policy/AADR_drug_poisoning_involving_OA_Heroin_US_2000-2014.pdf.

van Hooijdonk, Carolien, Mariël Droomers, Jeanne A. M. van Loon, Fons van der Lucht, and Anton E. Kunst. 2007. "Exceptions to the Rule: Healthy Deprived Areas and Unhealthy Wealthy Areas." *Social Science & Medicine* 64 (6): 1326–42. doi:10.1016/j.socscimed.2006.10.041.

Wagstaff, A. 2000. "Socioeconomic Inequalities in Child Mortality: Comparisons Across Nine Developing Countries." *Bulletin of the World Health Organization* 78 (1): 19–29.

[WHO] World Health Organization. 1986. *Cancer Pain Relief*. Geneva: World Health Organization.

[WHO] World Health Organization. 2009. "Life Expectancy at Birth (Years)." Accessed November 4. www.who.int/whosis/indicators/compendium/2008/2let/en/index.html.

[WHO] World Health Organization. 2018. *Global Health Observatory Data*. Geneva: World Health Organization.

[WHO] World Health Organization. 2019a. Accessed July 5, 2019. www.who.int/healthinfo/global_burden_disease/metrics_daly/en/.

[WHO] World Health Organization. 2019b. *Global Status Report on Road Safety 2018*. Geneva: World Health Organization.

Wilkinson, R. G., and K. E. Pickett. 2007. "The Problems of Relative Deprivation: Why Some Societies Do Better than Others." *Social Science & Medicine* 65 (9): 1965–78. doi:10.1016/j.socscimed.2007.05.041.

Williams, Allison. 2002. "Changing Geographies of Care: Employing the Concept of Therapeutic Landscapes as a Framework in Examining Home Space." *Social Science & Medicine* 55 (1): 141–54. doi:10.1016/S0277-9536(01)00209-X.

Williams, D. R., H. W. Neighbors, and J. S. Jackson. 2003. "Racial/Ethnic Discrimination and Health: Findings from Community Studies." *American Journal of Public Health* 93 (2): 200–8. doi:10.2105/ajph.93.2.200.

Wisner, B., P. M. Blaikie, T. Cannon, and I. Davis. 2004. *At Risk: Natural Hazards, People's Vulnerability and Disasters*. Routledge.

World Bank. 2019. "World Bank Open Data." Accessed August 22. https://data.worldbank.org/.

8

CULTURE AND IDENTITY

Humans are social beings, shaped by their cultural contexts. Just as socio-cultural circumstances influence the foods we eat or the languages we speak, they also affect our health and treatment of disease. Every individual has a unique personality and experiences, situated within a distinct constellation of social and cultural groupings, as defined by factors such as race, socioeconomic status, gender, and nationality. This combination of factors gives each of us a unique identity that influences how we respond to our circumstances.

In this chapter, we consider how **culture** and **identity** influence health. What drives what people view as healthy or unhealthy? How does culture affect how we treat disease? How do people's identities influence their likelihood of becoming ill or responding to treatment? We also consider the ways in which we identify people who are not in good health. How can ill health itself act to define an individual? Recognizing that such factors are highly case- and place-specific, we present a variety of case studies.

CULTURE

Culture refers to the beliefs and practices acquired from society and lived by a group of people (Helman 2007),

including factors such as religion, ethnicity, nationality, economic class, gender, and age. It is a nebulous term that represents both a way of life and a set of products created by people that reflect their values and beliefs (Mitchell 2000). This cultural framework is constantly evolving, subject to pressures from both within and outside the community (Kagawa Singer, Dressler, and George 2016). Culture is perhaps most commonly conceptualized as a place-based construct tied to particular countries or regions (e.g., Australian or Bengali culture), but other social groups may also have unique cultures. We may recognize a culture associated with an age cohort (e.g., Millennials), an ethnic group (e.g., Chinese Australians), or even an occupation (e.g., physicians). Sometimes it is useful to consider culture at small scales, such as a village, school, or household.

Study in health geography can include both how culture influences health and how health influences culture (Gesler and Kearns 2002), but most health geographers focus on how and why different cultural groups experience different health problems and outcomes. The reorientation of medical geography towards critical geographies of health in the 1990s broadened our understanding of the role of culture in health and encouraged researchers to directly engage with topics of diversity and inequality. This re-theorizing of culture emphasized its politicized nature, recognizing that powerful people play a dominant role in defining culture, and even what we consider to be "culture" (Mitchell 2000). For instance, this power-based framework pushes us to acknowledge the dominance of the European-American cultural lens through which science and medicine usually operate, marginalizing other cultural approaches to health and disease (Kagawa Singer, Dressler, and George 2016). Current research in culture and health considers the experiences of cultural groups marginalized by dominant power structures, including minority ethnic groups, migrants, women, **LGBTQ+** populations, and the those experiencing homelessness.

Although it is tempting to attribute health differences directly to cultural practices, cultural groups often differ along socioeconomic lines, which may contribute more to variations in health than cultural differences per se (Nazroo 1997). As we note in Chapter 7, socioeconomic status has a powerful influence on health and so it is always important to account for the potential **confounding** role of socioeconomic status before concluding that patterns are related to cultural practices. In other cases, it may not be the practices of cultural groups themselves that are critical but rather the way that dominant society treats a group and its practices. For instance, widespread and institutionalized racism can have significant impacts on the health of populations of color (Jones 2002). For

Figure 8.1 Local cuisine, Xi'an, China
Credit: Melissa Gould

Although many Westerners think of "Chinese food" as a single cuisine, China has huge regional variations in diet and food preparation practices.

this reason, Gesler and Kearns (2002, 13) warn that "culture should not be used as a source of explanation; rather it is something to be explained as it is continually being socially produced by people as they struggle to achieve power and meaning."

Culture and health

Many everyday behaviors have implications for health. For example, many communities have specific dietary practices, with cultural preferences and taboos related to what is eaten, how food is prepared, when people eat or fast, when food should be eaten, and how much should be eaten—all with coincidental health implications (Figures 8.1 and 8.2). In some cases, different community members may be expected to eat different things, with specific rules applied to groups such as children, invalids, and pregnant or nursing women. In some patriarchal societies, boys' nutrition is prioritized over girls', and so men may be served more nutrient-rich foods than women—biases that can be exacerbated at times of food shortage. These kinds of cultural biases have been held responsible for higher death rates among women during some famines (Neumayer and Plümper 2007).

Of course, ideas about diet are not static, and cultural preferences change over time. These changes are often particularly dramatic during profound cultural transitions such as emigration or colonization. To provide one example, changing dietary preferences associated with colonization have been associated with a rise in the deficiency disease beriberi, recorded in South and East Asia in the late nineteenth and early twentieth centuries. New ideas and technologies were rapidly being introduced into the region by the colonial regime, leading to changes in the foods available at local markets. Local people increasingly chose to eat industrially milled white rice instead of traditional hand-husked brown rice. The

Figure 8.2 Food market, Madrid, Spain
Credit: Melissa Gould

A tradition of curing meats is typical in parts of Southern Europe. Recent evidence suggests that diets high in processed meats may increase the risk of certain cancers. The Spanish tradition of coming together as a family to eat main meal is, by contrast, widely promoted as being protective of health.

white rice was not only easier to cook and had a longer shelf life than traditional varieties but had also quickly become associated with a prestigious "modern" way of life in a time of great uncertainty. Unfortunately, the aggressive milling and polishing process that produces white rice also removes the most nutritious parts of the rice grain—most significantly vitamin B-1 (thiamine). The loss of this essential vitamin was especially critical for poor consumers whose diet was heavily rice-based, leading to increased incidence of beriberi, which causes swelling of the legs, paralysis, shortness of breath, and cardiovascular problems (Arnold 2010). While the immediate biomedical cause of this disease was lack of

thiamine, a broader reading of the cultural landscape suggests that the significance of white rice as a source of prestige for communities suffering marginalization under a colonial regime may have underpinned this dietary shift.

Today, a dominant force shaping people's food preferences around the world is the transition from traditional diets to highly processed foods associated with a Western diet (Figure 8.3). Just like white rice in the foregoing example, these foods are often associated with affluence and modernization and are often a relatively cheap way that families can buy into the aspirational Western lifestyles they see on TV. Unfortunately, the

Figure 8.3 Coca-Cola delivery, Guatemala

Sugary soft drinks are a common sight throughout the Global South. Associated with modernization, as well as being tasty and cheap, processed foods are often extremely popular.

high palatability of these foods, which have been deliberately engineered to be as appealing to consumers as possible, is often achieved by adding fat, sugar, and salt at the expense of vitamins and micronutrients. The "obesity epidemic" that this dietary shift has fueled is rapidly spreading from affluent to middle-income and even low-income countries (Figure 8.4). In 2016, nearly 30 percent of Mexican adults were classified as obese, for instance, giving Mexico obesity rates comparable to the UK and Australia (Ritchie and Roser 2018).

Culture can also drive beliefs about which behaviors are healthy or unhealthy. For instance, the cultural image of smoking has shifted over time, with positive images of relaxation and social success in the 1960s and '70s incrementally replaced with negative ideas related to the health consequences of smoking by the 1980s and '90s. By the early 2000s, this image had shifted further as smokers began to be portrayed as a burden on society in terms of the problems created by passive smoking as well as the burden on public health systems of treating smoking-related diseases.

Beliefs around the healthiness of particular behaviors are shaped by both expert opinion and social norms, which may reinforce or contradict one another in complex ways. For instance, many commentators suggest that the rapid rise in cesarean sections that has occurred in many countries may be rooted in an "obstetric culture" of birth, which sees birth as a procedure that requires medical intervention. This medicalization of birth is also influenced by broader social norms around birth, which increasingly promote the hospital as the "safe" place for birth to occur. Although obstetric culture is based on supposedly objective biomedical understandings of birth, many researchers have identified what they perceive to be unnecessarily high rates of cesarean section (e.g., Plevani et al. 2017; Keirns 2015), suggesting that more women are being exposed to the potential risks of a significant surgical procedure than is necessary (Box 8.1).

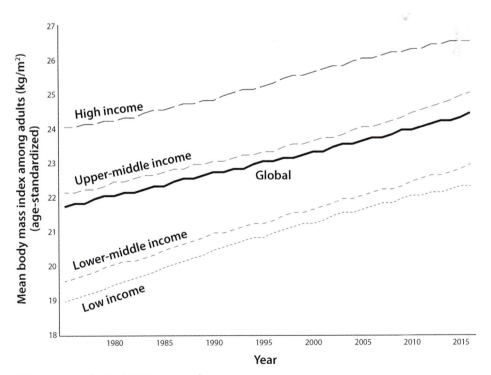

Figure 8.4 Mean age-standardized BMI over time by income group
Source: World Bank (2019)

Box 8.1

CESAREAN SECTION AND THE CULTURE OF BIRTH

Medical procedures such as cesarean sections are cultural practices, subject to both medical and broader social influences. Although cesareans (C-sections) provide a life-saving intervention for mother and child in many circumstances, many commentators are concerned about rapidly rising cesarean rates, generating a heated debate over the "optimal" C-section rate.

Since 1985, the WHO has suggested that cesareans should only be performed when medically necessary—for about 10 to 15 percent of births by their recommendation (WHO 2019a). Other researchers have suggested that improvements in maternal and fetal outcomes are recorded up to a C-section rate of 19 percent (Molina et al. 2015). The fact that cesarean rates range from just 4 percent in some parts of Africa to more than 40 percent of births in some Latin American countries suggests that "unacceptable disparities" exist in

access to the procedure (Wiklund et al. 2018) (Figure 8.5). Huge disparities also exist *within* many countries—in China, for instance, C-section rates range from around 4 percent in Tibet to more than 60 percent in Jilin province (Li et al. 2017). Low rates can usually be attributed to limited access to healthcare that leaves many women underserved and at risk from maternal and fetal injury or even death, and so increasing access to C-sections is of critical importance in many poor countries. Of greater interest to our cultural argument here, however, are places with high rates.

High C-section rates are a problem because they expose a larger number of women than is necessary to a potentially risky surgical procedure. High rates have been reported to increase neonatal intensive care admissions and even neonatal and maternal mortality (D'Alton and Hehir 2015). Some evidence suggests that babies born by C-section may be more

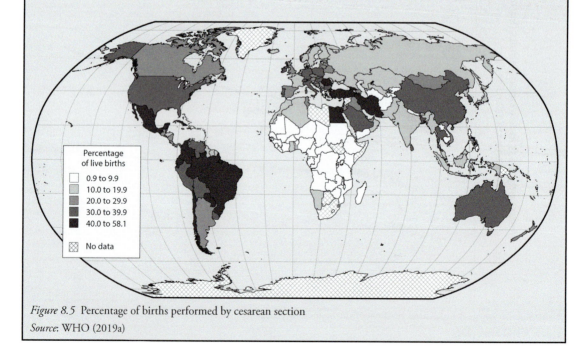

Figure 8.5 Percentage of births performed by cesarean section

Source: WHO (2019a)

prone to health conditions such as asthma, although the evidence remains inconclusive on many of these connections as well as whether these links are actually causal (Black et al. 2015). As an expensive procedure, "unnecessary" C-sections also divert healthcare resources away from other needs.

So, if obstetricians only considered best medical practice, we might expect that C-section rates should not greatly exceed 20 percent. In the Global North, where we might expect the greatest opportunity to follow best medical practice, rates are around 30 percent in many countries, including the US, Canada, and Australia. Researchers suggest that cultural context may help to explain this anomaly. Perhaps most influential, societal acceptance of C-sections as *the* way to treat complicated births leaves many doctors anxious about not performing them for fear of blame or even litigation in circumstances where there is any suggestion that a birth might have a negative outcome (Black et al. 2015). Efforts to manage doctors' and surgical workload schedules, and even personal financial benefit, may also contribute to high rates of C-sections in some places (Wiklund et al. 2018).

Broader societal expectations also influence cesarean rates. Some feminists have traditionally advocated for access to elective cesareans on the grounds that women should be able to avoid the pain and risk associated with childbirth, and even be able to schedule their birth at their convenience. In the early 2000s, a number of high-profile celebrity women were reported to have had cesareans purely for convenience, and the idea of elective cesareans became popular in some cultures, particularly in Latin America—some private clinics in Brazil, for instance, report C-section rates of more than 90 percent (Song 2004). In response, the popular press began to critique this trend, derisively referring to women choosing elective cesareans as "too posh to push," again articulating cultural expectations of womanhood and motherhood (*ibid.*).

In the end, the identification of an "optimal" C-section rate may be problematic owing to differences in healthcare systems and maternal health in different contexts. As d'Alton and Hehir (2015, 2239) conclude, "It is not whether the cesarean delivery rate is high or low that really matters, but rather whether appropriate performance of cesarean delivery is part of a system that delivers optimal maternal and neonatal care after consideration of all relevant patient and health system information."

Other studies have noted that scheduling of births may occur out of deep-seated cultural beliefs. Almond et al. (2015) investigated birth practices among Chinese Americans in California and found that 2.3 percent more births occurred to Chinese Americans on dates that included the number eight—a lucky number for many Chinese—than would be expected if births were distributed randomly. The authors suggest that some Chinese American parents may deliberately schedule procedures such as cesareans and birth inductions on these auspicious dates to provide good fortune for their children.

Individuals receive health-related messages from many professional and informal sources, and must make sense of these different messages, even when they contradict one another. This *intersection* of different cultural messages shapes our understandings of what is healthy. For example, a woman thinking about how to improve her diet might consider advice from family and friends, information from her doctor or a community health nurse, as well as the huge amounts of information available on the Internet. Her final decisions may therefore be driven by a mixture of professional, lay, and possibly even spurious information. Immigrants are in an especially challenging position because they live at the intersection of multiple distinct cultures. For example, Bradby (1997) found that Punjabi women living in Glasgow, Scotland, used two separate frameworks for understanding the relationship between food and health. The first, based on a Western biomedical view of the world, reduced food to its constituent parts (such as vitamins) that were perceived as either healthy or unhealthy. The second framework, based on South Asian cultural beliefs, used a holistic approach to consider the effect of the whole food on the body. In this cultural framework, health is an equilibrium that must be maintained, mediated by the careful consumption of the correct foods in the right context.

Differences in cultural frames of reference for what constitutes healthy or sanitary behavior can be contentious, as is illustrated by controversy over the practice of female genital cutting (sometimes known as "female circumcision"). The procedure involves the removal of parts of the female genitalia and is practiced as cultural tradition in some countries of Northern Africa, including Sudan, Somalia, and Egypt (Figure 8.6). In some communities, women who have not undergone the procedure may not be considered marriageable. The WHO estimates that approximately 200 million girls and women living today have undergone some form of the procedure (WHO 2018). Although the degree of cutting varies widely, with differing impacts on reproductive

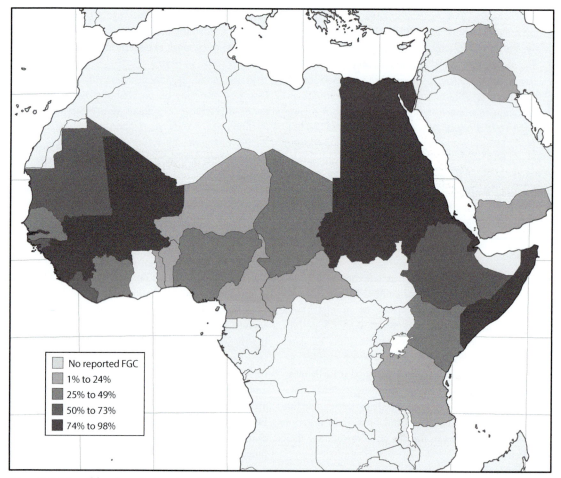

Figure 8.6 Map of female genital cutting, 2013

Source: UNICEF (2013)

This map shows the most recent estimate of the prevalence of female genital cutting (FGC), defined as the percentage of girls who are subjected to the practice.

health (Gruenbaum 2001), the health risks inherent in the procedure have led the WHO to declare the practice a violation of human rights (WHO 2018). In this context, it is sometimes referred to as "female genital mutilation."

The significance of labeling here is illustrative of widely differing attitudes to the practice—the clinical-sounding terms "clitoridectomy" and "infibulation" reflect the surgical specifics of the procedure; "female circumcision" sounds relatively innocuous, while "female genital mutilation" has been adopted by critics of the procedure and is highly emotive. Commentators have emphasized the power relations embedded in the practice—which is usually performed on girls under age 15 and as young as infancy—further complicating its significance. Many feminist scholars view the practice as an expression of men's attempts to control women's bodies and sexuality and thus a reflection of entrenched **patriarchy**. Western opposition to the practice, and colonial prohibitions of it, illustrate a different framework of power, with the practice becoming a site of struggle for autonomy within some cultures, even in situations where local actors might otherwise have opposed it (Gruenbaum 2001). In this context, the label of "cultural practice" has left many policymakers wary of confronting the issue for fear of being labeled racist, although many governments have now followed the WHO's lead in condemning the practice.

Maintaining a balance between respect for cultural traditions and regulating practices considered harmful to health doubtless involves treading a fine line. Since acknowledging the health impacts of a cultural practice usually involves singling out a community, it is often difficult to avoid charges of prejudice or discrimination. In the UK, for instance, a connection has been established between risk for **congenital** birth abnormalities and first-cousin marriages. The problem is well understood from a biomedical perspective: when close relatives marry, the chance of both partners carrying the genetic information for rare genetic abnormalities increases, raising the incidence of genetic conditions. This seemingly value-free biomedical explanation led several British Members of Parliament to suggest in the early 2000s that the issue should be addressed. Complications arose, however, because first-cousin marriages in the UK occur primarily in Britain's South Asian population, eliciting a vehement response from some Muslim community organizations, who saw raising the topic as an affront to their community. Indeed,

in 2008, when a British Member of Parliament raised the issue, the Muslim Public Affairs Committee asked for him to be removed from office (BBC News 2008). Nonetheless, the public health community has persisted in tracking the issue and seeking appropriate and sensitive interventions (e.g., Sheridan et al. 2013).

Cultural definitions of health

Culture not only influences health by encouraging or discouraging practices with health implications, but also constructs the very concept of "health." From a **postmodern** perspective, the biomedical definition of health is seen as just one among many understandings of health, albeit the dominant one. The biomedical approach is based on concepts such as **germ theory**, anatomical understandings of the functioning of the human body, and particularly the **positivist** idea that knowledge is derived from objective measurements. Doctors from the biomedical tradition, therefore, focus on measurable factors such as the presence or absence of pathogens, longevity, and physical size in assessing health (Figure 8.7). Other cultures may conceive of health differently. In many societies, physical health has traditionally been viewed as just one part of an integrated system, with other factors such as emotional and spiritual welfare, mental health, and even material wealth, also contributing to **well-being**. Such health systems are sometimes referred to as **holistic**.

In the case of the Matsigenka of the Peruvian Amazon, for instance, physical health was traditionally viewed as just one facet of well-being, with happiness, productivity, "goodness," balance, and social function also considered important (Izquierdo 2005). Although the Matsigenka's health was recorded to have improved in the early 2000s according to some biomedical indicators (such as levels of hemoglobin in the blood), interviews with the Matsigenka suggested that they perceived their own health to be in decline because of increased physical, mental, and emotional stress. The Matsigenka attributed this change in health status to contact with outsiders such as missionaries and oil workers, whom they perceived as bringing influences that conflicted with their own traditions (Izquierdo 2005).

How disease and health are defined can also have an impact on how disease is treated, or indeed whether it

Figure 8.7 Growth monitoring for a nutrition survey, Costa Rica

A health worker measures a child's height during a nutrition survey in Costa Rica. Height and weight are easily collected indicators that often contribute to biomedical approaches to monitoring health.

is treated at all. For instance, a study in rural Pakistan in the 1980s investigated why oral rehydration therapy (ORT) was rejected by mothers as a treatment for diarrhea, even though potentially life-saving ORT sachets were available to them for free. The researchers found that, in addition to not knowing how to use ORT, some mothers saw diarrhea as a natural part of childhood and therefore perceived no need to treat it (Mull and Mull 1988). Divergent understandings of health and disease can be a problem in affluent countries as well. For example, there has been a recent backlash against vaccination programs in the Global North, with some parents arguing that contracting diseases such as chicken pox and measles is a normal part of childhood.

Cultural conceptualizations of the body may also affect how we interpret symptoms and changes in the body, how we differentiate between sick and healthy bodies, what we view as an attractive body, and which bodily functions we view as socially acceptable or morally unclean (Helman 2007). In European cultures, for instance, slender bodies are seen as both attractive and healthy. In many pre-industrial societies, by contrast, being fat has traditionally been associated with health,

wealth, and fertility, in reaction to everyday experiences of hunger. This cultural preference for fatness—what Helman (2007) refers to as "cultural obesity"—has been identified in societies across the world. The Annang people of Nigeria, for instance, traditionally used "fattening rooms" to help women put on weight before their wedding (Brink 1995, 71). Similar traditions have been reported among Nauruans in the South Pacific (Pollock 1995). In some cultural contexts, slimness may even be perceived as a sign of poor health (Helman 2007); in Uganda, for instance, AIDS was traditionally known as "slim disease" (Serwadda et al. 1985). Despite these traditional views, the global diffusion of cultural images that associate slimness with beauty and health has led to the decline of cultural preferences for fatness in many regions. Although the prevalence of eating disorders is still highest in the Global North, eating disorders are on the rise in many other parts of the world, including parts of Latin America, Africa, and the Middle East, as well as in many immigrant communities in the West (Nasser 2009) (Figure 8.8). The Internet has been credited with a role in disseminating cultural expectations by reinforcing ideas about body image, greatly

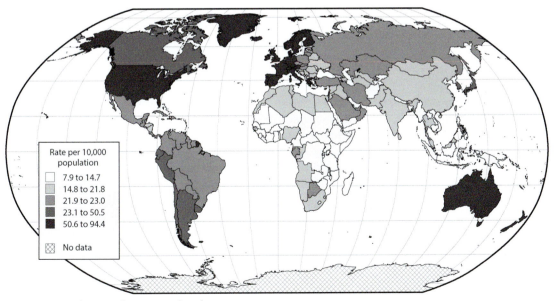

Figure 8.8 Population with an eating disorder

Source: Global Burden of Disease Collaborative Network (2017)

accelerating the globalization of notions of ideal bodies. While emphasizing slim, fit bodies can be a healthy message, it has also been credited with increasing body image disorders, as individuals hold themselves up to the impossible standards of images they see online. The Internet has also provided a forum in which users can find others with whom to share and reinforce extreme views on body image, leading to the proliferation of what have been termed "pro-eating disorder" websites, where users swap tips on extreme dieting techniques (Day and Keys 2008).

In all these examples, we see how the human tendency to associate with others who share a similar worldview—whether that worldview is based on ethnicity, gender, or even an affinity for extreme dieting—creates cultural practices with health implications. Cultural groups contain a diversity of perspectives, however, and belonging to several different cultural groups may lead to a collision of contradictory belief systems. How an individual navigates these different ideas is often as important to health outcomes as the cultural practices themselves, and so we now turn to an examination of aspects of identity that help us to understand the actions of individuals.

IDENTITY

Social geographers consider the everyday lived experiences of individuals as a valuable approach for understanding health phenomena. This **humanistic** approach calls attention to how our lived experiences are affected by our identity and position in society. Every individual has a unique personality and set of experiences from which their **identity** is constructed. Important aspects of an individual's identity reflect their affinity with particular cultural groups, based on factors such as gender, class, race, ethnicity, and sexuality.

People's individual identities evolve out of a complex and unique layering of these different cultural facets. For instance, imagine a woman growing up in Glasgow's Chinese community. She may consider herself to be simultaneously female, heterosexual, Scottish, British, Chinese, and Asian, and her own unique identity may assume aspects of all these cultural groups. She may also be vegan, a booklover, and own a pet snake, all of which

may bring expectations of behavior, as well as possible points of connection with others with similar views or lifestyle choices. Different aspects of identify may prove to be more or less important in different circumstances. As she lives outside China, her identity as Chinese may become significant to her (and others) as it becomes a way in which she differs from the majority population—it becomes a **marker of difference**. If she is working in a strongly gendered environment—perhaps she gets a job on an oil rig—her identity as female may be emphasized. Identities are, in this way, place-based, multi-faceted, and ever-changing.

It is important and sometimes challenging to avoid essentializing different groups in discussions of identity. **Essentialism** is the idea that there are certain characteristics that any member of a particular group naturally possesses: for instance, the notion that girls are inherently caring and nurturing or that boys are naturally rambunctious. Most scholars reject such simplistic associations and emphasize instead how concepts such as gender and race are both **socially constructed** and dynamic. In other words, cultural understandings about what it means to be female or white have been shaped by society, rather than by any inherent biological underpinning, and are constantly changing. In the context of gender, for instance, "there is, therefore, no universal 'woman,' but various possible enactments of 'womanhood' with certain strands of identity having greater salience according to specific contexts" (Dyck 1998, 106).

Undoubtedly, there are biological aspects to being female or male—different physical features and hormone levels, for instance—but here we are interested in how society defines, treats, and differentiates between men and women in ways that affect health. Many scholars articulate this distinction by reserving the term "**sex**" for the biological underpinnings of being male or female, and "**gender**" for the social aspects of how men and women behave in and are treated by society. Such a distinction allows us to acknowledge that some individuals do not identify as the same gender as their biological sex. These terms nonetheless still fall short in some circumstances. For instance, the traditional binary understanding of sex (where everyone can be assigned biologically male or female at birth) fails to recognize the experiences of individuals born with **intersex** traits—genitalia that are not clearly male or female.

How society responds to those with **intersex** traits exposes how uncomfortable people can be with individuals who do not conform to expectations of "normal" behavior or development. Experts suggest that 0.05 to 1.7 percent of the global population may be born with **intersex** traits (United Nations for LGBT Equality 2017), and yet many people are unaware of the phenomenon given a culture of secrecy around the topic. Doctors have traditionally taken the view that assigning a child an identity as male or female early on in life is crucial to their development, given society's strong cultural reliance on gender as a way to organize the world. In addition, having intersex traits may subject an individual to significant discrimination (United Nations for LGBT Equality 2017). As such, parents of babies identified as intersex have often been advised to surgically assign a sex to their child in infancy (typically female as this is usually the more straightforward surgery). Recently, however, with the growing movement to acknowledge and accept greater diversity in sex, gender, and sexuality issues, intersex is beginning to be viewed differently in some cultures. Mounting evidence of circumstances in which individuals have rejected the sex they were assigned at birth has led some commentators to argue that children with intersex traits should be allowed time to choose their own gender, or even not be asked to ascribe to either gender (United Nations for LGBT Equality 2017). In some countries, policy is slowly beginning to respond: in 2015, Malta became the first country to ban non-consensual modifications to sex characteristics; in 2018, Germany began allowing intersex identity on passports and birth certificates; and in 2019, Kenya recognized intersex as a census category (BBC 2019).

Despite this progress, individuals who do not conform to societal expectations often continue to experience significant discrimination. Entrenched ideas of "normal" and "abnormal," "sick" and "healthy," or "acceptable" and "unacceptable" can exclude individuals or entire groups. The concept of the "**Other**" refers to groups who are marginalized or undervalued because they fall outside the dominant culture. For instance, communities of color are frequently "othered" in white-dominated societies, while homosexuality is typically viewed as the "Other" in contrast to heterosexuality. Health-related appearances or behaviors can also "other" individuals through implicit assumptions constructed by society. As noted earlier, the idea that everyone should be unambiguously male or female leads to the othering of individuals with intersex traits. Similarly, the bodily appearance of people with a visible disability can become a **marker of difference** that influences how people treat them (Laws and Radford 1998). In this context, the notion of "diversity" (a positive construct that refers to variety within a society) becomes reinterpreted as "difference" or "otherness," with negative outcomes such as racism, sexism, and homophobia.

These forms of discrimination can be found throughout society. Even the medical profession is not immune to constructing or reinforcing discriminatory expectations. For example, in an Australian study, same-sex attracted women reported "interviewing" general practitioners carefully to assess their attitudes to same-sex attraction. This anxiety about potential discrimination had tangible impacts for the women in the study, who were less likely to seek the help of a doctor, less likely to use routine screening services, and more likely to delay visiting a doctor than their heterosexual counterparts (Edwards and van Roekel 2009).

Geographers have pointed out that identity often has an important spatial component, through individuals' desire to congregate with people who are similar and segregate those who are different. The process of othering therefore frequently occurs spatially, as the ability to create and maintain distinctions among groups depends on our ability to impose boundaries (Dear et al. 1997). This partitioning may be physical (e.g., dividing the sick from the healthy via quarantine) or social (e.g., discriminating between disabled and able bodies). For example, an individual who is unable to walk may be physically prevented from entering a building if access is barred by stairs, but they may also experience exclusion from social spaces as a result of prejudice against their disability.

Experiences of isolation—both physical and social—is an evolving area of research. Social isolation, often related to aging and increasing disability, has been associated with a wide variety of negative health outcomes, including heart disease, stroke, arthritis, Type II diabetes, and dementia, giving people with strong social networks a significant survival advantage (Holt-Lunstad, Smith, and Layton 2010; Valtorta et al. 2016; Brody 2017). Social isolation is a particularly important issue in aging societies, given that many people may live for decades after leaving the workforce. In the US, one third of people over age 65 and half of people over age 85 live alone, and the proportion of Americans who say they are lonely has increased from 20 percent to 40 percent over the past forty years (Khullar 2016). In combination with increasingly fluid family structures that have eroded some aspects of family support, many countries face an "epidemic" of isolation, with a vicious circle of loneliness and isolation producing risk factors for poor health, while poor health is itself a risk factor for isolation.

The "**life-course**" perspective seeks to acknowledge and incorporate many of these less tangible impacts on health such as isolation and discrimination, through positing that the experiences of an individual across the whole life-course potentially influence health outcomes later in life. This perspective focuses on people's day-to-day interactions with their social and physical environments, recognizing that the world is viewed through the lens of an individual's experiences and imagination, rather than as an objective reality. An individual's "**life-world**" is the sum of his or her lived experiences, with potential impacts on health. For instance, we might consider whether a recent personal event has led to trauma or how early experiences with poverty might be significant.

Geographers have noted how place can play a significant role in a person's life-world, influencing things like rootedness and sense of belonging. For example, Kornelsen et al. (2010) observe that place of birth has special significance for some First Nations women in Canada, tying them to a community identity. Giving birth locally, within reach of significant social relationships, was therefore considered important to achieving a sense of "belonging-through-birth" (*ibid.*). The impact of hospital closures, which have forced many women to travel beyond their community for maternity care, has been especially significant in this context.

Qualitative methods are especially valuable for studying the influence of culture and identity on health. Scholars commonly use **in-depth interviews** and **focus groups** to explore the particularity of experience

that underlies these aspects of health. **In-depth interviews** allow researchers to prompt respondents to relate their experiences and ideas on a topic that can then be explored further via follow-up questions. The open-ended nature of this format allows respondents to direct the conversation towards topics that they deem to be important. A **focus group** uses a similar approach but incorporates a small group of participants, with the idea that interactions among respondents may be important or may push the conversation along lines that would not be reached by a conversation between an interviewer and a lone interviewee.

In Kornelsen et al.' s (2010) study, for instance, the researchers used in-depth interviews and a written survey to explore the individual experiences of women in a First Nation community from the Heiltsuk nation in Canada. Recognizing that their respondents' experiences as members of the Heiltsuk nation could be significant to their experiences around birth, the researchers sought advice from a community-based advisory committee to guide their study and included a member of the local community as part of their research team. Efforts to make research more **participatory**, whereby community members from the study site are actively involved in a study's development and production, are increasingly acknowledged as strengthening the quality of research on culture and identity.

Identity and ill health

Distinctions between "able" and "disabled" bodies are found in almost all societies (Helman 2007), with stigma and discrimination common to many social understandings of illness and disability (Person et al. 2009). People with disabilities may have difficulty finding work, accommodation, a marriage partner, and status in many societies. For instance, in a study of lymphatic filariasis—a condition that can result in significant disfigurement through the swelling of limbs—women in the Dominican Republic and Ghana reported being denied access to education and employment, as well as being isolated and criticized by friends, community members, and even caregivers (Person et al. 2009). Similarly, early responses to the HIV/AIDS epidemic in many communities included efforts to push individuals living with HIV/AIDS out of mainstream society. In some cases, people lost their jobs, had their leases terminated, or were even placed in sanatoria (Baldwin 2005) (Box 8.2).

Not all cultural constructions of disability are negative, although most emphasize disability as the defining feature of an individual's identity, with potentially

---

Box 8.2

STIGMA AND HIV/AIDS IN THE US

In the US, social stigma around HIV/AIDS evolved from the initial identification of the disease in homosexual populations and the disease's subsequent connection to other marginalized populations, such as intravenous drug users. The prevailing moral discourse implied that HIV was a punishment for those "considered deviant or promiscuous, contributing to a less-than-sympathetic attitude towards those afflicted" (Keeler 2007, 615). By contrast, some groups with the infection were framed as innocent victims (such as hemophiliacs or babies born to HIV-positive mothers), reinforcing the blame assigned to those who had contracted the disease through supposedly deviant or promiscuous behaviors.

In the early 1980s, public health experts attempted to make sense of the emerging epidemic by identifying high-risk groups. In the US, these efforts led to the identification of an HIV "4-H Club"—heroin addicts, hemophiliacs, homosexuals, and Haitians—as supposedly high-risk populations. While identifying high-risk groups may make sense from a biomedical perspective, the social implications of this approach were devastating, as *all* gay men, *all* Haitians, and even *all* African Americans were perceived as potential sources of contagion.

The inclusion of Haitians as a high-risk group was particularly confusing. Despite knowing little about routes of transmission of the disease, public health officials identified a cluster of cases in the US Haitian community that did not seem to fit prevailing understandings of the disease. They consequently identified Haitians as a high-risk group despite little evidence of anything specific to this community that placed it at high risk. Indeed, Farmer (1992) suggests that the apparent high risk of this group may have been partly a result of simply underestimating the size of the US's Haitian community. The

fact that Haitians represented the "Other" in US imaginations—not only black but also mythologized by unfamiliar Voodoo practices—was probably instrumental in the rapid acceptance of Haitians as carriers of disease. Before long, theories emerged that suggested that AIDS had been brought to the US from Haiti, despite Haitian researchers' reports that many aspects of the HIV epidemic appeared just as new to Haiti as they were to the US (Farmer 1992). Prejudice against the Haitian community became overt, leading to harassment of Haitian children at school, refusal to hire Haitians into jobs, failure of Haitian businesses, and eviction of Haitian families from their homes (Sabatier and Tinker 1988).

As Treichler (1999, 11) summarizes, "the AIDS epidemic . . . is simultaneously an epidemic of a transmissible lethal disease and an epidemic of meanings or signification." The use of labels not only stigmatizes particular populations but may also have accelerated the spread of the disease among individuals who took few precautions because they perceived themselves as outside "at risk" groups. Over time, the discriminatory and confusing nature of labeling "at risk" groups has become clear, and public health discourse increasingly prefers the term "risk behaviors" instead of "risk groups" to avoid implying connections between a disease and a specific population.

Attention to the structural constraints that contribute to infection has shifted blame even further from particular communities or individuals (e.g., Barnett and Whiteside 2002; Farmer 2001; Rhodes et al. 2005). This structural perspective suggests that political, economic, and social constraints limit the opportunities of some people to avoid infection, for instance, by precluding them from obtaining the information that they need to understand their risk of infection.

negative implications. In traditional Hmong culture, for instance, epilepsy is regarded as a serious condition, and yet people with epilepsy are often respected as shaman (Fadiman 1998). In rural Korea, some blind men have traditionally acted as fortune-tellers and diviners, while some blind people among the Tiv of Nigeria are believed to have powers of "second-sight" (Levinson and Gaccione 1997). These case studies draw attention to the fact that the meaning of disability is highly place-specific. Meanings associated with disease may also change over time, as is illustrated by the recent rapid rise in diagnoses of attention deficit hyperactivity disorder (ADHD) in many Western countries. Before the recognition of ADHD as a disorder in the 1980s, children who exhibited the typical symptoms of the condition—impulsiveness, hyperactivity, and inattention—were commonly labeled as badly behaved.

Providing an official diagnosis for children with ADHD encouraged therapeutic rather than punitive approaches to be used to try to help children with ADHD to operate successfully in social settings, encouraging some parents to actively seek a diagnosis for their child. However, a careful analysis of the history of ADHD illustrates how power differentials between dominant and marginalized groups in society continue to drive how and if ADHD is diagnosed and treated (Box 8.3).

Health geographers have begun to consider how experiences of ill health or disability are socially constructed within particular cultural frameworks, leading to an entire sub-discipline of geography around "geographies of disability." Here the plural term "geographies" is used to emphasize the multiple ways in which disability is envisioned and enacted, reinforcing the significance of individual experiences.

Box 8.3

ADHD IN SOCIAL CONTEXT

The incidence of attention deficit hyperactivity disorder (ADHD) increased rapidly during the 1990s in the US and some other Western countries. Between 1994 and 1999, the production of Ritalin, a drug used to treat ADHD, rose by 800 percent, more than 90 percent of which was used in the US (Breggin 2007). What does a geographic analysis of this epidemic tell us about the intersection of culture, identity, and health?

Few people dispute the legitimacy of symptoms of ADHD in many children, including impulsive behavior and an inability to concentrate in a classroom setting. The US *Diagnostic and Statistical Manual of Psychiatric Disorders* formally recognizes the condition, but diagnosis remains challenging because of a lack of definitive diagnostic indicators, and there is continued disagreement over whether ADHD should be viewed as a medical condition or a socially

constructed label for children with behavioral problems Wright 2012). While we do not aim to answer this controversial question here, considering the social framing of ADHD can illustrate the importance of understanding social context to health.

Rates of diagnosis and treatment of ADHD vary significantly from place to place. In one study, researchers analyzed per capita consumption of Ritalin across the US (Bokhari, Mayes, and Scheffler 2005). In both state- and county-level analyses, the authors concluded that high Ritalin usage is associated with places with higher per capita incomes, lower unemployment rates, better access to healthcare, and more private schooling. At a regional scale, patterns related to race also became apparent. The five states with the highest consumption of Ritalin were in the Northeast, where 80 percent of the student population is white. In contrast, only 40 percent of the student population is white in the five states with the lowest consumption rates. The researchers also found an achievement gap between high-use and low-use states; students in states with high consumption of Ritalin achieve better proficiency in writing, for instance. Bokhari, Mayes, and Scheffler (2005) argue that these data cast doubt on biomedical interpretations of the disease that suggest that ADHD is the product of congenital abnormalities because the most privileged communities with the smallest likelihood of birth injuries and congenital disability report the highest incidence of the disease. Instead, they compel us to consider social influences on diagnosis and treatment of the disease.

What social factors could explain the finding that children living in more privileged circumstances are more likely to receive Ritalin for ADHD? We could hypothesize that more privileged populations have better access to treatment, and that there is a huge unmet demand for treatment in low-consumption states, and this is probably a contributory factor. Other scholars have argued that changing meanings associated with ADHD provide a further critical explanation (see Rafalovich 2001). In the past, children from marginalized communities were often considered "deviant" or "hyperactive" if they displayed typical symptoms of ADHD—descriptions that middle-class parents actively avoided for their children. As the disability rights movement attempted to destigmatize behavioral problems, however, a diagnosis of ADHD has become increasingly socially acceptable, and treatments and resources have become associated with the diagnosis. In this context, it is understandable that parents would be more likely to actively seek a diagnosis to make sense of their child's behavioral problems.

Geographies of disability

The bodily and social experiences of people with disabilities has become an important, if understudied, topic within geography. Disability geography has a relatively short history as a distinct sub-discipline, dating only from the early 1990s (Worth 2008). During this time, however, disability geography has experienced a major theoretical shift. Early work on disability focused on spatial patterns in the incidence of disabling conditions and service provision, as well as access and mobility for disabled people (Chouinard, Hall, and Wilton 2010, 13). Although the value of this work should not be understated, critical geographers have pointed out that much of this research focused on physical aspects of disability at the expense of considering the disabling nature of the socio-spatial environment. In this context, it is helpful to distinguish between "impairment"—representing the physical reality of an individual's situation (e.g., a missing limb)—and "disability"—the disadvantages imposed by society on an individual with a physical impairment (Oliver 1990). From this perspective, part of the disability of being blind, for instance, is in living in a world designed for the sighted (Butler and Parr 1999). This conceptualization challenges the medical construction of disability and focuses attention instead on constraints imposed by social and physical environments (Wright 2015).

Acknowledging these new ideas, a "first wave" of critical research in disability studies emerged in the late 1990s, focusing on "disability as a socially and spatially produced form of exclusion and oppression [and] recognizing the capacity of disabled people to challenge or transform such disabling social relations and spatial structures" (Chouinard, Hall, and Wilton 2010, 3). Many geographic studies of disability focused on exclusion from social spaces. For instance, in a study of the lives of individuals with intellectual impairments in Toronto, Laws and Radford (1998, 99–100) concluded that their respondents experienced "low levels of interaction with the so-called normal world, small action spaces, attenuated life worlds, and precarious finances, all point[ing] to a life on the outer fringes of the daily round." Others have demonstrated that social space is maintained between dominant and "othered" populations by considering "hierarchies of acceptance," rankings or preferences expressed by the general population about the relative acceptability of different groups of people. Dear et al. (1997) report that people with mental illness are commonly ranked at the bottom of these hierarchies (least socially acceptable), on a par with alcoholics, drug users, and ex-convicts. Some geographic work has considered the exclusions and experiences of people with mental illness, although the topic remains understudied (Moon 2000; Philo 1987).

The relationship between disability and space continues to be re-theorized in increasingly nuanced ways. For instance, in response to the dramatic shift towards social aspects of disability, some scholars have pointed out that physical aspects of disability are now being neglected. There is now recognition of the importance of the intersection of physical and social aspects of disability, accepting that bodies are physical entities that can be damaged, but that it is also important to frame this in socio-spatial context (Hall 2000; Hansen and Philo 2007; Moon 2000). Shakespeare (2006) argues that it is useful to think about the *interaction* between social and physical aspects of disability, recognizing that "people are disabled by society, and by their bodies." Notions of "disabled" and "able-bodied" are also being re-theorized as a continuum rather than as polar opposites, with the experience of disability seen as relative to particular times and places (Worth 2008).

Chouinard et al. (2010) recognize three additional ways in which disability studies are evolving. First, the meaning of disability has been broadened to encourage examination of additional groups that are othered by society in ways similar to those with disabilities, including older people (e.g., Milligan 2006), people

who are overweight (e.g., Longhurst 2010), and people with chronic illnesses (Wright 2015; del Casino 2001; Crooks and Andrews 2009). Second, geographers are considering ways in which technology can influence the lives of people with disabilities, acknowledging the paradox that technology can both enable and constrain them. For instance, Skelton and Valentine (2010) discuss how the Internet has opened new spaces for accessing information for people with hearing impairments, and yet does nothing to improve their access to the hearing world. Third, geographic work on disability continues to emphasize the political nature of "disability" and the need for political engagement. Giving people with disabilities a voice through methods such as participant observation and interviews and engaging with policy in realms such as social exclusion and employment are two ways forward.

KEY CULTURAL INFLUENCES ON HEALTH: CASE STUDIES OF GENDER AND RACE

In the remainder of the chapter, we consider two aspects of culture and identity that are commonly studied as important shapers of health experiences: gender and race. We build our discussion around case studies in recognition of the highly context-specific nature of these factors.

Gender

As we noted earlier, social theorists argue that many of the differences between men and women are socially constructed. In other words, men and women have different experiences because each group behaves differently and is treated differently owing to gendered social norms and attitudes. For instance, there is a **gendered division of labor** in societies around the world, with some jobs expected to be filled by men and others by women, irrespective of the actual abilities of men and women. For instance, catching fish is a traditionally male job in many cultures (implicit in the term fisher*man*), while women are often restricted to processing, trading, or cooking fish. This is not because women are physically incapable of fishing but is borne out of cultural expectations of gender roles, reinforced by social structures that influence access to fish through who owns fishing equipment or who has permission to fish particular waters. The implications of gendered expectations are often unequal—in this example, fishing is usually more profitable compared with fish processing and trading, leaving women economically disadvantaged. In affluent and poor countries alike, numerous studies have shown that women are overrepresented in lower-paid jobs, particularly in the caring professions (e.g., nurse, teacher) and in support roles (e.g., secretarial staff), while men dominate higher-status jobs and are generally better paid, even when men and women are performing the same job.

In making sense of gender inequalities, we can consider gender from both a structural and an individual perspective. The notion of **structural violence** refers to the ways in which social structures and institutions prevent people from meeting basic needs and is often considered to be gendered. **Patriarchy**, the structuring of society around men as dominant figures, can be viewed as a structural constraint that limits the **agency** of women. Patriarchy may operate in obvious ways: for instance, women are more likely than men to be the victims of sexual violence. Many of the impacts of gendered social structures are subtler, however. For example, in societies where women are undervalued, women and girls may be less likely to be given medical care when they are sick (Bhan et al. 2005). There is also a gendered mortality gap associated with environmental disasters, with women dying at higher rates than men during some disasters. This gap is narrower in countries with higher socioeconomic status, suggesting that these inequalities are largely a social rather than biological phenomenon (Neumayer and Plümper 2007).

To avoid **essentializing** women, we must accept that not all women are affected by patriarchy in the same way, however. For instance, it is often reported that social and economic structures force women into commercial sex, leaving them vulnerable to HIV infection, and yet few women turn to prostitution, even when living in poverty. Gendered analyses must acknowledge these differences to avoid depicting women as victims of circumstance. Often, the most marginalized women in society face the gravest consequences of patriarchy, particularly if they have few options for living independently of men. In this way, gender can act in concert with other aspects of social identity to generate greater vulnerability, a theme we explore further later.

Some feminist scholars have argued that we should also focus on how the lived experiences and identity of individual women and men are shaped by gender. Culturally situated conceptualizations of gender influence time- and place-specific understandings of what particular people, or groups of people, can or should do. For instance, women are viewed as nurturing in many cultures. Although this is an apparently positive stereotype, it often serves to validate the idea that women and girls should bear the greatest burden of providing care for family members. By contrast, young men are often viewed as risk-takers, with different implications for health. Indeed, it is important to remember that social structures and inequalities can harm men's health in some instances, leading some scholars to investigate the ways in which masculinity, too, can be viewed as a precarious state. For instance, among cohorts born since 1942 in wealthy countries, men aged between 20 and 50 are more than twice as likely to die as women in any particular year (Rigby and Dorling 2007). Although "no single medical, biological or 'lifestyle' explanation can account for why young men's mortality is improving so much more slowly than women's across all the richer nations of the world," the emancipation of women and related changes in the relative social position of men may have played a role (Rigby and Dorling 2007, 163). Males are also consistently more likely to experience violent death (Payne 2006). Worldwide, males have a suicide rate almost twice that of women and are almost four times as likely to die from homicide (WHO 2019b; United Nations Office on Drugs and Crime 2019). Overall, men are more likely than women to die from accidental death, non-intentional violence, and intentional violence. This is perhaps partly due to "men's own self-destructive and risky behavior to prove their heterosexual male identity" (Riska 2003, 74), but it is not simply a matter of men living up to societal expectations about aggression that underlies these differences. For instance, occupational differences in accidental deaths may be related to constructions of masculinity that depict men as "breadwinners." Consequently,

---

Box 8.4   *continued*

increases. Indeed, Mojola (2011) argues that the "fish-for-sex" economy appears to be relatively new to the region, perhaps a response to declining fish stocks as a result of pollution and eutrophication of the lake, combined with growing populations that put greater pressures on the remaining fish. Thus, it is arguably the intersection of ecological and economic challenges with patriarchal structures of society that help explain women's high rates of HIV infection in the Lake Victoria region (*ibid.*).

Not all women traders participate in "fish-for-sex" practices, of course—generally speaking it is only the most economically marginalized among the female traders who turn to transactional sex, once again illustrating the significance of intersections of vulnerabilities (Camlin, Kwena, and Dworkin 2013). Many of the women who participate in "fish-for-sex" transactions are recent migrants who have been forced to move following shocks such as widowhood, divorce, family conflict, and gender-based violence, as women with established familial relationships in the region typically have access to fish without having to turn to transactional sex (*ibid.*).

Fish trading is seen by many women as a chance to improve their life circumstances in challenging conditions—a way of earning a living with low start-up costs and limited training. For those who manage to become established, fish trading can provide a measure of economic security and social status, enabling women to provide for their dependents. By contrast, the "fish-for-sex" economy is widely stigmatized by local communities and held responsible for fueling the HIV epidemic (Béné and Merten 2008; Camlin, Kwena, and Dworkin 2013; Fiorella et al. 2015). As such, most women participate reluctantly. Efforts to provide women with alternative means to establish themselves economically in the region are critical to reducing the vulnerability of women so that they are not compelled to turn to risky behaviors that fuel HIV transmission. As fish stocks continue to decline, increasing women's access to resources within the fishery industry may be hard to achieve, suggesting that longer-term strategies of ecological stabilization and alternative economic opportunities may be essential (Fiorella et al. 2015).

Broader understandings of gender politics and cultural norms can influence decision-making about sexual activity even in consensual relationships. In a comparative study of adolescent attitudes to sex in Kenya and Sweden, for example, women in both countries were found to be expected to assume the leading role in ensuring moral standards and safety, leaving women in the difficult situation of being expected to take on this responsibility and yet requiring sexual equality for this role to be effectively fulfilled (Ahlberg, Jylkäs, and Krantz 2001). Much of the pressure for women to behave responsibly came from other women, reflecting how individuals can internalize and reproduce gender inequalities.

These case studies complicate our understanding of the ways in which identity, culture, and socioeconomic structures intersect with respect to health. The biological reality of sex cannot provide a satisfactory framework to explain health differences between men and women but must be supplemented with an understanding of the inequalities and constraints society constructs around gender and the way that these constraints intersect with other vulnerabilities. This has important implications for how we address health issues like HIV/AIDS, suggesting that efforts to control disease require a more ambitious and nuanced approach than can be offered by strictly biomedical solutions. Instead, it is critical to consider the structural inequalities and constructions of identity that leave some individuals more vulnerable to infection than others.

Race and ethnicity

Race-based disparities in health have been identified in countries around the world, among indigenous groups, immigrant minorities, and even in some cases among majority populations facing discrimination, such as black South Africans. For example, in the US, where there are particularly profound racial disparities in health, blacks have higher all-cause mortality rates than whites in all age groups under 65 years, as well as higher rates of chronic disease at younger ages (Cunningham et al. 2017). In Australia, Aboriginal and Torres Strait Islander people have a life expectancy almost ten years shorter than Australia's non-indigenous population. A similar situation exists in New Zealand, where there is a six- or seven-year gap in life expectancy between Maori and non-Maori populations (New Zealand Ministry of Health 2013). Although the issue is best studied in the Global North, similar health inequalities exist in many countries of the Global South too: for example, Paulino et al. (2019) explore the reasons for poorer maternal health outcomes among indigenous populations in Guatemala, Mexico, Peru, and Bolivia.

So, how can we explain how race might play a role in health outcomes? As with gender, we must consider both biological and socially constructed aspects of race. In the past, many race-related differences in health outcomes were assumed to have a biological basis, and recent genetic analysis has confirmed that, in some cases, genetic factors play a role. For instance, researchers have identified a genetic predisposition to obesity in some Pacific Islander populations (Minster et al. 2016) and a genetic basis for susceptibility to asthma among African Americans (White et al. 2016).

Despite these examples, many scholars have argued that race has more meaning as a socially constructed phenomenon, with far more cultural, social, and political meaning than biological foundation (Cooper 1984; Marks 2001). Indeed, some researchers have used genetics to highlight misconceptions about the biological basis of race, noting that biological differences *within* many racial groups are far greater than differences *among* races (see Rosenberg et al. 2002; Marks 2001). For this reason, some people prefer the term **ethnicity** to race, referring to groupings that have less to do with skin color or other biological variations than with cultural similarities (although the terms "race" and "ethnicity" are often used interchangeably).

One common social explanation for race-based variations in health suggests that socioeconomic status acts as a significant **confounding** variable. We know that socioeconomic status is one of the most robust determinants of health across the world, and that many racial and ethnic divisions in society correlate to some degree with affluence, with minority groups often poorer than majority populations. It is therefore reasonable to consider whether poor health in a minority group has more to do with marginal economic position than with culture or other social circumstances. Research across a broad range of topics suggests that socioeconomic status does indeed frequently explain some proportion of racial differences in health outcomes. However, a growing body of evidence suggests that a relationship between race/ethnicity and certain health outcomes remains even after socioeconomic status has been accounted for. Much research now supports the idea that racism, and the inherent stress individuals experience through being marginalized by other social groups, may play an important role in the onset and development of a variety of health conditions, including diabetes, heart disease, hypertension, and pre-term birth.

Racism may also influence the way in which health conditions are viewed or the resources devoted to tackle health problems. One powerful example is drug addiction. In their discussion of the US opioid crisis, Netherland and Hansen (2017) point out that drug use in inner-city communities of color is still commonly viewed as a moral failing to be addressed with punishment, while it was only when opioid addiction became common in white communities that researchers began to redefine addiction as a medical problem using evidence such as brain scans. They argue that this reframing of drug use from moral failing to medical problem, although well intentioned, has failed to acknowledge the race-based premises which underlie many of the assumptions made about drug use and abuse in the US.

As noted earlier, an individual's identity is multifaceted, and it is often when identity becomes a **marker of difference** that it becomes especially significant. Although immigrant populations often report superior health compared with native-born populations when they first arrive in their host country—the so-called healthy immigrant effect—their health status can deteriorate rapidly after just a generation. Factors such as institutionalized racism, the psychological impact of differences in status in society, and a lack of feelings of connection to the new society may explain immigrants' and minorities' health disparities beyond differences in socioeconomic status. All these factors may also affect indigenous populations, who may feel that their cultures have been overwhelmed or displaced.

This does not mean that all individuals or communities respond in the same way to these pressures, however. For instance, in a UK study, white and South Asian single mothers were found to have high rates of mental illness compared with the UK's Caribbean community, in which lone parenthood was not associated with mental illness (Lloyd 1998). Although this may suggest that the UK's Caribbean community has high levels of resiliency, the answer may not be as simple as this. In a subsequent study, Edge and Rogers (2005) found that black Caribbean women reported less pregnancy-associated depression than white women, despite being more likely to have risk factors for depression such as being single mothers and living in poverty. Intriguingly, black women in the study acknowledged many symptoms of depression, but most rejected the diagnosis of "postnatal depression" because they associated depression with weakness and an inability to cope with adversity. Instead, black women emphasized how they were tackling their troubling situation and the ways in which they were "dealing with it." This was associated with the self-ascribed identity of being a "Strong Black Woman," an idea borne out of the need to succeed in a place with prejudiced and racist histories. Although this appears to be a powerfully positive response to adversity, research in the US suggests that lower rates of common mental illnesses in the African American community may result from a trade-off between mental and physical health (Williams et al. 1997). In this case, strategies to cope with racism may provide a buffer against the development of psychological conditions, but this resilience takes a significant *physical* toll, leaving minority populations vulnerable to a range of ailments.

Generalizing about race and health within or between minority groups is therefore highly problematic, and so here we look at two case studies to illustrate specific aspects of the issue. In the first case study, we consider how ethnic identity can influence an individual's health status through pressures related to being marginalized within society. In the second, we extend our discussion of experiences of immigration, asking whether some aspects of race could protect health.

African American women and pre-term labor in the US

In the US, infant mortality rates among African American women were over eleven per thousand in 2016 (equivalent to infant mortality rates in Argentina, Turkey, and China that year), compared with less than five per thousand among white mothers (CDC 2017; PRB 2016). These figures for 2016 perpetuate a long-standing divergence in infant health outcomes by race in the US (Figure 8.9), which has been growing since the 1950s (David and Collins 2007).

Most of this difference in birth outcomes is due to a larger proportion of black infants being born at very low birth weight (David and Collins 2007), with three times as many black infants as white infants weighing less than 1,500 g (3lb 5oz) at birth (Iyasu et al. 1992). Poorer birth outcomes persist for blacks in the US even after socioeconomic status has been controlled for.

To investigate whether health differences could be explained by genetic differences between black and white populations, David and Collins (1997) compared pre-term labor patterns among US whites, US blacks, and African immigrants in the US. They found that rates of pre-term delivery for African immigrants to the US were more similar to US whites than to US blacks, suggesting that genetic characteristics of black Africans cannot explain US blacks' greater likelihood of delivering their babies prematurely (Figure 8.10). Furthermore, after only one generation in the US, the daughters of African immigrants were found to give birth to babies that were, on average, lighter than their mothers had been at birth, suggesting that something about living in the US had influenced birth outcomes (David and Collins 1997).

The authors propose a variety of explanations for these findings. They note, for instance, that African American women are more likely than white women to have partners who were incarcerated during their pregnancy, are more likely to be exposed to violence in their

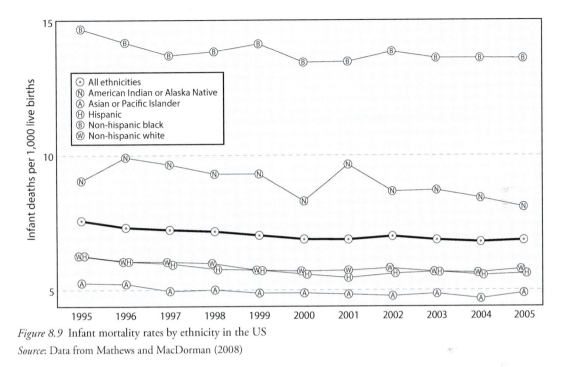

Figure 8.9 Infant mortality rates by ethnicity in the US
Source: Data from Mathews and MacDorman (2008)

neighborhoods, and are more likely to report higher levels of life stress (David and Collins 2007). African American women have also reported stress from coping with racism (Collins et al. 2004; Mustillo et al. 2004). The experience of chronic racism is increasingly being identified as a significant factor that leads to poor health outcomes in minority populations. Jones (2002) identifies three levels at which racism operates. *Institutionalized racism* refers to the structures, social norms, and institutions that provide different ethnic groups with different opportunities. *Personally mediated racism* is prejudice derived from assumptions about a person's abilities and motives on the basis of his or her ethnicity, and manifests itself in lack of respect, devaluation, and suspicion. Finally, *internalized racism* is the acceptance of negative stereotypes by members of a stigmatized group, leading to feelings of helplessness and hopelessness. Jones argues that all three types of racism have significant impacts on health. In the context of identity, *personally mediated* and *internalized racism* are especially significant factors,

explaining why African Americans continue to experience negative health outcomes even after accounting for structural factors such as poor access to healthcare (*institutionalized racism*).

The impact of feelings of difference, rejection, and even inadequacy are thought to build over an extended period. As a result, Lu and Halfon (2003) argue it is important to consider a time frame longer than just the pregnancy in research on pre-term labor. They advocate taking a **life-course perspective** that considers the cumulative impact of a lifetime of stresses on a mother's health. In this context, the notion of **weathering** is important. Weathering suggests that marginalized groups such as African American women experience something akin to accelerated aging, "as a consequence of the cumulative impact of repeated experience with social, economic, or political exclusion," including exposure to environmental hazards, ambient stressors, and psychosocial stress (Geronimus 2001, 133). Consequently, African American women are more likely to

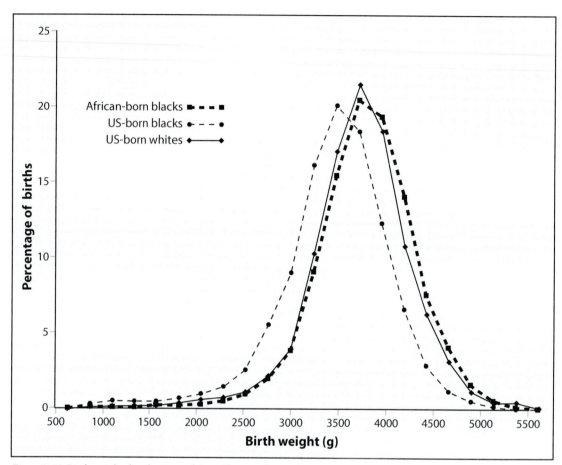

Figure 8.10 Birth weight distributions of three Illinois sub-populations

Source: David and Collins

experience chronic conditions or premature death at an earlier age. Weathering may also be exacerbated by unhealthy behaviors adopted to cope with day-to-day stresses.

For many African American women, poor health begins during child-bearing years, potentially leading to problematic birth outcomes. Indeed, African American women in their twenties and thirties actually have *worse* birth outcomes than teenage white mothers (Geronimus 2001), suggesting that the African American mothers (who would be expected to have optimal birth outcomes at that age) are influenced by something that is even

more significant than the social challenges of teenage motherhood. Even well-educated, professional African American women experience poorer birth outcomes than their white counterparts. Indeed, the huge effort that many professional African American women must exert to overcome social barriers could be yet one more pressure that contributes to weathering.

Although much of the research on the impact of race on pregnancy outcomes has emerged from the US, similar disparities have been identified in Europe. In one analysis of the pregnancy outcomes of native and immigrant women in twelve European countries from 1966 to 2004, immigrant women had a 43 percent higher risk of low birth weight, 24 percent higher risk of pre-term delivery, 50 percent higher risk of perinatal mortality, and 61 percent higher risk of congenital malformations (Bollini et al. 2009, 452). These unfavorable outcomes persist across generations, despite immigrant communities having access to Western healthcare in their host countries. The authors identify factors such as immigration, stress, disruption of social networks, low socioeconomic status, and discrimination within the healthcare sector as possible reasons for this disparity.

Immigration has different impacts on health in different countries, however. In Norway and Sweden, for instance, there is little difference in birth outcomes between native and immigrant populations, while greater disparities are found in Italy and the UK (Bollini et al. 2009). Countries with the strongest integration policies for immigrants display the smallest differences in birth outcomes. Social and psychological problems associated with poverty, racism, and social exclusion may help explain this finding (*ibid.*). This evidence suggests that greater efforts to help integrate immigrant communities could be a means to reduce the psychological and physical impacts of discrimination.

The Latino Paradox

Although we tend to think of minority status leading to health problems, this is not always the case. The **Latino Paradox** refers to the fact that Latino populations in the US appear to show similar, and sometimes better, health indicators than the majority non-Hispanic white population, despite the fact that Latino populations have below-average socioeconomic status. One recent CDC assessment of the Latino Paradox notes that Hispanic populations in the US are twice as likely to live below the poverty line and four times less likely to have completed high school than white populations—both risk factors for poor health. Hispanics are also far less likely to have health insurance and be slightly more likely to delay or fail to seek medical attention because of cost (CDC 2015). Despite these risk factors for poor health, US Latino populations had a 24 percent lower all-cause death rate than white populations in 2013, as well as lower death rates in a variety of specific causes of death,

including cancer and heart disease. Death rates were higher among Latinos for only a small number of causes of death, including diabetes, chronic liver disease, and homicide (*ibid.*).

This health advantage was first noted in the 1980s and labelled the "Hispanic paradox" (Markides and Coreil 1986). Initially, scholars explored the possibility that Latino migrants arriving in the US were healthier than average (the "healthy immigrant" hypothesis) or that some migrants were returning to their home countries in old age and therefore not being recorded in US mortality statistics (so-called salmon bias), but these factors are insufficient to explain lower mortality rates in US Latino populations (Turra and Elo 2008; Abraído-Lanza et al. 1999). Hispanic populations are also younger than non-Hispanic white groups in the US, by 15 years on average (CDC 2015), but age-adjusted rates allow us to control for this.

So, what is it about Latino communities that can explain these positive health indicators? One factor that has consistently proved to be significant is smoking status. Lower smoking rates in Latino populations are estimated to account for at least 50 percent of their longer life expectancy (Blue and Fenelon 2011). This leaves space for other factors, however. One common explanation suggests that strong family and community networks in Latino populations provide a protective effect on health by providing support for vulnerable community members, as well as a shielding effect against some of the more negative aspects of American culture (Eschbach et al. 2004). Diet could be another contributory factor. Latino immigrants tend to eat more fruits, vegetables, and vitamins, as well as less fat, than their white American counterparts, although this benefit erodes over time as immigrants adopt more Westernized diets (Montez and Eschbach 2008).

Indeed, **acculturation** to American ways of life—through changes to diet and other cultural practices—leads to a weakening of the Latino Paradox over time. US-born Latino populations have considerably worse health indicators than their immigrant forebears in factors such as obesity, hypertension, heart disease, and cancer (CDC 2015). Residence in the US appears to have negative health effects through worsening diet; raised likelihood of substance abuse, alcohol use, and smoking; and increased feelings of low self-esteem and declining aspirations among youth (Lara et al. 2004). Mental health shows a complex relationship with immigration and minority status. For instance, one meta-study shows that adults who migrate may experience improvements in depressive symptoms and anxiety over time as they settle in after the stress of immigration; by contrast, Latinos who arrive as children or adolescents often experience poorer mental health over time, as well as escalating problems with conduct and aggression (Teruya and Bazargan-Hejazi 2013).

Although some evidence supports the legitimacy of a "Latino Paradox," there are many challenges in assessing the phenomenon. In particular, defining a unified "Latino" population can be challenging, particularly given the diversity of places and cultures represented among the Latino population, and yet many data sets are not stratified by country of origin (The Lancet 2015; Lara et al. 2004). Grouping Latino populations together may therefore hide important details. For instance, Puerto Ricans lack the positive health indicators typical of the paradox, possibly because smoking occurs at far higher rates in the US Puerto Rican population than among most other Hispanic immigrant populations (CDC 2015). Some scholars have even argued that the Latino Paradox may not exist at all because of misclassification errors and other data problems (Smith and Bradshaw 2006).

Furthermore, many aspects of the Latino Paradox are reflected in a more general **healthy immigrant phenomenon**, which recognizes that many immigrant groups—not just Latinos—are healthier than native-born populations when they first arrive in the US and other affluent countries (Teruya and Bazargan-Hejazi 2013). The significance of a unique "Latino culture" on health is therefore questionable. The sheer complexity of Latinos' experiences, as well as the distinctions that must be made between recent Latino immigrants and more established Latino populations, call into question whether Latino culture is most significant in influencing health status (*ibid.*). Perhaps instead, the Latino Paradox is simply the best-studied example of a broader healthy immigrant phenomenon.

We can nonetheless draw several lessons from this case study. First, some recent immigrant groups—many Latino populations among them—appear to bring some healthy cultural practices to the US, particularly related to diet and smoking. Second, strong, cohesive family and community ties may provide a protective barrier against ill health. Over time, however, many of these benefits erode. Communities begin to acculturate to practices of the new host country, including different diet and smoking behaviors, often with negative health implications. In addition, mental health is often negatively affected by migration—by the stress associated with the migration itself, and then by the reality of economic and cultural marginalization, racism, and limited economic mobility that faces many new immigrants. In many ways, therefore, the Latino Paradox and healthy immigrant phenomenon can be read as a sad reflection of the challenges of Western lifestyles as much as a positive view of minority groups' cultural patterns on health.

CONCLUSION

Culture and identity are powerful influences on our everyday lives—this influence extends to the realm of health. Our beliefs about ourselves and others, and about our relative position in society, are integral to our health, how we think about improving it, and how we perceive those who are not in good health. The social construction of these ideas is value laden, with certain sectors of society wielding far greater power to define themselves and those around them. In the next chapter, we turn to explore the importance of power, considering how power becomes entrenched in institutions and social mechanisms and how these power dynamics have significant implications for health.

DISCUSSION QUESTIONS

1   How is health defined differently by different cultural or social groups? How would you define "health"?

2   Treichler (1999, 11) argues that "the AIDS epidemic . . . is simultaneously an epidemic of a transmissible lethal disease and an epidemic of meanings or signification." In what ways have these signifiers led to discrimination and stigma?

What other diseases show strong evidence of the influence of socially constructed meanings?

3 Cultural practices are often viewed as distinct from other human actions as somehow inalienable. How can we define a cultural practice? Should groups be allowed to continue cultural practices even if they are detrimental to health?

4 Are there specific ways that healthcare services or public health initiatives could be more sensitive to the identities and cultural contexts of minority groups? Can you think of examples of policies or programs that are particularly sensitive or insensitive to issues of culture or identity?

5 Is the distinction between "disability" and "impairment" a significant one? Why or why not? How might characteristics of an individual's environment be significant in turning an impairment into a disability?

SUGGESTED READING

Assari, S. 2017. "Why Is It so Hard to Close the Racial Health Gap in the US?" *The Conversation.* https://theconversation.com/why-is-it-so-hard-to-close-the-racial-health-gap-in-the-us-69012.

Crooks, V. A. 2007. "Exploring the Altered Daily Geographies and Lifeworlds of Women Living with Fibromyalgia Syndrome: A Mixed-Method Approach." *Social Science & Medicine* 64: 577–88.

Forster, K. 2017. "Revealed: How Dangerous Fake Health News Conquered Facebook." *The Independent.* www.independent.co.uk/life-style/health-and-families/health-news/fake-news-health-facebook-cruel-damaging-social-media-mike-adams-natural-health-ranger-conspiracy-a7498201.html.

Izquierdo, C. 2005. "When 'Health' Is Not Enough: Societal, Individual and Biomedical Assessments of Well-Being Among the Matsigenka of the Peruvian Amazon." *Social Science & Medicine* 61: 767–83.

Netherland, J., and H. Hansen 2016. "White Opioids: Pharmaceutical Race and the War on Drugs That Wasn't." *Biosocieties* 12 (2): 217–38.

VIDEO RESOURCES

"Unnatural Causes: Is Inequality Making Us Sick." 2008. California Newsreel. Unnaturalcauses.org.

REFERENCES

Abraído-Lanza, A. F., B. P. Dohrenwend, D. S. Ng-Mak, and J. B. Turner. 1999. "The Latino Mortality Paradox: A Test of the 'Salmon Bias' and Healthy Migrant Hypotheses." *American Journal of Public Health* 89 (10): 1543–48. doi:10.2105/ajph.89.10.1543.

Ahlberg, Beth Maina, Eila Jylkäs, and Ingela Krantz. 2001. "Gendered Construction of Sexual Risks: Implications for Safer Sex Among Young People in Kenya and Sweden." *Reproductive Health Matters* 9 (17): 26–36. doi:10.1016/S0968-8080(01)90005-9.

Almond, D., C. P. Chee, M. M. Sviatschi, and N. Zhong. 2015. "Auspicious Birth Dates Among Chinese in California." *Economics & Human Biology* 18: 153–59. doi:10.1016/j.ehb.2015.05.005.

Arnold, David. 2010. "British India and the 'Beriberi Problem', 1798–1942." *Medical History* 54 (3): 295–314. doi:10.1017/s0025727300004622.

Asiki, G., J. Mpendo, A. Abaasa, C. Agaba, et al. 2011. "HIV and Syphilis Prevalence and Associated Risk Factors Among Fishing Communities of Lake Victoria, Uganda." *Sexually Transmitted Infections* 87 (6): 511–15. doi:10.1136/sti.2010.046805.

Baldwin, P. 2005. *Disease and Democracy: The Industrialized World Faces AIDS.* University of California Press.

Barnett, T., and A. Whiteside. 2002. *AIDS in the Twenty-First Century: Disease and Globalization.* Basingstoke: Palgrave Macmillan.

BBC. 2019. "Kenya Census to Include Male, Female and Intersex Citizens." Accessed August 24, 2019. www.bbc.com/news/world-africa-49127555.

BBC News. 2008. "Birth Defects Warning Sparks Row." Accessed July 6, 2019. http://news.bbc.co.uk/2/hi/uk_news/7237663.stm.

Béné, Christophe, and Sonja Merten. 2008. "Women and Fish-for-Sex: Transactional Sex, HIV/AIDS and Gender in African Fisheries." *World Development* 36 (5): 875–99. doi:10.1016/j.worlddev.2007.05.010.

Bhan, G., N. Bhandari, S. Taneja, S. Mazumder, and R. Bahl. 2005. "The Effect of Maternal Education on Gender Bias in Care-Seeking for Common Childhood Illnesses." *Social Science & Medicine* 60 (4): 715–24. doi:10.1016/j.socscimed.2004.06.011.

Black, M., S. Bhattacharya, S. Philip, J. E. Norman, and D. J. McLernon. 2015. "Planned Cesarean Delivery at Term and Adverse Outcomes in Childhood Health." *JAMA* 314 (21): 2271–79. doi:10.1001/jama.2015.16176.

Blue, L., and A. Fenelon. 2011. "Explaining Low Mortality Among US Immigrants Relative to Native-Born Americans: The Role of Smoking." *International Journal of Epidemiology* 40 (3): 786–93. doi:10.1093/ije/dyr011.

Bokhari, F., R. Mayes, and R. M. Scheffler. 2005. "An Analysis of the Significant Variation in Psychostimulant Use Across the U.S." *Pharmacoepidemiology Drug Safety* 14 (4): 267–75. doi:10.1002/pds.980.

Bollini, P., S. Pampallona, P. Wanner, and B. Kupelnick. 2009. "Pregnancy Outcome of Migrant Women and Integration Policy: A Systematic Review of the International Literature." *Social Science & Medicine* 68 (3): 452–61. doi:10.1016/j.socscimed.2008.10.018.

Bradby, H. 1997. "Glaswegian Punjabi Women's Thinking About Everyday Food." In *Food, Health, and Identity*, edited by P. Caplan. Routledge.

Breggin, P. 2007. *Talking Back to Ritalin: What Doctors Aren't Telling You About Stimulants and ADHD*. Da Capo Press, Incorporated.

Brink, P. 1995. "Fertility and Fat: The Annang Fattening Room." In *Social Aspects of Obesity*, edited by I. de Garine and N. J. Pollock. Gordon and Breach Publishers.

Brody, J. 2017. "The Surprising Effects of Loneliness on Health." *New York Times*, December 11. www.nytimes.com/2017/12/11/well/mind/how-loneliness-affects-our-health.html.

Butler, R., and H. Parr. 1999. *Mind and Body Spaces: Geographies of Illness, Impairment and Disability*. Routledge.

Camlin, C. S., Z. A. Kwena, and S. L. Dworkin. 2013. "Jaboya vs. Jakambi: Status, Negotiation, and HIV Risks Among Female Migrants in the 'Sex for Fish' Economy in Nyanza Province, Kenya." *AIDS Education and Prevention* 25 (3): 216–31. doi:10.1521/aeap.2013.25.3.216.

[CDC] Centers for Disease Control and Prevention. 2015. "Vital Signs: Leading Causes of Death, Prevalence of Diseases and Risk Factors, and Use of Health Services Among Hispanics in the United States—2009–2013." *Morbidity and Mortality Weekly Report* 64 (17): 10.

[CDC] Centers for Disease Control and Prevention. 2017. "Table 11. Infant Mortality Rates, by Race: United States, Selected Years 1950–2016." Accessed July 5, 2019. www.cdc.gov/nchs/data/hus/2017/011.pdf.

Chouinard, V., E. Hall, and R. Wilton. 2010. *Towards Enabling Geographies: 'Disabled' Bodies and Minds in Society and Space*. Ashgate Gower.

Collins, James W., Jr., Richard J. David, Arden Handler, Stephen Wall, and Steven Andes. 2004. "Very Low Birthweight in African American Infants: The Role of Maternal Exposure to Interpersonal Racial Discrimination." *American Journal of Public Health* 94 (12): 2132–38. doi:10.2105/ajph.94.12.2132.

Connell, R. W. 2000. *The Men and the Boys*. University of California Press.

Cooper, R. 1984. "A Note on the Biologic Concept of Race and Its Application in Epidemiologic Research." *American Heart Journal* 108 (3 Pt 2): 715–22. doi:10.1016/0002-8703(84)90662-8.

Crooks, V. A., and G. J. Andrews. 2009. *Primary Health Care: People, Practice, Place*. Ashgate.

Cunningham, T. J., J. B. Croft, Y. Liu, H. Lu, et al. 2017. "Vital Signs: Racial Disparities in Age-Specific Mortality Among Blacks or African Americans—United States, 1999–2015." *Morbidity and Mortality Weekly Report* 66 (17): 444–56.

D'Alton, Mary E., and Mark P. Hehir. 2015. "Cesarean Delivery Rates: Revisiting a 3-Decades-Old Dogma.

Editorial." *JAMA* 314 (21): 2238–40. doi:10.1001/jama.2015.15948.

David, R. J., and J. W. Collins. 1997. "Differing Birth Weight Among Infants of US-Born Blacks, African-Born Blacks, and US-Born Whites." *New England Journal of Medicine* 337 (17): 1209–14. doi:10.1056/nejm199710233371706.

David, R. J., and J. W. Collins. 2007. "Disparities in Infant Mortality: What's Genetics Got to Do with It?" *American Journal of Public Health* 97 (7): 1191–97. doi:10.2105/ajph.2005.068387.

Day, Katy, and Tammy Keys. 2008. "Starving in Cyberspace: A Discourse Analysis of Pro-Eating-Disorder Websites." *Journal of Gender Studies* 17 (1): 1–15. doi: 10.1080/09589230701838321.

Dear, M., R. Wilton, S. Lord Gaber, and L. Takahashi. 1997. "Seeing People Differently: The Sociospatial Construction of Disability." *Environment and Planning D: Society and Space* 15 (4): 26.

del Casino, V. 2001. "Enabling Geographies? Non-Governmental Organizations and the Empowerment of People Living with HIV and AIDS." *Disability Studies Quarterly* 21 (4).

Dyck, I. 1998. "Women with Disabilities and Everyday Geographies." In *Putting Health into Place: Landscape, Identity, and Well-Being*, edited by R. A. Kearns and W. M. Gesler. Syracuse University Press.

Dyck, I. 2003. "Feminism and Health Geography: Twin Tracks or Divergent Agendas?" *Gender, Place & Culture* 10 (4): 361–68. doi:10.1080/0966369032000153331.

Edge, D., and A. Rogers. 2005. "Dealing with It: Black Caribbean Women's Response to Adversity and Psychological Distress Associated with Pregnancy, Childbirth, and Early Motherhood." *Social Science & Medicine* 61 (1): 15–25. doi:10.1016/j.socscimed.2004.11.047.

Edwards, J., and H. van Roekel. 2009. "Gender, Sexuality and Embodiment: Access to and Experience of Healthcare by Same-Sex Attracted Women in Australia." *Current Sociology* 57 (2): 18.

Eschbach, K., G. V. Ostir, K. V. Patel, K. S. Markides, and J. S. Goodwin. 2004. "Neighborhood Context and Mortality Among Older Mexican Americans: Is There a Barrio Advantage?" *American Journal of Public Health* 94 (10): 1807–12. doi:10.2105/ajph.94.10.1807.

Fadiman, A. 1998. *The Spirit Catches You and You Fall Down: A Hmong Child, Her American Doctors, and the Collision of Two Cultures*. Farrar, Straus and Giroux.

Farmer, P. 1992. *AIDS and Accusation: Haiti and the Geography of Blame*. University of California Press.

Farmer, P. 2001. *Infections and Inequalities: The Modern Plagues*. University of California Press.

Fiorella, Kathryn J., Carol S. Camlin, Charles R. Salmen, Ruth Omondi, et al. 2015. "Transactional Fish-for-Sex Relationships Amid Declining Fish Access in Kenya." *World Development* 74: 323–32. doi:10.1016/j.worlddev.2015.05.015.

Geronimus, A. T. 2001. "Understanding and Eliminating Racial Inequalities in Women's Health in the United States: The Role of the Weathering Conceptual Framework." *Journal of the American Medical Women's Association* 56 (4): 133–36, 149–50.

Global Burden of Disease Collaborative Network. 2017. "Global Data Health Exchange: Institute for Metrics and Evaluation Data." Accessed August 24, 2019. http://ghdx.healthdata.org/ihme_data.

Gesler, W. M., and R. A. Kearns. 2002. *Culture/Place/Health*. Routledge.

Gruenbaum, E. 2001. *The Female Circumcision Controversy: An Anthropological Perspective*. University of Pennsylvania Press, Incorporated.

Hall, Edward. 2000. " 'Blood, Brain and Bones': Taking the Body Seriously in the Geography of Health and Impairment." *Area* 32 (1): 21–29.

Hansen, N., and C. Philo. 2007. "The Normality of Doing Things Differently: Bodies, Spaces and Disability Geography." *Tijdschrift voor Economische en Sociale Geografie* 98 (4): 493–506. doi:10.1111/j.1467-9663.2007.00417.x.

Helman, C. G. 2007. *Culture, Health and Illness*. 5th ed. Taylor & Francis.

Holt-Lunstad, Julianne, Timothy B. Smith, and J. Bradley Layton. 2010. "Social Relationships and Mortality Risk: A Meta-Analytic Review." *PLoS Medicine* 7 (7): e1000316. doi:10.1371/journal.pmed.1000316.

Iyasu, S., J. E. Becerra, D. L. Rowley, and C. J. Hougue. 1992. "Impact of Very Low Birthweight on the Black-White Infant Mortality Gap." *American Journal of Preventive Medicine* 8 (5): 271–77.

Izquierdo, C. 2005. "When 'Health' Is Not Enough: Societal, Individual and Biomedical Assessments of Well-Being Among the Matsigenka of the Peruvian Amazon." *Social Science & Medicine* 61 (4): 767–83. doi:10.1016/j.socscimed.2004.08.045.

Jones, Camara. 2002. "Confronting Institutionalized Racism." *Phylon* 50 (1/2): 7–22. doi:10.2307/4149999.

Kagawa Singer, M., W. Dressler, and S. George. 2016. "Culture: The Missing Link in Health Research." *Social Science & Medicine* 170: 237–46. doi:10.1016/j.socscimed.2016.07.015.

Keeler, Rebecca L. 2007. "Analysis of Logic: Categories of People in U.S. HIV/AIDS Policy." *Administration & Society* 39 (5): 612–30. doi:10.1177/0095399707303637.

Keirns, Carla. 2015. "Watching the Clock: A Mother's Hope for a Natural Birth in a Cesarean Culture." *Health Affairs* 34 (1): 178–82. doi:10.1377/hlthaff.2014.0563.

Khullar, D. 2016. "How Social Isolation Is Killing Us." *New York Times.* Accessed August 24. https://www.nytimes.com/2016/12/22/upshot/how-social-isolation-is-killing-us.html

Kornelsen, J., A. Kotaska, P. Waterfall, L. Willie, and D. Wilson. 2010. "The Geography of Belonging: The Experience of Birthing at Home for First Nations Women." *Health and Place* 16 (4): 638–45. doi:10.1016/j.healthplace.2010.02.001.

Lara, Marielena, Cristina Gamboa, M. Iya Kahramanian, Leo S. Morales, and David E. Hayes Bautista. 2004. "Acculturation and Latino Health in the United States: A Review of the Literature and its Sociopolitical Context." *Annual Review of Public Health* 26 (1): 367–97. doi:10.1146/annurev.publhealth.26.021304.144615.

Laws, G., and J. Radford. 1998. "Place, Identity, and Disability." In *Putting Health into Place: Landscape, Identity, and Well-Being*, edited by R. A. Kearns and W. M. Gesler. Syracuse University Press.

Levinson, D., and L. Gaccione. 1997. *Health and Illness: A Cross-Cultural Encyclopedia.* ABC-CLIO.

Li, H. T., S. Luo, L. Trasande, S. Hellerstein, C. Kang, et al. 2017. "Geographic Variations and Temporal Trends in Cesarean Delivery Rates in China, 2008–2014." *JAMA* 317 (1): 69–76. doi:10.1001/jama.2016.18663.

Lloyd, K. 1998. "Ethnicity, Social Inequality, and Mental Illness—In a Community Setting the Picture Is Complex." *British Medical Journal* 316 (7147): 1763. doi:10.1136/bmj.316.7147.1763.

Longhurst, R. 2010. "The Disabling Affects of Fat: The Emotional and Material Geographies of Some Women Who Live in Hamilton, New Zealand." In *Towards Enabling Geographies: 'Disabled' Bodies and Minds in Society and Space*, edited by V. Chouinard, E. Hall, and R. Wilton. Ashgate.

Lu, M. C., and N. Halfon. 2003. "Racial and Ethnic Disparities in Birth Outcomes: A Life-Course Perspective." *Maternal and Child Health Journal* 7 (1): 13–30.

Lupton, D. 2003. *Medicine as Culture: Illness, Disease and the Body in Western Societies.* Sage Publications.

Markides, K. S., and J. Coreil. 1986. "The Health of Hispanics in the Southwestern United States: An Epidemiologic Paradox." *Public Health Reports* 101 (3): 253–65.

Marks, J. M. 2001. *Human Biodiversity: Genes, Race, and History.* Transaction Publishers.

Mathews, T. J., and M. F. MacDorman. 2008. "Infant Mortality Statistics from the 2005 Period Linked Birth/Infant Death Data Set." *National Vital Statistics Reports* 57 (2): 1–32.

Milligan, Christine. 2006. "Caring for Older People in the 21st Century: 'Notes from a Small Island'." *Health & Place* 12 (3): 320–31. doi:10.1016/j.healthplace.2004.12.002.

Minster, R. L., N. L. Hawley, C. T. Su, G. Sun, et al. 2016. "A Thrifty Variant in CREBRF Strongly Influences Body Mass Index in Samoans." *Nature Genetics* 48 (9): 1049–54. doi:10.1038/ng.3620.

Mitchell, D. 2000. *Cultural Geography: A Critical Introduction*. Wiley.

Mojola, Sanyu A. 2011. "Fishing in Dangerous Waters: Ecology, Gender and Economy in HIV Risk." *Social Science & Medicine* 72 (2): 149–56. doi:10.1016/j.socscimed.2010.11.006.

Molina, G., T. G. Weiser, S. R. Lipsitz, M. M. Esquivel, et al. 2015. "Relationship Between Cesarean Delivery Rate and Maternal and Neonatal Mortality." *JAMA* 314 (21): 2263–70. doi:10.1001/jama.2015.15553.

Montez, J. K., and K. Eschbach. 2008. "Country of Birth and Language Are Uniquely Associated with Intakes of Fat, Fiber, and Fruits and Vegetables Among Mexican-American Women in the United States." *Journal of the American Dietetic Association* 108 (3): 473–80. doi:10.1016/j.jada.2007.12.008.

Moon, G. 2000. "Risk and Protection: The Discourse of Confinement in Contemporary Mental Health Policy." *Health and Place* 6 (3): 239–50.

Mull, J. Dennis, and Dorothy S. Mull. 1988. "Mothers' Concepts of Childhood Diarrhea in Rural Pakistan: What ORT Program Planners Should Know." *Social Science & Medicine* 27 (1): 53–67. doi:10.1016/0277-9536(88)90163-3.

Mustillo, S., N. Krieger, E. P. Gunderson, S. Sidney, et al. 2004. "Self-Reported Experiences of Racial Discrimination and Black-White Differences in Preterm and Low-Birthweight Deliveries: The CARDIA Study." *American Journal of Public Health* 94 (12): 2125–31. doi:10.2105/ajph.94.12.2125.

Nasser, Mervat. 2009. "Eating Disorders Across Cultures." *Psychiatry* 8 (9): 347–50. doi:10.1016/j.mppsy.2009.06.009.

Nazroo, J. Y. 1997. *Ethnicity and Mental Health: Findings from a National Community Survey*. Policy Studies Institute.

Netherland, J., and H. Hansen. 2017. "White Opioids: Pharmaceutical Race and the War on Drugs That Wasn't." *Biosocieties* 12 (2): 217–38. doi:10.1057/biosoc.2015.46.

Neumayer, E., and T. Plümper. 2007. "The Gendered Nature of Natural Disasters: The Impact of Catastrophic Events on the Gender Gap in Life Expectancy." *Annals of the Association of American Geographers* 97 (3): 56.

New Zealand Ministry of Health. 2013. "Life Expectancy." www.health.govt.nz/our-work/populations/maori-health/tatau-kahukura-maori-health-statistics/nga-mana-hauora-tutohu-health-status-indicators/life-expectancy.

Oliver, M. 1990. *The Politics of Disablement: A Sociological Approach*. Basingstoke: Palgrave Macmillan.

Opio, Alex, Michael Muyonga, and Noordin Mulumba. 2013. "HIV Infection in Fishing Communities of Lake Victoria Basin of Uganda—A Cross-Sectional Sero-Behavioral Survey." *PLoS One* 8 (8): e70770. doi:10.1371/journal.pone.0070770.

Paulino, Nancy Armenta, María Sandín Vázquez, and Francisco Bolúmar. 2019. "Indigenous Language and Inequitable Maternal Health Care, Guatemala, Mexico, Peru and the Plurinational State of Bolivia." *Bulletin of the World Health Organization* 97 (1): 59–67. doi:10.2471/BLT.18.216184.

Payne, S. 2006. *The Health of Men and Women*. Wiley.

Person, B., L. K. Bartholomew, M. Gyapong, D. G. Addiss, and B. van den Borne. 2009. "Health-Related Stigma Among Women with Lymphatic Filariasis from the Dominican Republic and Ghana." *Social Science & Medicine* 68 (1): 30–38. doi:10.1016/j.socscimed.2008.09.040.

Philo, C. 1987. "'Not at Our Seaside': Community Opposition to a Nineteenth Century Branch Asylum." *Area* 19 (4): 297–302.

Plevani, C., M. Incerti, D. Del Sorbo, A. Pintucci, et al. 2017. "Cesarean Delivery Rates and Obstetric Culture—An Italian Register-Based Study." *Acta Obstetricia et Gynecologica Scandinavica* 96 (3): 359–65. doi:10.1111/aogs.13063.

Pollock, N. J. 1995. "Social Fattening Patterns in the Pacific: The Positive Side of Obesity—A Nauru Case Study." In *Social Aspects of Obesity*, edited by I. de Garine and N. J. Pollock. Gordon and Breach Publishers.

[PRB] Population Reference Bureau. 2016. 2016 World Population Data Sheet.

Rafalovich, Adam. 2001. "The Conceptual History of Attention Deficit Hyperactivity Disorder: Idiocy, Imbecility, Encephalitis and the Child Deviant, 1877–1929." *Deviant Behavior* 22 (2): 93–115. doi:10.1080/016396201750065009.

Rhodes, T., M. Singer, P. Bourgois, S. R. Friedman, and S. A. Strathdee. 2005. "The Social Structural Production of HIV Risk Among Injecting Drug Users." *Social Science & Medicine* 61 (5): 1026–44. doi:10.1016/j.socscimed.2004.12.024.

Rigby, Janette E., and Danny Dorling. 2007. "Mortality in Relation to Sex in the Affluent World." *Journal of Epidemiology and Community Health* 61 (2): 159–64. doi:10.1136/jech.2006.047381.

Riska, E. 2003. "Gendering the Medicalization Thesis." In *Gender Perspectives on Health and Medicine: Key Themes*, edited by M. T. Segal, V. Demos and J. J. Kronenfeld. Emerald Group Publishing Limited.

Ritchie, H., and M. Roser. 2018. "Causes of Death." *OurWorldInData.org*. Accessed February 4, 2019. https://ourworldindata.org/causes-of-death.

Rosenberg, Noah A., Jonathan K. Pritchard, James L. Weber, Howard M. Cann, et al. 2002. "Genetic Structure of Human Populations." *Science* 298 (5602): 2381–85. doi:10.1126/science.1078311.

Sabatier, R., and J. Tinker. 1988. *Blaming Others: Prejudice, Race, and Worldwide AIDS*. Panos Institute.

Selik, Richard M., Susan Y. Chu, and James W. Buehler. 1993. "HIV Infection as Leading Cause of Death Among Young Adults in US Cities and States." *JAMA* 269 (23): 2991–94. doi:10.1001/jama.1993.03500230073032.

Serwadda, D., R. D. Mugerwa, N. K. Sewankambo, A. Lwegaba, et al. 1985. "Slim Disease: A New Disease in Uganda and Its Association with HTLV-III Infection." *The Lancet* 2 (8460): 849–52. doi:10.1016/s0140-6736(85)90122-9.

Shakespeare, T. 2006. *Disability Rights and Wrongs*. Taylor & Francis.

Sheridan, Eamonn, John Wright, Neil Small, Peter C. Corry, et al. 2013. "Risk Factors for Congenital Anomaly in a Multiethnic Birth Cohort: An Analysis of the Born in Bradford Study." *The Lancet* 382 (9901): 1350–59. doi:10.1016/S0140-6736(13)61132-0.

Skelton, T., and G. Valentine. 2010. "It's My Umbilical Cord to the World . . . the Internet: D/deaf and Hard of Hearing People's Information and Communication Practices." In *Towards Enabling Geographies: 'Disabled' Bodies and Minds in Society and Space*, edited by V. Chouinard, E. Hall, and R. Wilton. Ashgate Gower.

Smith, David P., and Benjamin S. Bradshaw. 2006. "Rethinking the Hispanic Paradox: Death Rates and Life Expectancy for US Non-Hispanic White and Hispanic Populations." *American Journal of Public Health* 96 (9): 1686–92. doi:10.2105/AJPH.2003.035378.

Song, S. 2004. "Too Push to Push?" Accessed August 23, 2019. http://content.time.com/time/magazine/article/0,9171,610086,00.html.

Teruya, Stacey A., and Shahrzad Bazargan-Hejazi. 2013. "The Immigrant and Hispanic Paradoxes: A Systematic Review of Their Predictions and Effects." *Hispanic Journal of Behavioral Sciences* 35 (4): 486–509. doi:10.1177/0739986313499004.

The Lancet. 2015. "The Hispanic Paradox." *The Lancet* 385 (9981): 1918. doi:10.1016/S0140-6736(15)60945-X.

Treichler, P. A. 1999. *How to Have Theory in an Epidemic: Cultural Chronicles of AIDS*. Duke University Press.

Turra, Cassio M., and Irma T. Elo. 2008. "The Impact of Salmon Bias on the Hispanic Mortality Advantage: New Evidence from Social Security Data." *Population Research and Policy Review* 27 (5): 515–30. doi:10.1007/s11113-008-9087-4.

UNAIDS. 2016. "HIV and Gender." Accessed July 14, 2019. www.uncares.org/content/hiv-and-gender.

UNICEF. 2013. *Female Genital Mutilation/Cutting: A Statistical Overview and Exploration of the Dynamics of Change*. United Nations Children's Fund (UNICEF).

United Nations for LGBT Equality. 2017. "Fact Sheet: Intersex." Accessed October 25, 2019. https://www.unfe.org/wp-content/uploads/2017/05/UNFE-Intersex.pdf.

United Nations Office on Drugs and Crime. 2019. "Global Study on Homicide 2019." Accessed August 22, 2019. www.unodc.org/unodc/en/data-and-analysis/global-study-on-homicide.html.

Valtorta, Nicole K., Mona Kanaan, Simon Gilbody, Sara Ronzi, and Barbara Hanratty. 2016. "Loneliness and Social Isolation as Risk Factors for Coronary Heart Disease and Stroke: Systematic Review and Meta-Analysis of Longitudinal Observational Studies." *Heart* 102 (13): 1009–16. doi:10.1136/heartjnl-2015-308790.

Weiser, S., K. Leiter, D. Bangsberg, L. Butler et al. 2007. "Food Insufficiency is Associated with High-risk Sexual Behavior among Women in Botswana and Swaziland. *PLOS Medicine* 4 (10): e260. https://doi.org/10.1371/journal.pmed.0040260.

White, M. J., O. Risse-Adams, P. Goddard, M. G. Contreras, et al. 2016. "Novel Genetic Risk Factors for Asthma in African American Children: Precision Medicine and the SAGE II Study." *Immunogenetics* 68 (6–7): 391–400. doi:10.1007/s00251-016-0914-1.

[WHO] World Health Organization. 2018. "Female Genital Mutilation." Accessed July 6, 2019. www.who.int/news-room/fact-sheets/detail/female-genital-mutilation.

[WHO] World Health Organization. 2019a. "Caesarean Sections Should Only Be Performed When Medically Necessary Says WHO." Accessed August 31, 2019. www.who.int/reproductivehealth/topics/maternal_perinatal/cs-statement/en/.

[WHO] World Health Organization. 2019b. "Suicide Rates (per 100 000 Population)." Accessed August 24, 2019. www.who.int/gho/mental_health/suicide_rates_male_female/en/.

Wiklund, I., A. M. Malata, N. F. Cheung, and F. Cadee. 2018. "Appropriate use of Caesarean Section Globally Requires a Different Approach." *The Lancet* 392 (10155): 1288–89. doi:10.1016/s0140-6736(18)32325-0.

Williams, D. R., Yu Yan, J. S. Jackson, and N. B. Anderson. 1997. "Racial Differences in Physical and Mental Health: Socio-economic Status, Stress and Discrimination." *Journal of Health Psychology* 2 (3): 335–51. doi:10.1177/135910539700200305.

World Bank. 2016. "Reviving Lake Victoria by Restoring Livelihoods." Accessed July 22, 2019. www.worldbank.org/en/news/feature/2016/02/29/reviving-lake-victoria-by-restoring-livelihoods.

World Bank. 2019. "World Bank Open Data." Accessed August 22, 2019. https://data.worldbank.org/.

Worth, Nancy. 2008. "The Significance of the Personal Within Disability Geography." *Area* 40 (3): 306–14.

Wright, Gloria Sunnie. 2012. "ADHD Perspectives: Medicalization and ADHD Connectivity." Australian Association for Research in Education.

Wright, J. D. 2015. *International Encyclopedia of the Social & Behavioral Sciences*. Elsevier Science.

Wyatt, G. 1995. "Transaction Sex and HIV Risks: A Woman's Choice?" In *HIV Infection in Women: Setting a New Agenda*. Washington, DC.

illness, and death when faced with hazards such as famines or earthquakes (e.g., Watts and Bohle 1993). Political economy approaches are now increasingly being used to frame discussions about access to healthcare and other health services (e.g., Klodawsky, Aubry, and Farrell 2006). All these studies rest upon the notion that underlying political and economic structures have a powerful and important influence on health.

Political ecology

Recall that **cultural ecology** argues that cultural practices influence how individuals and groups interact with their environments. If we combine this idea with a **political economy** approach, we find many examples of situations in which political and economic structures influence human interactions with their natural environment, giving rise to the field of **political ecology**. Developing from early work on issues such as land degradation (Blaikie and Brookfield 1987), political ecology considers the ways in which *political* and *economic* factors, such as capitalist economic structures, colonial histories, and social relations of production, influence *ecological* access, transformation, and management (see Robbins 2004). The critical approach of political ecology tends to view the relationship between humans and the natural environment as an exploitative and despoiling one, in which limited resources from the biosphere are channeled to the benefit of a privileged few (Baer 1996). Both the environment and marginalized social groups are the losers in the context of powerful societal structures, particularly the global economy.

A growing body of literature considers how ideas of political ecology can be applied to the study of health. In this framework, health and disease can be seen as the result of both *politics*—influencing who gets sick and who stays healthy, how healthcare is provided, and how knowledge of health is created and disseminated—and *ecology*, which sees health and disease as rooted in our biophysical environment (Jackson and Neely 2015). This approach acknowledges the value of studying human and biophysical domains as integrated systems, with careful place-based assessments critical to understanding the complexity of human-environment systems (King and Crews 2013). As with political economy

approaches, scholars of political ecology tend to frame their work through engagement with **critical geography** and social theory in exploring political and economic patterns (Jackson and Neely 2015). Geography, situated at a critical point of intersection between ecology and social theory, provides the theoretical expertise to make these sorts of connections, and there have been repeated calls for geographers to make more use of political ecology in health studies (Jackson and Neely 2015; King 2010; Mayer 1996).

An early focus of the political ecology literature was infectious disease, again particularly HIV/AIDS. For instance, Barnett and Blaikie (1992) considered the impacts of AIDS in Africa on land and environments, emphasizing how normal interactions between social and environmental systems can be disrupted in the context of the disease. Kalipeni and Oppong (1998) investigated the implications of Africa's refugee crisis for the emergence and reemergence of infectious diseases. Baylies (2002) considered how AIDS intersected with drought in Zambia, negatively influencing household production. Drimie (2003) traced the loss of assets and livelihoods associated with AIDS to loss of land rights, particularly for women and orphans, illustrating how environmental issues can intersect with social vulnerability.

Another major focus has been health-related **exposures**. Scholars have argued that some communities are exposed to **geogens** at higher rates than others, with health implications. For example, Hanchette (2008) explored high rates of lead poisoning in parts of the US state of North Carolina and concluded that the dislocation of African American populations associated with segregation and desegregation increased their risk of lead exposure because many African Americans moved into older housing vacated by suburbanizing white communities. Sultana (2006) used a gendered political ecology framework to examine the social mechanisms through which women have been disadvantaged by Bangladesh's arsenic crisis. She points out that the contamination of water with arsenic has led to a greater burden of work for women, who must often procure water from distant wells to avoid contamination, as well as bear the burden of caring for sick family members. Women may even be divorced or perceived as unmarriageable if they develop visible symptoms of arsenic poisoning. She goes on to argue that a complex intersection of geology, development planning, and social structures have produced "contaminated citizens," who are contaminated both literally (by the arsenic itself) and metaphorically (owing to social stigma associated with arsenic-related disease) (Sultana 2011). More recently, Nyantakyi-Frimpong, Arku, and Inkoom (2016) argue that the high intensity of agrochemical usage among urban farmers in Ashaiman, Ghana, is shaped by factors related to access to and control over land and water. For a farmer with only a short lease on a plot of land, heavy application of fertilizer is a rational strategy to ensure a return on investment by getting crops to harvest as quickly as possible, even though many farmers recognize the potential health problems associated with agrochemicals and many have reported symptoms consistent with pesticide poisoning. A gendered division of labor leave women doing many of the tasks that lead to direct contact with pesticides, such as weeding and harvesting, revealing a further layer of gendered power dynamics.

As these varied examples illustrate, political ecology differs from traditional disease ecology by emphasizing the political nature of environmental decisions and interactions. In this regard, it can also provide a channel for previously unheard voices and interrogate conflicting discourses related to the environment (King 2010). For instance, in a study of respiratory health in Houston, Texas, Harper (2004) concludes that local understandings of respiratory health differ significantly from formalized public health knowledge. This has political significance as public health messages from college-educated elites are targeting disproportionately low-income populations of color. As Harper (2004, 203) comments, "given a history of social tensions between the classes and races, the message to clean the house, rather than being viewed as authoritative and helpful, may well be politely acknowledged and then ignored." She concludes that the voices that she highlights through her qualitative techniques are "at the very heart of a political ecology of health approach to healthcare in that they illuminate the ways in which differing voices reflect changing cultural perceptions as well as resistance to dominant explanatory models" (Harper 2004, 317).

POLITICS AND HEALTH

The study of politics offers another key means for conceptualizing power structurally. Narrowly defined, *politics* refers to the ways that groups of people make decisions to formulate policy across a range of geographic scales (although in broader terms "politics" may refer to a host of behaviors, practices, norms, institutions, and systems of control that affect power relations). The most obvious way in which politics and health are intertwined is through government policies that directly affect health and healthcare. Although the United Nations recognizes access to medical care as a human right, noting that "everyone has the right to a standard of living adequate for the health and wellbeing of himself and of his family, including food, clothing, housing and medical care and necessary social services" (United Nations 1948), a lack of government investment in healthcare leaves many poor populations with little access. Rich countries also continue to struggle with healthcare provision, particularly as aging populations and the development of new therapies mean that healthcare costs have escalated rapidly in recent years.

Because funding is finite, political choices are always also economic choices, which often leads to controversy. While most people would like access to top-notch healthcare, there is strong disagreement about how and if society should organize and pay for healthcare. As a result, different societies have negotiated different ways to organize healthcare. In **socialized medicine** systems, the government pays for healthcare for the whole population through revenue generated from taxes, as occurs in the UK, forming a highly centralized system. At the other extreme are systems in which healthcare is primarily provided by the private sector, often with some government oversight, such as in the US.

Socialized medical systems are usually praised for offering the best equality of access to healthcare across a whole population, although the rationing of medical treatment, role of general practitioners as gatekeepers who control access to specialist medicine, and use of waiting lists may significantly limit choices and opportunities for patients. Significant inequalities are therefore reported even in socialized health systems. In the early 2000s, for instance, the breast cancer drug Herceptin hit the headlines in the UK after women in some regions of the UK were offered the drug for early-stage breast cancer, while women in other regions were reportedly being denied it, owing to regional policy variations among local health authorities (Box 9.1). Similar spatial inequalities have been noted with respect to organ donation in the US, where waiting times for organs often differ across regions (Smith 2017).

Although socialized healthcare systems are often criticized for "rationing care"—whereby certain treatments are deemed too costly or to provide too little benefit to be offered to patients—it is important to recognize that even private medical systems are forced to make similar value judgments with respect to which treatments will be offered. Insurance companies can increase the cost of premiums to cover escalating healthcare costs, but they must continually balance the cost of the care they cover with the cost of insurance premiums for the consumer in order to remain competitive in the healthcare market. This ultimately requires similar decisions over how to "ration" care.

Generally, greater inequalities exist in healthcare in market-based systems because those in greatest need of healthcare are often those least able to pay for it. As a result, health systems often provide the most benefit to affluent populations who need them least (Gwatkin, Bhuiya, and Victora 2004). By contrast, the poor are less likely to use health services and have been shown to benefit less from even basic health services (Gwatkin et al. 2007). In market-based systems, public health

---

Box 9.1

FUNDING HERCEPTIN IN THE UK

Within the UK's healthcare system, the National Health Service (NHS), policymakers are instrumental in deciding how funds are spent since demand for healthcare inevitably exceeds supply. Whether the improvement in outcomes generated by a particular drug or procedure justifies its cost is a key question in such circumstances. The UK formed the National Institute for Health and Care Excellence (NICE) to independently appraise new medical treatments and recommend whether, and in what circumstances, they should be used.

In 2005, Herceptin was available through the NHS for the treatment of advanced breast cancer. In April 2005, with an announcement from Roche Pharmaceuticals that some women with *early-stage* breast cancer would also benefit from the drug, patients with early-stage breast cancer began to actively lobby, and even threatened to sue their healthcare trusts, for access to the drug (Wilson et al. 2008). Although the drug was not licensed for use in early-stage breast cancer when the media debate began, and indeed there were some significant medical concerns associated with its use (Ferner and McDowell 2006), an emotive discussion soon revolved around the idea that women were being denied a potentially life-saving drug on the grounds of cost (Wilson et al. 2008). Much was made of the fact that women in some health authorities had been given access to the drug while those living in other parts of the country were being denied access (BBC 2002)—the so-called postcode lottery. Government policies were pitted against human-interest stories of women with breast cancer.

The situation was not straightforward, however. In the case of Herceptin, the cancer charity CancerBackup, which promoted access to the drug as part of its patient advocacy, was actually sponsored by the drug's manufacturer (Wilson et al. 2008). Indeed, patient advocacy groups frequently receive support from drug companies, providing far more compelling media stories to support the provision of a particular drug than any drug company press release ever could (Ferner and McDowell 2006). Teasing apart "objective science" from political and economic goals is challenging in such circumstances. NICE subsequently endorsed the use of Herceptin in most patients with early-stage breast cancer, in addition to its use for advanced breast cancer.

interventions may even initially exacerbate health inequalities because the rich typically have better access, as Victora et al. (2000) report in Brazil. This disparity only diminishes once a government makes deliberate efforts to extend the reach of public health programs to poor populations. Even in wealthy countries, affluent populations will have better access to healthcare if government does not make specific efforts to subsidize access to the poor, leading to huge inequalities where this is not a political priority. In the US, for instance, a patchwork of government and private healthcare systems have traditionally left large numbers uninsured and having to pay out-of-pocket for healthcare. We extend our discussion of healthcare in Chapter 10.

Many governments also play a role in promoting health through legislation such as regulation of water pollution or occupational safety. Other policies focus on controlling individual behaviors by, for instance, enforcing the use of seatbelts or motorcycle helmets, prohibiting drinking and driving, or attempting to influence diet or smoking behaviors. This kind of legislation can have significant health impacts, apparent in the significant decrease in road traffic deaths that typically occurs once governments begin to legislate for road safety best practice, as we previously explored in Box 7.1.

Some health-related legislation also has a spatial dimension because it can drive where particular behaviors are sanctioned. For instance, geographers have argued that whether alcohol consumption is considered to be a problem depends on *where* it occurs, with particularly salient public-private and rural-urban divides (Jayne, Valentine, and Holloway 2008). The ultimate example of place-based policies is probably anti-smoking legislation, however. In light of overwhelming evidence of the negative health impacts of cigarettes for smokers, many countries adopted legislation that defines where people are allowed to smoke, with power exercised "through the selective criminalization of smoking on a territorial basis" (Poland 1998, 209). These policies are premised on the idea that smoking is a polluting activity that has a negative impact on air quality, harming the health of others who share that space. There is some evidence that the health impacts of *passive* smoking may have been overstated by these sorts of campaigns (Kabat 2008), particularly in outdoor spaces where smoke is quickly dissipated (Figure 9.1). As such, the goal of smoking bans in outdoor spaces could be read as making smoking socially unacceptable, rather than because of genuine risks to the public from passive smoking. Anti-smoking campaigners would argue that this is justifiable if it leads more smokers to quit, but critical geographers have pointed out that the impacts of anti-smoking legislation are not experienced equally by all individuals. Instead, the impact is greatest in low socioeconomic classes, where smoking is more commonly practiced, potentially further stigmatizing an already marginalized group (Thompson, Pearce, and Barnett 2009; Poland 1998).

The degree to which governments are willing to attempt to influence individual behaviors varies widely. In general, it is mostly in the Global North that governments have had the "luxury" to try to regulate health-related behaviors, once broader public health problems such as sanitation have been addressed. There is also a political divide between welfare-focused governments that view efforts to influence individual behaviors as a logical extension of public health and **libertarian** political thought, which sees this as government overreach. For instance, in many European governments, as well as Australia, Canada, and New Zealand, which typically embrace a strong welfare role, heavy cigarette taxes and aggressive anti-smoking campaigns have been accepted with relatively little controversy. Twenty years ago, Canada became the first government to mandate graphic warnings of the dangers of smoking on cigarette packages, with a list of countries quickly following suit, including European Union members, India, and New Zealand (BBC 2019) (Figure 9.2). In 2012, Australia went a step further, mandating the use of "plain packaging," which also restricts the use of branding and promotional information. Australia today has some of the most stringent legislation for cigarette packaging, with all aspects of the packet engineered to make cigarettes as unappealing as possible, including the color (brown), the font, and the use of 75 percent of the front and 90 percent of the back of the package for graphic images of smoking-related diseases (BBC 2019). The US, by contrast, with its stronger tradition of support for individual and corporate rights, has so far stopped short of requiring graphic images. In 2011, the Food and Drug Administration (FDA)

Figure 9.1 Smoking restrictions near a children's playground, Rotorua, New Zealand

Smoking bans initially began in enclosed spaces such as bars and restaurants, where passive smoke inhalation posed a legitimate health risk for staff and other customers. Where smoking bans have been extended to outdoor spaces, their impact is arguably more significant in terms of reinforcing negative images around smoking than in protecting the health of non-smokers.

attempted to require graphics indicating the negative consequences of smoking, but intense lobbying by the cigarette industry brought the issue to the federal appeals court where it was quashed on the grounds of freedom of speech. The US still does not meet WHO best practice for cigarette packaging, although the FDA is making renewed efforts to extend the mandating of health warnings on packets (BBC 2019).

As these examples illustrate, the formal power structures of government can have significant impacts on

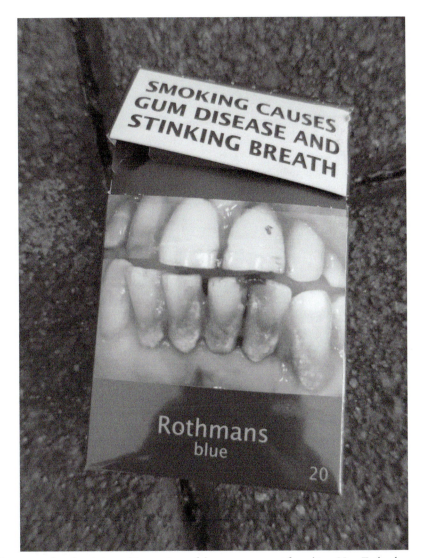

Figure 9.2 Cigarette packet showing graphic warnings of the consequences of smoking, New Zealand

This cigarette packet illustrates efforts to discourage smoking through use of graphic images, drab coloring, explicit health warnings, and a lack of explicit branding.

health. Not all power relations are explicit, however. With recent attention to social and cultural theory, geographers have focused more on implicit power relations that stem from broad social inequalities. We now turn to these critical approaches to power.

CRITICAL APPROACHES TO POWER

Critical geographers and other scholars have emphasized that power differentials characterize many (some would go so far as to argue *all*) interactions related to health.

For instance, a power differential exists between doctors and patients in both formal (i.e., codified in policy) and informal ways. Doctors have the legal power to prescribe medications and perform medical procedures to which patients otherwise do not have access. In a less formal manifestation of power, society often gives professional medical knowledge much greater value than lay or informal knowledge. By this line of reasoning, healthcare can disempower individuals by subjecting them to governmental and professional oversight and intervention. On the other hand, patients can also be empowered by healthcare, for instance, through actively pursuing biomedical solutions to health problems in order to tap into the power that they perceive to be associated with "modern" or "technological" therapies.

Critical geographers take power to be a fundamental facet of geographic study. Here we explore scholarly work that exemplifies three geographic aspects of health in a critical framework: segregation, power and the body, and surveillance.

Segregation and health

In the previous chapter, we considered ways in which geographic research has considered the socio-spatial boundaries that segregate those "othered" by ill health (Laws and Radford 1998; Dear et al. 1997). For instance, Takahashi (1998) describes how local opposition to facilities serving marginalized individuals—such as the mentally ill or persons living with HIV—has increased since the 1980s, associated with a resurgence of the rejection of difference. Although policymakers may overtly emphasize the "normalization" of such populations, structures of power in society-at-large maintain spatial and social boundaries, systematically marginalizing people identified by their difference. Other populations such as the homeless, itinerant populations, and refugees experience similar socio-spatial exclusions (Curtis 2004). These exclusions may themselves have health ramifications, leading to a vicious cycle in which exclusion compromises health, and ill health then leads to further exclusion. For instance, there is a well-documented relationship between homelessness and mental health problems (*ibid.*)—mental health problems are often a risk factor for homelessness,

while homelessness forms a significant barrier to getting psychiatric or psychological care.

Another group facing socio-spatial exclusions is people with infectious diseases, who have traditionally been stigmatized in public health discourse as sources of contagion and a threat to others in society (Lupton 2003). In this situation, segregation may be exercised in formal ways. **Quarantine** is a clear spatial manifestation of the idea that "dangerous" bodies should be removed from public spaces in the name of health. For example, isolation of people with leprosy remained a common practice in many countries well into the twentieth century. Many "leper colonies" were built in remote areas, often on islands, reinforcing the notion of isolation as a means of removing contagion from society. The small island of Chacachacare in Trinidad and Tobago operated as a "leper colony" from the 1920s to the 1980s; the Kalaupapa Peninsula on the island of Molokai, Hawaii—bounded by ocean and steep cliffs—was used to isolate people with leprosy between 1865 and 1969. Today, with the advent of an effective treatment, the spread of leprosy has been dramatically reduced. Nonetheless, the term "leper" remains in common parlance to refer to someone to be avoided, reflecting continuing psycho-social discrimination. Acknowledging the power of labeling, the name "Hansen's disease" is now sometimes used instead of "leprosy" to avoid these negative connotations.

Physical and social segregation of infectious individuals is not merely a historical phenomenon, however. More recently, penal and civil codes have been adopted to try to prevent behaviors that could spread HIV and to hold transgressors liable for their actions (Baldwin 2005). Although convictions are rare, in 2010 a singer was prosecuted in Germany for having unprotected sex with several partners without informing them that she was HIV-positive, and a British man was jailed in 2018 for deliberately infecting other men with HIV (BBC 2018).

Power and the body

Related to an increasing focus on individuals and their unique experiences has been a re-theorizing of the "body" within social studies disciplines. Until recently, scholarly work on the human body was almost entirely the domain of the natural sciences, with the body viewed as a biological entity. Increasingly, however, the body is being understood as socially constructed by particular forms of knowledge and discourses. "[B]odies, then, are not born; they are made" (Haraway 1999). One key way in which the body has been "made" is through biomedicine's monopoly on explaining how the body functions, excluding alternative forms of knowledge derived from complementary or traditional forms of medicine (Lupton 2003). To provide another example, a key focus of disability geographers has been how society constructs understandings of disabled bodies and the implications of these understandings in terms of how we expect disabled people to look and act.

The writings of Michel Foucault have been especially relevant to the study of the body and power. Foucault argued that the body is the ultimate point at which political states exercise ideological control, and that it is therefore a focal point of state-mandated discipline. State systems, such as education, law, and medicine, ultimately operate at the scale of the body, disciplining bodies that violate established boundaries and keeping bodies economically and politically productive. Medicine then serves to label these bodies as "deviant" or "normal," "unhygienic" or "hygienic" (Foucault 1979), influencing how we believe people should look and act.

Bodies, and the spaces they occupy, have thus been framed as sites of both oppression and resistance (see Imrie and Imrie 1996; Dorn and Laws 1994). As geographers, we can consider the ways in which these expectations can be place-based, influencing what we expect particular bodies to do in particular places. For instance, Mahon-Daly and Andrews (2002) used participant-observation techniques to examine why some women in the US and the UK do not breastfeed despite publicly available information about the medical benefits of the practice. Through their qualitative research, they found that many women continue to be embarrassed by the act of breastfeeding and so women seek "liminal spaces," places used as "transitional zones, which women move in to breastfeed and out of to reintegrate with society and 'normal' daily activities" (Mahon-Daly and Andrews 2002, 70). Women's perception that they must conform to social norms when breastfeeding is situated in a social context in which breasts are constructed as sexualized parts of the body, requiring women to carefully regulate where and when they feed their babies. Ultimately, this discourages a healthy behavior for women and their children.

Healthcare spaces can also exert power over bodies. Foucault (1973) argues that by moving bodies out of their everyday spaces and into special spaces like hospitals, people become subject to the *medical gaze* where they can be compared and treated scientifically. Such understandings of healthcare settings highlight the way in which bodies are increasingly monitored by modern medicine. This *surveillance* of bodies in particular places exemplifies an additional way in which space intersects with health and healthcare.

Public health and surveillance

Whereas public health has traditionally considered symptoms and conditions that already exist (e.g., through draining malarial swamps or building sanitation systems), today much public health is involved with future conditions that may or may not appear (Brown and Duncan 2002). Quarantine, vaccination, anti-smoking campaigns, even get-fit campaigns can all be read as encouraging activities that support population health by minimizing projected future disease risks, but at the cost of individual choice over day-to-day activities. For critics, this "surveillance medicine" targets not only the sick, but also the healthy, with the idea that individuals and populations "[hang] precariously between health and illness" (Armstrong 1995, 396). In such cases, power relations are often made invisible as these practices are voluntarily imposed, with individuals self-regulating their behavior to conform to societal expectations of what it is to be "healthy" (Grosz 1990). While supporters would argue that individuals are benefitting from both better health and the control of healthcare costs at a societal scale, those ascribing to more libertarian ideals argue that an individual's health decisions should be theirs and theirs alone. The anti-vaccination movement provides an excellent example of a community arguing for their right to make their own individual decisions in this way. In this respect, public health always treads a fine line between supporting the public good and compromising individual liberties.

The turn from clinical medicine towards public health also has spatial implications, bringing healthcare out of hospitals and into the community, as disease has been reinterpreted in terms of the "risk" posed by everyday activities (Armstrong 1995). For instance, with respect to heart disease, this approach advocates making changes to the health landscape and the behaviors of the people who live there. Could the construction of more cycle paths encourage people to get more exercise? Could creating smoke-free spaces encourage people to give up smoking, with positive implications for heart health? In the case of skin cancer, the behavior of asymptomatic individuals becomes a target via public health campaigns that encourage sun safety at school (Collins, Kearns, and Mitchell 2006) (Figure 9.3). While public health campaigns such as providing cycleways and improving sun safety for kids are, of course, positive in numerous ways, here we illuminate some of the veiled power differentials that critics have identified.

The major critique is that staying healthy becomes a moral obligation for responsible individuals to pursue (Collins, Kearns, and Mitchell 2006). Disease is no longer an unavoidable and unforeseen event but is instead viewed as a calculable risk that can be avoided if the "correct"

Figure 9.3 Sunscreen over a children's playground, New Zealand

Sun safety is taken seriously in Australia and New Zealand, where damage to the ozone layer has led to an increased risk of skin cancer. The government-based healthcare systems of Australia and New Zealand have a vested interest in preventive health campaigns to avoid society having to bear future healthcare costs.

actions are taken, such as maintaining a healthy diet and attending regular health screenings. Particularly amid rapidly escalating healthcare costs, such moral obligations may even be framed in terms of avoiding becoming a burden on society. In this framework, a person who smokes or becomes obese is no longer only influencing their own health but may be detrimental to the future financial health of society. This often adds another source of stigma to communities of low socioeconomic status where many of these behaviors are frequently more common.

Medical interventions have also begun to influence parts of people's lives that were not previously considered "unwell" (Armstrong 1995). Critics suggest that this has often undermined the value of non-professional and lay practices. Processes such as birth, reproduction, care-giving, and death, which have traditionally been overseen by lay people, have been subsumed by biomedical practice (Rosenfeld and Faircloth 2009). For example, in many societies, lay midwives have been largely replaced by professionally trained midwives and doctors. This is a positive change if it means that lay midwives receive more training, perhaps concurrently also raising their status in society. However, we often see an educated elite simply take over activities that were previously in the hands of lay individuals, thereby eroding their status and employment opportunities. Feminist scholars have suggested that these changes also have a significant gendered aspect, given traditional gendered divisions of labor (Box 9.2).

The importance of power in these philosophical starting points resides in the potential of health-related

---

Box 9.2

POWER AND PLACES OF BIRTH

In the US, people have historically given birth at home, attended by self-taught or apprentice-trained midwives. Over the course of the nineteenth and twentieth centuries, however, "traditional" midwifery declined in the face of a growing number of hospitals and an increasingly formalized medical profession. By the 1940s, the majority of women were birthing their babies in hospitals and by 1970 hospital birth was the norm, with 90 percent of women giving birth in hospitals (Rooks 1999).

Feminist readings of the demise of traditional forms of midwifery in the US have argued that these changes had a significant gendered dimension. As (Beckett 2005, 253) notes, moving birth into hospitals was related to the:

> political machinations of the emerging medical profession and the impact of the profession's propaganda on women's beliefs and preferences. According to these analysts, doctors used their growing political and cultural authority to redefine childbirth as a dangerous, pathological event, to denigrate and eliminate midwives, and to fuel the perception

that middle- and upper-class women were less able to withstand the challenges of childbirth.

By this reading, the male domination of new health spaces, notably hospitals, undermined the traditional home-based role of women in midwifery, disempowering women in the process.

The power dynamics are not completely straightforward, however. Beckett (2005) emphasizes that such a reading can actually further disempower women by implying that they lacked agency in this transition. She suggests that an alternative feminist reading should recognize the influence of women in also *claiming access* to hospital births—for instance, as women sought modern forms of pain relief. Furthermore, for other populations, a *lack of access* to hospital births was significant. With white, well-educated women at the vanguard of the move to hospital births, primarily low-income women of color continued to give birth at home by the mid-twentieth century. Power relations related to places of birth are, therefore, complex—not only gendered, but also including significant race- and class-based aspects.

continued

Box 9.2 *continued*

Although less than 1 percent of births in the US today occur in homes, commentators have noted a return to home birth among certain populations. Some home births occur at home unplanned or owing to a lack of alternative options for poor and very young women (Michie 1998), clearly a disempowering situation. There has nonetheless been an increase in *planned* home births, particularly among well-educated white women. According to Michie (1998), these "upper-middle-class college-educated women . . . plan a home birth because of a feminist or proto-feminist critique of medical ideology," using their empowered position to reclaim the home as a space for birth. In other words, these women are turning to home births as a way to regain control over the birthing experience through rejection of a space associated with loss of control: the hospital. Among this group, there is often a fear that the "natural" process of birth will be overwhelmed by unnecessary technological and professional intervention in the hospital, as well as a desire to seek out a space with a more comfortable, relaxed atmosphere (Hazen 2017).

Emerging from this image of the home as a space of comfort and empowerment, there has been a recent effort to make hospital spaces more "homelike" and appealing in the US, particularly as hospitals increasingly compete for medical "consumers." Fannin (2003) documents the rise of these "homelike" hospital settings, where the trappings of modern technology are veiled by aspects of home—attractive quilts, comfortable beds, rocking chairs. Despite these changes, Fannin argues that the traditional balance of power remains the same, with women still disempowered by the hospital experience in many cases. Instead, such changes represent mere window-dressing, with the control afforded to women over their birthing space usually only over superficial elements such as lighting and music. Indeed, "placing birth in the homelike space of the hospital birthing room situates the birthing woman within both domestic and institutional space—equally ambivalent sites with respect to women's political agency" (Fannin 2003, 518). The domestication of the hospital carries with it further ambiguous meaning in the context of class, as it interprets an ideal birth in the context of white, middle-class, heterosexual domestic values, as seen in the particular representation of "home" that is enacted via the hospital décor (Fannin 2003; Michie 1998). This idealization of home presupposes that home carries meanings of comfort and security for all, when the reality for many women, particularly women of lower socioeconomic classes, may be far from this ideal (Michie 1998).

policies and decisions to empower certain people while disempowering others. As is often the case, marginalized groups often end up disempowered in such circumstances. In a further examination of power, we now turn to the topic of fertility, which offers illustrations of the role of both explicit policies and implicit power imbalances in influencing health.

FERTILITY POLICY

Fertility policy sits at a critical intersection of power and health. Most governments are keenly interested in fertility because of its influence on population size, with ramifications related to economic productivity, resource consumption, and national defense. Fertility policy is, therefore, a common way in which governments intervene in individual health decisions through policies that promote or deter births, as well as proscriptions on appropriate and inappropriate birth-related technologies such as abortion and contraception. Increasingly, non-governmental organizations have also become involved in fertility policy, either through funding fertility campaigns or campaigning for specific fertility outcomes. These interventions have generated social concerns around who controls fertility policy and how this affects less powerful communities, including issues of race, nationality, and gender. For those invoking postmodern understandings of power and the "body," fertility policy can even be read as an attempt to control certain bodies, particularly women's.

Policies related to fertility are among the most highly contested worldwide, regardless of a population's affluence, culture, or religion (Knudsen and Hartmann 2006). Fertility is controversial for several reasons. First, fertility is subject to a wide range of cultural and religious beliefs, which many argue are inviolable. Second, fertility patterns vary substantially across geographic space and among population groups, and so efforts to control fertility affect certain populations more than others. This has significance in terms of race, class, and other issues of identity, if more powerful groups or individuals advocate policies to influence birth rates in poorer or minority communities. Third, issues of reproduction and fertility are clearly gendered, with women often bearing the brunt of the physical, emotional, and economic toll of their reproductive choices (or lack of choice). Finally, with its obvious connections to population size, fertility policy impinges on the distribution of resources, again leading to arguments about the true motives of many fertility campaigns. In this respect, fertility policy highlights tensions between individual and community goals. While having twelve children or no children may be your personal preference, your choice may not be consistent with what would most benefit the community at large. This is frequently a key point of disagreement in politics: how do we balance individual rights with the community good?

Until the latter half of the twentieth century, government attempts to manipulate fertility tended to focus on trying to increase population size (**pro-natalism**) in order to reinforce political or economic power. For instance, in Russia in the post-World Wars period, women were awarded medals for bearing many children, culminating in the "Order of Mother Heroine" for mothers of ten children. Special honors were also awarded to mothers of large families in France in the 1920s and Nazi Germany (Hoffmann 2000). In Romania, under Ceausescu's dictatorship from 1965 to 1989, contraception and abortion were banned in all but a few circumstances to try to increase birth rates, with disastrous consequences for maternal and infant health (Gold 2014). In East Germany, social policy in the 1960s and

'70s was explicitly designed to encourage multi-child families and the incorporation of women into the labor force. Policies included the provision of low-cost daycares, child benefits, paid maternity leave, and shorter working hours (Ochel and Osterkamp 2007).

By the 1960s, however, considerable effort began to be expended towards trying to reduce population growth rates, in response to a widely publicized crisis of population explosion (e.g., Ehrlich 1971). It is on these population policies that we focus.

Family planning campaigns

Communities have attempted to control fertility for thousands of years, but it was the advent of effective family planning methods such as the contraceptive pill in the twentieth century that brought fertility issues to the fore, opening debates about how contraceptive technologies should be used and who has the right to make individual and community fertility decisions. Although some people reject contraceptive technologies out of religious or ethical objections, here we focus on the geographic issue of inequalities among communities regarding the way that fertility policy is applied. In this context, the history of **family planning** has been interpreted as a battle between those who want to expand reproductive choice for women and couples and those who have used reproductive technologies as a way to control the fertility of other groups.

A considerable body of literature describes situations in which family planning has been applied in ways that restrain reproductive freedom rather than enhance it. For instance, in the early part of the twentieth century, many health and welfare officials in the US encouraged the use of contraception as a critical part of welfare programs in the belief that alcoholism, sexual promiscuity, and poverty were hereditary, and so births among "undesirable" populations should be discouraged. Many women, particularly poor women of color, were believed to be incapable of making rational reproductive choices or effectively using contraceptives, and so it was argued that government should make fertility decisions on their behalf. This involved efforts to promote sterilization or the insertion of intra-uterine devices (IUDs) as a condition for receiving welfare payments (Schoen 2006). In other cases, women—often from minority groups—were sterilized without giving fully informed consent to the procedure, often not understanding that the procedure was permanent, for instance (Gold 2014). At its worst, involuntary sterilizations have been sanctioned by law in some countries as part of **eugenic** policies, whereby reproduction in some populations is deemed not to be in the interests of the community. Although eugenics societies were common across much of Europe and the US in the early part of the twentieth century, Nazi Germany implemented the most notorious campaigns, targeting groups including Jews, people with mental illness, and criminals.

In Europe and the US, other women were concurrently arguing that access to new and effective reproductive technologies was being unnecessarily restricted. In the first half of the 1900s, many healthcare providers refused to distribute contraceptive technologies to unmarried women, and married women were often required to obtain their husband's consent. Consequently, women sometimes turned to illicit or humiliating means to gain access to contraceptive technologies. In the US, for instance, some women were reported to have applied for eugenic sterilization on the grounds that they were feeble-minded because they had no access to elective sterilization (Schoen 2006).

The concept of power provides a unifying framework to explain this contradictory history of family planning, with "sexual, class, and racial conflicts shap[ing] negotiations over reproductive control" (Schoen 2006, 5). In the US, for instance, certain groups—especially poor and minority groups—have frequently been assumed to be inadequate parents and unable to control their sexuality, requiring government intervention to control their fertility. As a result, great efforts were made by many researchers and public health officials—who tended to be white and male—to develop contraceptives that were out of the control of female patients (*ibid.*). For white middle-class women, by contrast, contraceptive technologies were often perceived as undesirable, even "dirty," leaving these women underserved. In this way, "assumptions about the links between sexuality, class, and race shaped public perceptions of women's sexual behavior, policy debates surrounding issues of sexuality and reproduction, the formulation of reproductive policies, and the delivery of services to patients" (Schoen 2006, 5).

Although we must put such concerns in historic context—most modern Western governments now offer access to family planning in ways that more actively promote reproductive choice—concerns continue to resurface about how family planning is implemented. Roberts (1999), for instance, describes how the advent of the contraceptive "Norplant" in the 1990s provided a new opportunity for the fertility of poor black women to be controlled by policymakers and public health officials in the US. Norplant is a contraceptive device that is implanted under the skin; it releases a synthetic hormone into the bloodstream, preventing pregnancy for up to five years. The technology was initially marketed as a breakthrough for women because, once inserted, the device effectively prevents pregnancy for several years. It is reversible, its effectiveness is rivaled only by sterilization, and it can be used without assistance from the woman's partner—points that are often cited as enhancing women's reproductive freedom. As an additional contraceptive option, it was indeed an innovative breakthrough. Roberts (1999) argues, however, that the way the device was perceived and used by policymakers in the US was highly problematic. Roberts' critique focuses on the power differential between the professionals and policymakers who have promoted Norplant and the poor women who have been its primary users.

An initial point of controversy was the way that Norplant was preferentially promoted to teens and women on welfare, arguably as a cheap way to prevent these groups from having children. Every US state quickly made Norplant available on Medicaid (the US's health program for people with low income), and several states set aside money to fund the contraceptive for poor women without Medicaid coverage. While this in itself is not necessarily problematic, other state legislation went further in encouraging the use of Norplant among targeted groups by proposing to offer financial bonuses for implantation of the device, or even requiring use of Norplant as a condition of receiving welfare benefits (Roberts 1999; Gold 2014). In other cases, women were offered reduced prison sentences if they agreed to have the implants (Gold 2014). The point at which incentives become coercive is a critical issue within this discussion.

Another key issue in the Norplant debate is the fact that this method of contraception requires considerable specialist intervention. Healthcare practitioners must not only insert the device, but also remove it. Women have reported a variety of ways in which this has led to coercion, ranging from healthcare providers attempting to persuade women not to have the device removed early (even in situations in which it was causing serious side effects), to women finding that their insurance or Medicaid would cover insertion of the device but not its early removal (Roberts 1999). After lawsuits related to side effects combined with falling demand, Norplant was removed from the UK market in 1999 and distribution discontinued in the US in 2002.

Today, so-called long-acting reversible contraceptives (LARCs), which include intra-uterine devices and hormonal implants, are being tentatively embraced once again on the grounds that they can offer advantages. Given that many unplanned pregnancies occur when contraceptives are used inconsistently or incorrectly, LARCs offer a potentially powerful tool as, once implanted, their effectiveness is not subject to human error (Gold 2014). They are therefore becoming increasingly popular once again as a method of birth control in high-income countries and have begun to increase in popularity in many parts of the low-income world. In Africa, injectable contraceptives are now the commonest form of birth control. Whether LARCs end up increasing reproductive freedom or are instead used as a weapon of reproductive control is largely a matter of policy.

International population programs

Power dynamics related to fertility policy at the national level are reflected also at the international scale. In reaction to rapid global population growth from the 1950s onwards, many Western commentators argued in the 1960s and '70s that population expansion should be controlled to avoid a resource crisis, environmental degradation, and even political chaos. In response to such concerns, family planning programs were widely implemented in low-income countries where most population growth was occurring.

Criticism of these policies has focused on the idea that many of them were implemented with the goal of meeting demographic targets rather than improving reproductive choice. Commentators often make a distinction between fertility programs that focus on improving access to birth control, thereby enhancing individual agency to control fertility, and policies of "population control" that have a political agenda—usually the reduction of a particular population's birth rate for political, economic, or ecological reasons. Just as affluent policymakers have been criticized for exerting power over poorer, minority populations in the West, so critics of global population control policies argue that elite, often Western-educated policymakers are influencing, or even determining, the reproductive choices of poor women of color in the low-income world. In this vein, Roberts (1999, 143) argues that "[population] incentive programs have tried to substitute mass sterilization for the equitable distribution of wealth in Third World countries, sacrificing the health and dignity of poor women of color in the process."

The significance of population control is critical in considering global population policies. Subtle differences in campaigns can dramatically affect whether they are empowering or coercive. For instance, many fertility campaigns have traditionally relied on sterilization as the primary method of contraception. Sterilization is a permanent and cost-effective way to prevent births, making it an ideal choice if the goal is to achieve the largest possible reduction in fertility at the population level. Female sterilization remains the commonest form of birth control, used by 19 percent of married or in-union women worldwide (Department of Economic and Social Affairs 2015, 2). From the perspective of individual choice, however, sterilization is less often an optimal method. In particular, the permanency of sterilization removes future reproductive options. There are also inherent surgical risks, particularly for women, for whom the procedure is more complicated than for men. The fact that many sterilization programs have nonetheless focused on women is often interpreted as a further reflection of power dynamics. In many traditional cultures, men are unwilling to undergo sterilization for fear of compromising their masculinity, and so rates of male sterilization remain very low, except in a few countries such as Canada and the UK. As a result, sterilization campaigns have often targeted women, who are typically more willing to undergo the procedure despite the risks. Family planning programs that focus heavily on sterilization have therefore often come under fire for being overly concerned with demographic targets rather than enhancing reproductive choice.

The meaning of particular reproductive technologies is highly case specific, however, leading some scholars to argue that there is no such thing as a "good" or "bad" method of contraception, only technologies that are applied appropriately or inappropriately. Assuming a humanistic perspective, we can consider the appropriateness and meaning of particular contraceptive technologies and policies for individuals. For instance, in a study of the meaning of sterilization for women in a poor community in Brazil, Dalsgaard (2004) concluded that individual women often purposefully selected surgical sterilization as a means to control their own fertility. However, a closer analysis of the life-worlds of these women showed that this choice of sterilization was often an effort to counter significant structural constraints, suggesting that efforts to seek sterilization were not necessarily as empowering as they might initially have appeared (Box 9.3).

Box 9.3

FEMALE STERILIZATION IN NORTHEAST BRAZIL

In a study of female sterilization in a low-income community of Recife, Brazil, Dalsgaard (2004) argues that high rates of sterilization are connected to women's desire to maintain their own autonomy in a "wider context of poverty, disrespect and constrained agency." Assuming a humanistic approach, Dalsgaard emphasizes the need to include individual experiences within studies of reproductive behavior, highlighting the fact that behavior is dependent on what is at stake for particular individuals in specific situations.

Among her study participants, Dalsgaard found that responsible motherhood was interpreted as having only as many children as could be provided for. Given their marginalized situation, women struggled to define their own identity, but being a "responsible mother" was one aspect they could control through careful fertility choices. In this context, choosing sterilization reflected a woman's agency to make wise choices in a difficult situation. Furthermore, its permanent nature was attractive in some respects, representing certainty in an uncertain world. By contrast, "neither contraceptive pills nor IUDs were the right solution; they both represented too much of the uncertainty, fluidity and movement that people were striving to leave behind" (Dalsgaard 2004, 167).

Women's agency was nonetheless constrained by economic uncertainty and social problems, summed up by Dalsgaard's respondents' frequent assertions that "we don't have the conditions for bringing up children" (Dalsgaard 2004). Power relations between healthcare workers and low-income women in the study also influenced reproductive choices. Clinical healthcare was one of the main ways in which low-income women left their domestic lives and encountered upper- and middle-class values. Viewing sterilization as a connection to the affluence and "modern" lifestyles that they associated with professional healthcare workers further encouraged many women to accept sterilization. However, interactions with these professionals often left women feeling inferior, raising questions over the kind of empowerment that women received from such reproductive "choices" (Dalsgaard 2004).

Dalsgaard ultimately argues that while sterilization in itself is not necessarily problematic, the fact that the participants in her study often turned to sterilization to address unmet needs in their lives—and not because it was the optimal family planning strategy for them—can be read as a reflection of the structural constraints acting upon them. In this way, high rates of sterilization in this community can be viewed as a paradox, simultaneously serving as a manifestation of power structures and as a mechanism through which women exercise agency to gain control over their lives.

In other cases, critics have argued that low-income countries have been used as laboratories to test new contraceptive technologies. In these contexts, women were not offered reproductive choices or counseled to select the contraceptive technique that best suited them but were instead signed up for trials of whichever pharmaceutical product required study subjects. One of the most infamous cases occurred in Puerto Rico, whose colonial relationship with the US gave the more powerful nation considerable opportunity to abuse reproductive rights. A variety of contraceptive products, including contraceptive foams, pills, and IUDs, were tested on Puerto Rican women in the 1950s. Although many women participated voluntarily, viewing the novel forms of contraception as an opportunity to control fertility and to participate in the modernization of their country, for many women this was their *only* opportunity to gain access to reproductive technologies and so these programs were far from offering reproductive *choice* to women. Many of the products being tested were ineffective or had unpleasant and dangerous side effects, leaving women highly disillusioned (Schoen 2006).

Over time, the rhetoric surrounding population concerns has changed, particularly in the context of a series of United Nations population conferences that shifted global emphasis from demographic targets towards greater concern for reproductive health and gender issues. Many countries' population policies now include language designed to protect women's rights and emphasize broader reproductive health goals, although changes on the ground are often still superficial. In many cases, funding continues to depend on political whim, requiring that reproductive health programs meet donors' concerns—again reinforcing the power dynamics inherent in fertility policy. For instance, the US has maintained a stringent stance with regard to abortion, denying federal funding to family planning organizations with any connection to abortion, even those that provide only abortion counseling or lobby governments on abortion issues (Bishop 2004). This policy was again reinforced by the Trump administration when, on his third day in office, Trump ordered that US foreign aid could not be given to organizations that offered abortion services, mirroring legislation from the Reagan era (Crane, Daulaire, and Ezeh 2017). Organizations that attend to broader issues of women's health and safe births, and sometimes even HIV/AIDS prevention strategies, may thus be precluded from US funding because of their parallel involvement with abortion services.

The challenge remains: how can policymakers craft population policies that recognize individual human rights and yet promote the greater good of society? Although identifying an "ideal" policy is difficult, we can learn from the experience of past population policies. China's family planning program provides a particularly instructive case because it is one of the most stringent examples of a national effort to reduce fertility.

China's population policies

With the distinction of being the most populous country in the world (although India will likely soon surpass it), China's population policies have a major global impact. China has experimented with some of the most draconian fertility policies ever implemented, lending its story special relevance to the study of geographies of power.

Between 1949 and 1953, China had pro-natalist policies, providing financial incentives for couples to have many children and outlawing abortion, sterilization, and the production of contraceptives (Zhenming 2000). The Communist Party under Mao Zedong saw distribution of resources as more problematic than overall population size, but an increasing recognition of problems resulting from population growth led to the adoption of family planning as a key national policy by the mid-1950s. The serious famines associated with the economic and social policies of China's "Great Leap Forward" from 1958 to 1961 further focused concern on population growth. At first, the goal was to make family planning methods and abortion available, but policies were soon broadened to incorporate overt fertility limitation as well.

China's third family planning campaign of 1971, based around the pillars of late marriage and childbearing (*Wan*), birth spacing (*Xi*), and fertility limitation (*Shao*), epitomized the quota system that typified this new approach. In cities, couples were limited to two children and encouraged to delay marriage until age 25 for women and 28 for men. In rural areas, three children were allowed, and marriage was permitted at 23 for women and 25 for men. In both situations, births

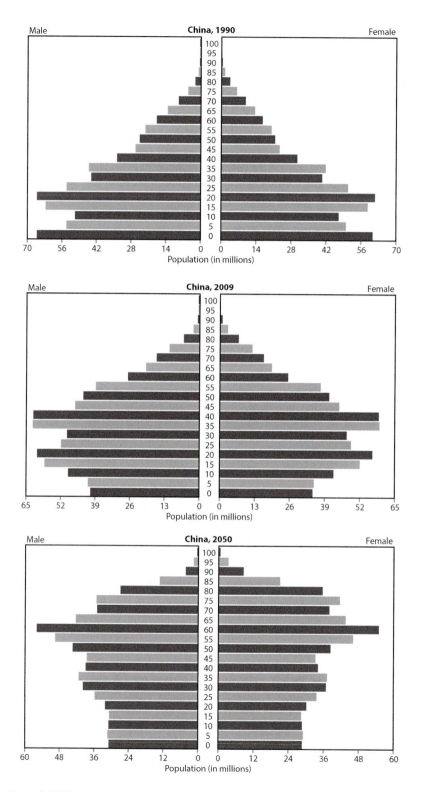

Figure 9.4 China, population pyramids for 1990, 2009, and 2050

Source: US Census Bureau (2018)

Grosz, E. 1990. "Inscriptions and Body-Maps: Representations and the Corporeal." In *Feminine/Masculine and Representation*, edited by T. Threadgold and A. Cranny-Francis, Sydney Association for Studies in Society and Culture. Allen & Unwin.

Gwatkin, D. R., Abbas Bhuiya, and Cesar G. Victora. 2004. "Making Health Systems More Equitable." *The Lancet* 364 (9441): 1273–80. doi:10.1016/S0140-6736(04)17145-6.

Gwatkin, D. R., S. Rutstein, K. Johnson, E. Suliman, et al. 2007. "Socio-Economic Differences in Health, Nutrition, and Population Within Developing Countries: An Overview." *Nigerian Journal of Clinical Practice* 10 (4): 272–82.

Hanchette, Carol L. 2008. "The Political Ecology of Lead Poisoning in Eastern North Carolina." *Health & Place* 14 (2): 209–16. doi:10.1016/j.healthplace.2007.06.003.

Haraway, D. 1999. "The Biopolitics of Postmodern Bodies: Determinations of Self in Immune System Discourse." In *Feminist Theory and the Body: A Reader*, edited by J. Price and M. Shildrick. Routledge.

Harper, J. 2004. "Breathless in Houston: A Political Ecology of Health Approach to Understanding Environmental Health Concerns." *Medical Anthropology* 23 (4): 295–326. doi:10.1080/01459740490513521.

Hartmann, B. 1987. *Reproductive Rights and Wrongs: The Global Politics of Population Control and Contraceptive Choice*. Harper & Row.

Hazen, H. 2017. "'The First Intervention Is Leaving Home': Reasons for Electing an Out-of-Hospital Birth Among Minnesotan Mothers." *Medical Anthropology Quarterly* 31 (4): 555–71. doi:10.1111/maq.12358.

Hoffmann, David L. 2000. "Mothers in the Motherland: Stalinist Pronatalism in Its Pan-European Context." *Journal of Social History* 34 (1): 35–54.

Huang, W., X. Lei, and Y. Zhao. 2016. "One-Child Policy and the Rise of Man-Made Twins." *The Review of Economics and Statistics* 98 (3): 10.

Imrie, P. R. F., and R. I. R. Imrie. 1996. *Disability and the City: International Perspectives*. Sage Publications.

Jackson, Paul, and Abigail H. Neely. 2015. "Triangulating Health: Toward a Practice of a Political Ecology of Health." *Progress in Human Geography* 39 (1): 47–64. doi:10.1177/0309132513518832.

Jayne, Mark, Gill Valentine, and Sarah L. Holloway. 2008. "The Place of Drink: Geographical Contributions to Alcohol Studies." *Drugs: Education, Prevention and Policy* 15 (3): 219–32. doi:10.1080/09687630801969465.

Kabat, G. C. 2008. *Hyping Health Risks: Environmental Hazards in Daily Life and the Science of Epidemiology*. Columbia University Press.

Kalipeni, E., and J. Oppong. 1998. "The Refugee Crisis in Africa and Implications for Health and Disease: A Political Ecology Approach." *Social Science & Medicine* 46 (12): 1637–53.

Kassebaum, Nicholas J., Rafael Lozano, Stephen S. Lim, and Christopher J. Murray. 2017. "Setting Maternal Mortality Targets for the SDGs—Authors' Reply." *The Lancet* 389 (10070): 697–98. doi:10.1016/S0140-6736(17)30339-2.

King, B. 2010. "Political Ecologies of Health." *Progress in Human Geography* 34 (1): 38–55. doi:10.1177/0309132509338642.

King, B., and K. A. Crews. 2013. "Human Health at the Nexus of Ecologies and Politics." In *Ecologies and Politics of Health*, edited by B. King and K. A. Crews. Taylor & Francis.

Klodawsky, Fran, Tim Aubry, and Susan Farrell. 2006. "Care and the Lives of Homeless Youth in Neoliberal Times in Canada." *Gender, Place & Culture* 13 (4): 419–36. doi:10.1080/09663690600808577.

Knudsen, L. M., and B. Hartmann. 2006. *Reproductive Rights in a Global Context: South Africa, Uganda, Peru, Denmark, United States, Vietnam, Jordan*. Vanderbilt University Press.

Kornelsen, J., A. Kotaska, P. Waterfall, L. Willie, and D. Wilson. 2010. "The Geography of Belonging: The Experience of Birthing at Home for First Nations Women." *Health and Place* 16 (4): 638–45. doi:10.1016/j.healthplace.2010.02.001.

Laws, G., and J. Radford. 1998. "Place, Identity, and Disability." In *Putting Health into Place: Landscape, Identity, and Well-Being*, edited by R. A. Kearns and W. M. Gesler. Syracuse University Press.

Lupton, D. 2003. *Medicine as Culture: Illness, Disease and the Body in Western Societies*. Sage Publications.

Mahon-Daly, Patricia, and Gavin J. Andrews. 2002. "Liminality and Breastfeeding: Women Negotiating Space and Two Bodies." *Health & Place* 8 (2): 61–76. doi:10.1016/S1353-8292(01)00026-0.

Mayer, Jonathan D. 1996. "The Political Ecology of Disease as One New Focus for Medical Geography." *Progress in Human Geography* 20 (4): 441–56. doi:10.1177/030913259602000401.

Michie, H. 1998. "Confinements: The Domestic in the Discourses of Upper-Middle-Class Pregnancy." In *Making Worlds: Gender, Metaphor, Materiality*, edited by S. H. Aiken. University of Arizona Press.

Nyantakyi-Frimpong, Hanson. 2017. "Agricultural Diversification and Dietary Diversity: A Feminist Political Ecology of the Everyday Experiences of Landless and Smallholder Households in Northern Ghana." *Geoforum* 86: 13.

Nyantakyi-Frimpong, Hanson, Godwin Arku, and Daniel Kweku Baah Inkoom. 2016. "Urban Agriculture and Political Ecology of Health in Municipal Ashaiman, Ghana." *Geoforum* 72: 38–48. doi:10.1016/j.geoforum.2016.04.001.

Ochel, W., and R. Osterkamp. 2007. "Fertility Policy in Germany." *Pharmaceuticals Policy and Law* 9: 9.

OECD [Organisation of for Economic Co-operation and Development]. 2018. "Family Benefits Public Spending." Accessed October 27, 2019. https://data.oecd.org/socialexp/family-benefits-public-spending.htm.

Poland, B. 1998. *Putting Health into Place: Landscape, Identity, and Well-Being*, edited by R. A. Kearns and W. M. Gesler. Syracuse University Press.

[PPB] Population Reference Bureau. 2010. "Did South Korea's Population Policy Work Too Well?" Accessed September 1, 2019. www.prb.org/koreafertility/.

Qirjo, D. 2017. "Enticing the Stork: Can We Evaluate Pro-Natal Policies Before Having Children?" *Institute of Economic Affairs Monographs* 36 (2).

Rishworth, A., and J. Dixon. 2019. *Reproducing Inequalities: Examining the Intersection of Environment and Global Maternal and Child Health*, edited by M. R. England, M. Fannin, and H. Hazen, *Reproductive Geographies: Bodies, Places and Politics*, Routledge.

Robbins, P. 2004. *Political Ecology: A Critical Introduction*. Wiley.

Roberts, D. E. 1999. *Killing the Black Body: Race, Reproduction, and the Meaning of Liberty*. Vintage.

Rooks, J. 1999. *Midwifery and Childbirth in America*. Temple University Press.

Rosenfeld, D., and C. Faircloth. 2009. *Medicalized Masculinities*. Temple University Press.

Scharping, T. 2013. *Birth Control in China 1949–2000: Population Policy and Demographic Development*. Taylor & Francis.

Schoen, J. 2006. *Choice and Coercion: Birth Control, Sterilization, and Abortion in Public Health and Welfare*. University of North Carolina Press.

Schulz, Leslie O., and Lisa S. Chaudhari. 2015. "High-Risk Populations: The Pimas of Arizona and Mexico." *Current Obesity Reports* 4 (1): 92–98. doi:10.1007/s13679-014-0132-9.

Schwalfenberg, Gerry K. 2010. "Indigenous Health." *Canadian Medical Association Journal* 182 (6): 592. doi:10.1503/cmaj.110-2042.

Smith, J. M. 2017. "The Gross Inequality of Organ Transplants in America." *The New Republic*. https://newrepublic.com/article/145682/gross-inequality-organ-transplants-america.

[SPLASH]. The Social Policy and Law Shared Database (SPLASH). 2014. "Family Policies: Sweden (2014)." Accessed October 27, 2019. https://splash-db.eu/policydescription/family-policies-sweden-2014/.

Sultana, F. 2006. "Gendered Waters, Poisoned Wells in Bangladesh." In *Fluid Bonds: Views on Gender and Water*, edited by K. Lahiri-Dutt. National Institute for Environment and Australian National University and Stree.

Sultana, F. 2011. "Producing Contaminated Citizens: Toward a Nature-Society Geography of Health and Well-being." *Annals of the Association of American Geographers* 102 (5): 1165–72.

Takahashi, L. 1998. "Concepts of Difference in Community Health." In *Putting Health into Place: Landscape, Identity, and Well-Being*, edited by R. A. Kearns and W. M. Gesler. Syracuse University Press.

Thévenon, Olivier, and Anne H. Gauthier. 2011. "Family Policies in Developed Countries: A 'Fertility-Booster' with Side-Effects." *Community, Work & Family* 14 (2): 197–216. doi:10.1080/13668803.2011.571400.

Thompson, Lee, Jamie Pearce, and Ross Barnett. 2009. "Nomadic Identities and Socio-Spatial Competence: Making Sense of Post-Smoking Selves." *Social & Cultural Geography* 10 (5): 565–81. doi:10.1080/14649360902974431.

Turshen, M. 1984. *The Political Ecology of Disease in Tanzania*. Rutgers University Press.

United Nations. 1948. "The Universal Declaration of Human Rights." United Nations. Accessed July 15. www.ohchr.org/EN/UDHR/Documents/UDHR_Translations/eng.pdf.

United Nations. 2019. "World Population Prospects: 2019." Accessed September 1, 2019. https://population.un.org/wpp/.

United Nations Population Division. 2019. *World Population Prospects: The 2019 Revision*. Geneva: United Nations Population Division.

United Nations Population Division, and East-West Center. 2015a. *Can the Republic of Korea Afford Continuing Very Low Fertility?* edited by United Nations Department of Economic and Social Affairs. New York: United Nations Department of Economic and Social Affairs, Population Division.

United Nations Population Division, and East-West Center. 2015b. *The Influence of Family Policies on Fertility in France*, edited by United Nations Department of Economic and Social Affairs. New York: United Nations Department of Economic and Social Affairs, Population Division.

United Nations Population Division, and East-West Center. 2015c. *Not so Low Fertility in Norway—A Result of Affluence, Liberal Values, Gender-Equality Ideals and the Welfare State*, edited by United Nations Department of Economic and Social Affairs. New York: United Nations Department of Economic and Social Affairs, Population Division.

United States Census Bureau. 2018. "International Programs." Accessed May 27, 2010. https://www.census.gov/programs-surveys/international-programs/about/idb.html.

Victora, C. G., J. P. Vaughan, F. C. Barros, A. C. Silva, and E. Tomasi. 2000. "Explaining Trends in Inequities: Evidence from Brazilian Child Health Studies." *Lancet* 356 (9235): 1093–98. doi:10.1016/s0140-6736(00)02741-0.

Wang, B. 2018. *Next Big Future*. Accessed September 1, 2019. www.nextbigfuture.com/2018/11/free-medical-fertility-and-bribes-could-boost-japans-population-in-2100-by-50.html.

Wang, F., B. Gu, and Y. Cai. 2016. "The End of China's One-Child Policy." Brookings. Accessed July 19, 2019. www.brookings.edu/articles/the-end-of-chinas-one-child-policy/.

Watts, Michael J., and Hans G. Bohle. 1993. "The Space of Vulnerability: The Causal Structure of Hunger and Famine." *Progress in Human Geography* 17 (1): 43–67. doi:10.1177/030913259301700103.

Wilson, Paul M., Alison M. Booth, Alison Eastwood, and Ian S. Watt. 2008. "Deconstructing Media Coverage of Trastuzumab (Herceptin): An Analysis of National Newspaper Coverage." *Journal of the Royal Society of Medicine* 101 (3): 125–32. doi:10.1258/jrsm.2007.070115.

Yi, Zeng. 2007. "Options for Fertility Policy Transition in China." *Population and Development Review* 33 (2): 215–46. doi:10.1111/j.1728-4457.2007.00168.x.

Yongping, J., and P. Xizhe. 2000. "Age and Sex Structures." In *The Changing Population of China*, edited by X. Peng and Z Guo. Wiley.

Zhenming, X. 2000. "Population Policy and the Family Planning Programme." In *The Changing Population of China*, edited by Xizhe Peng and Guo Zhigang. Wiley.

10

GEOGRAPHIES OF HEALTHCARE

Healthcare includes the diagnosis, treatment, and prevention of disease. Because health is integral to human well-being, the provision of adequate healthcare is often a fundamental societal goal. The importance, and arguably the ethical imperative, of ensuring the best possible health to all people is codified in the constitution of the World Health Organization, which states: "The enjoyment of the highest attainable standard of health is one of the fundamental rights of every human being without distinction of race, religion, political belief, economic or social condition" (WHO 1946). Nonetheless, the subject of healthcare raises complex social, political, philosophical, and economic questions. How should we treat illness? Who should provide healthcare? How much healthcare do people have a right to, and what procedures should we consider a privilege? Is government responsible for ensuring access to healthcare for everybody or is it appropriate to rely on market forces? As societies and cultures have taken different approaches to these questions, the study of healthcare is embedded in social context.

Geographers use a variety of social and spatial approaches to analyze healthcare issues. The study of healthcare *provision* considers where and how healthcare is provided across different communities. Often this work focuses on examining inequalities. Geographers have traditionally focused on spatial methods such as **location-allocation modeling**, which aims to determine the best location for healthcare facilities based on factors such as travel time between facilities and their users. Inequalities are addressed through the study of healthcare *access* (the availability of healthcare) and healthcare *utilization* (the actual use of healthcare services). Studies of healthcare access and utilization consider not only the spatial distribution of facilities, but also sociocultural and structural barriers to healthcare, such as language, cost, or eligibility for service.

We begin our discussion by describing historical developments in philosophies of healing. Building on our discussion of culture and politics in previous chapters, we then consider how different societies provide healthcare, noting in particular the dominance of Western biomedicine and concurrent marginalization of **complementary** and **alternative medical systems**. We then turn to healthcare access and utilization, before exploring alternative geographies of healing.

FOUNDATIONS OF MEDICAL THOUGHT AND PRACTICE

The practice of medicine originated more than 4,000 years ago, when people began to systematically address symptoms of disease (Magner 1992). During this initial development, medical practice was often tied to religious beliefs. In ancient India, for instance, Vedic medicine evolved from Hindu mythology. In Vedic medical practice, disease results from sin or demons, requiring the curative power of rituals, charms, and herbal remedies. Surgical techniques and medical practice could only be effective if combined with appropriate religious rituals (*ibid.*). Ayurvedic medicine, which is still practiced today, may have developed out of Vedic medicine and other early historical developments around Hinduism. The primary objective of Ayurvedic medicine is to promote the maintenance of health rather than the treatment of disease. Typical of many traditional medical systems, Ayurvedic medicine is closely linked with moral concerns, with *karma*—the good or bad deeds done in current and previous lives—used to explain disease. Ayurvedic healing combines both spiritual and medical approaches, including religious mantras and rituals, diet and drugs, and psychic therapy (*ibid.*).

Chinese medicine also has a long history. Historical documents demonstrate that Chinese practitioners understood the role of the heart and the circulation of blood long before such concepts were incorporated into Western medical practice. Traditional Chinese medicine is based on the concepts of *yin* and *yang*—symbolizing respectively female and male, darkness and light, cold and heat—dualistic oppositions that form the basis for everything in the universe and the origin of life and death. Disease results when *yin* and *yang* are no longer balanced, and treatment is dedicated to restoring this harmony. Chinese medicine also emphasizes connections between the human body and the environment. Practitioners seek to restore harmony by restoring elements of a patient's environment and culture. Specific treatments generally follow five methods: living in harmony with the universe, diet, acupuncture, drugs, and the treatment of bowels, blood, and breath (Magner 1992).

In Western civilization, the first organized thought about medicine originated in ancient Greece. During this time, important foundations were laid as medical practitioners began to systematically investigate the causes of disease through observation. Although historians debate his precise role, the Greek philosopher Hippocrates (circa 460–360BC) is commonly recognized for rejecting superstition as an explanatory framework for disease, emphasizing instead the need for careful observation in generating medical knowledge. Of significance to health geographers, Hippocrates showed an interest in the role of interactions between people and their environment in causing disease:

> Whoever wishes to investigate medicine properly, should proceed thus: in the first place to consider the seasons of the year, and what effects each of them produces for they are not at all alike, but differ much from themselves in regard to their changes. Then the winds, the hot and the cold, especially such as are common to all countries, and then such as are peculiar to each locality. We must also consider the qualities of the waters, for as they differ from one another in taste and weight, so also do they differ much in their qualities. In the same manner, when one comes into a city to which he is a stranger, he ought to consider its situation, how it lies as to the winds and the rising of the sun; for its influence is not the same whether it lies to the north or the south, to the rising or to the setting sun.
>
> *On Airs, Waters, and Places*, Hippocrates,
> 400 BCE, translated by Francis Adams

In the original form of medicine developed by Hippocrates, the human body is a microcosm of the universe—somewhat reminiscent of ideas in traditional Chinese medicine. The body contains four *humors*: blood, phlegm, black bile, and yellow bile, which are paired with elements of the universe: earth, air, fire, and water. Ill health is caused by an excess or deficiency of these humors, and treatment involves replacing the deficiency or purging the excess (Helman 2007). Galen, a Greek physician working in Rome during the second century, elaborated on these ideas, using experimentation and observation to develop his theories. Galen's systematic approach to medicine, with its emphasis on **empiricism**, was particularly influential in the development of Islamic medicine, which flourished as European medicine stagnated during the Dark Ages. Ideas of experimentation and observation were subsequently reintroduced to European medicine from Islamic texts in the Middle Ages.

Humoral conceptions of health are still practiced today in many communities as part of **folk medicine**—medical beliefs passed down through the generations as general knowledge within a community. In Latin America, for instance, folk medicine continues to revolve around the idea of balancing "hot" and "cold" elements (Helman 2007). Although humoral conceptions of health have largely disappeared from the West, vestiges of these beliefs remain. For example, the notion of catching a cold is an allusion to the idea that temperature is important to one's health. In English folk beliefs about colds and chills, for instance, the idea of counteracting the penetration of environmental cold or dampness is reflected in the practice of administering heat via warm food or drink (*ibid.*).

As the scientific method developed during the Renaissance, so did scientific approaches to the study of the human body. These again emphasized the importance of observation and measurement for understanding physical processes, rejecting supernatural or spiritual foundations of health. By the end of the nineteenth century, **germ theory** had gained widespread acceptance,

paving the way for modern approaches to medicine and the emergence of the now dominant paradigm of **biomedicine**. Biomedical approaches explain disease and prescribe treatment based on biological evidence and hypothesis testing. The biomedical approach to healthcare led to rapid and revolutionary changes in human health and is now the dominant medical system across much of the world, although it does not exist unchallenged.

ALTERNATIVE MEDICAL SYSTEMS

The very notion of "alternative" medicine reflects the dominance of Western biomedicine, since the term is commonly applied to any healthcare practices that fall outside the biomedical model. Although a wide variety of alternative medical approaches exist across the globe, most government-provided healthcare focuses on biomedicine, and alternative systems are often systematically undervalued in comparison. Nonetheless, traditional forms of medicine remain an important source of healthcare for many populations, particularly in remote rural areas, where **folk medicine** may be the primary source of healthcare (Figure 10.1). In some countries, however, non-Western medical systems are recognized as important systems of mainstream professional care.

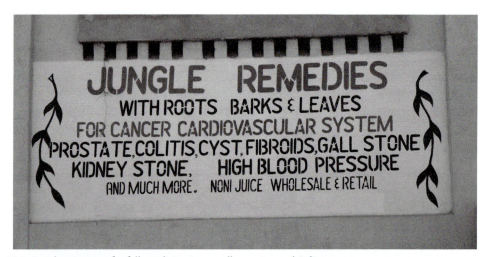

Figure 10.1 An advertisement for folk medicine in a small town in rural Belize

In a *professional* medical system, a widely accepted body of knowledge and practice is passed from practitioner to practitioner via a formalized training system. For instance, traditional Chinese medicine remains a mainstay of the formal Chinese healthcare system, and its practitioners are found in Chinese communities throughout the world. Similarly, Ayurvedic medicine is a professional medical system that is still used in the Indian subcontinent and parts of the Arab world.

There has also been a resurgence of interest in **complementary and alternative medicine (CAM)** in the West, often in recognition of the limits of biomedicine. CAM therapies are united by the belief that the body has the power to heal itself and that healing should involve the mind, body, and spirit, often emphasizing the need for individualized treatments and a strong partnership between patient and practitioner (Tabish 2008). CAM includes approaches as diverse as massage, yoga, herbal remedies, meditation, homeopathy, and faith healing, and may include elements of both professional and non-professional medical systems. Acupuncture, for instance, is commonly used in the West to treat conditions such as chronic pain and nausea, often by patients with little understanding of the underlying philosophy of traditional Chinese medicine. Despite growing acceptance of CAM in the West, a cultural bias often persists against the use of alternative therapies. This may explain why many individuals report turning to social networks for advice on complementary medicine, rather than seeking the advice of their primary doctor (CORDIS 2012).

Nonetheless, the use of alternative medicine in the West has grown dramatically over the past thirty years. In Europe, for instance, one recent study calculated that 26 percent of the population of the countries studied had used CAM during the past year, although use rates varied widely by country (Kemppainen et al. 2018). Similarly, 79 percent of Canadians reported in 2016 that they had used an alternative therapy at some point in their lives (Esmail 2017). As a result, spending on CAM is significant, with a recent US report suggesting that 59 million individuals had made at least one expenditure on a complementary health approach in 2015, representing more than $30 billion in out-of-pocket costs (Nahin, Barnes, and Stussman 2016). In Europe, spending on CAM probably tops 100 million euros (CORDIS 2012). Meanwhile, China is looking to develop a medical tourism industry based on traditional Chinese medicine (Cyranoski 2018).

Different medical systems therefore frequently coexist—a phenomenon known as **medical pluralism**. In some countries, such as India and China, Western biomedicine and alternative, professional systems of care are widely respected and even closely integrated. In China, for instance, many doctors practice both traditional Chinese medicine and Western biomedicine, and patients may be referred from one system to the other. The combination of modern medicine with traditional forms of healing is known as **integrated medicine**. In other countries, such as Germany, the US, and New Zealand, Western biomedicine is the dominant medical system, but other practitioners such as herbalists, aromatherapists, and chiropractors may practice. In most of the Global North, patients must pay privately for alternative therapies, although some treatments may be offered by some government-sponsored healthcare systems. More mainstream CAM therapies such as acupuncture or massage may even be offered alongside biomedicine in the same healthcare facilities, suggesting a degree of biomedical endorsement of the treatment through the sharing of space (Figure 10.2). People with conditions that have proved difficult to treat using biomedical approaches, such as chronic pain or depression, are particularly likely to turn to alternative medicine (CORDIS 2012).

Openness to alternative treatments should not imply uncritical acceptance of all therapies. The use of an ineffective treatment clearly has an opportunity cost in terms of time, energy, and expense. At their worst, alternative medical systems involve potentially damaging or even life-threatening activities or substances. For instance, ephedra (known as Ma Huang in China)—an herb traditionally used in Chinese medicine to treat short-term respiratory conditions—was marketed in the US as an energy enhancer and then dieting aid in the early 2000s. As a "natural" herbal remedy, it was popularly viewed as a safe option, but it was subsequently connected to 16,000 reports of adverse health impacts and 155 deaths (Harvard Medical School 2004). In 2004, the US Food and Drug Administration (FDA) banned supplements that included ephedra, concluding that there was little

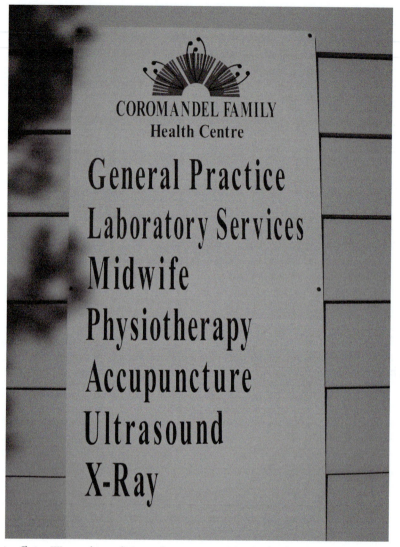

Figure 10.2 Clinic offering Western biomedicine and acupuncture, New Zealand

Although complementary medicine tends to be marginalized by biomedicine in the West, clinics in many countries are beginning to offer some CAM treatments as patients demand more alternatives to care. In this case, acupuncture is listed as just one more treatment option alongside more mainstream biomedical therapies, indicating its increasing acceptance in the West.

evidence of ephedra's effectiveness beyond short-term weight loss, and that any benefits were outweighed by an increased risk of heart problems and stroke. Ephedra was the first dietary supplement to be banned in the US, although its use in traditional Chinese remedies was excluded from the ban (National Center for Complementary and Integrative Health 2016). Ironically, alternative therapies may not be given the same level of scrutiny as biomedical treatments before coming to market. In the US, for instance, dietary supplements are subject to a different set of regulations than drugs, allowing many alternative remedies to be marketed without full evidence of their safety or effectiveness.

The potential benefits and dangers of many CAM treatments therefore remain open to question (Tabish 2008). One common critique of alternative therapies is that their efficacy has not been demonstrated with **randomized controlled clinical trials**, making it unclear whether any perceived benefits are due to the treatment itself or to the placebo effect. The **placebo effect** refers to the therapeutic benefits that can be generated when an individual believes that s/he is receiving a treatment, even if the treatment has no objective curative effect. Many studies have found that the placebo effect can be powerful, underscoring the significance of psychological aspects of healing. It is standard practice in clinical drug trials for a placebo, such as a sugar pill, to be administered to the control group so that both the control and study groups believe they are receiving some form of treatment. Control groups often report health improvements, despite having received no active ingredients.

Some alternative therapies have also been criticized for failing to demonstrate a plausible biological mechanism through which the treatment could work. For instance, homeopathy, which relies on administering extremely dilute preparations of substances with curative properties, has been criticized because the concentrations of the active ingredient are too small to be effective according to biomedical understandings of the body. This concern was voiced during a UK parliamentary inquiry into homeopathy, which came to the conclusion that the National Health Service (NHS) should cease funding homeopathy because the evidence indicated that it was not efficacious beyond the placebo effect and that homeopathic approaches were scientifically implausible (British House of Commons 2010). A subsequent Australian inquiry came to the same conclusion, adding that homeopathy could be dangerous if patients delayed biomedical treatments while they pursued homeopathic remedies (Ernst 2015).

Supporters of CAM argue that their treatments work, even if we cannot explain them with scientific methods. Some proponents argue that Western biomedical methods may not be an effective way to test CAM treatments because scientific approaches fail to capture the holistic way in which many alternative therapies work. Furthermore, if people report improvements in their health following a CAM treatment, why should we prevent its use? Opponents of CAM argue that, beyond safety concerns, the danger lies in delaying more effective treatments as well as the economic cost of pursuing treatments of doubtful efficacy.

Despite the challenge of proving the effectiveness of alternative therapies using positivist approaches, there is growing interest in the clinical investigation of CAM. In 1998, the *Journal of the American Medical Association* published an entire edition on alternative treatments, "in response to demand from doctors who wanted scientifically sound data on treatments and products that so many patients were asking about" (Grady 1998). Of the seven studies reported in the special issue, four provided evidence in support of the effectiveness of alternative treatments and three did not. Since then, a growing literature has explored CAM therapies using experimental approaches, resulting in support for some CAM therapies. Most notably, artemisinin—a product used in traditional Chinese medicine—has been found to be a powerful treatment for malaria (Cyranoski 2018). Other evidence suggests that hypnosis and relaxation techniques may reduce anxiety, panic disorders, and insomnia (WHO 2002a); yoga can reduce asthma attacks (WHO 2002a); acupuncture can relieve postoperative nausea and vomiting (Lee and Fan 2009); probiotics can prevent upper respiratory tract infections (Hao et al. 2011); and zinc supplements can prevent pneumonia in young children (Lassi, Moin, and Bhutta 2016). Evidence on most CAM therapies remains inconclusive, however. In one meta-analysis of the efficacy of CAM therapies in pediatrics, the authors found that less than 4 percent of studies found evidence to fully support the therapy, 20 percent offered conditional support, and 6 percent found evidence that the therapy should *not* be used, while an overwhelming 70 percent of the studies reviewed were found to be inconclusive (Meyer et al. 2013). The authors call for the use of randomized controlled clinical trials to substantiate claims made with respect to CAM.

Ironically, the success of alternative therapies in the Western world has led to concerns about access to these therapies in other parts of the world. In 2002, the global market for "traditional therapies" was US$60 billion per year and growing (WHO 2002a). The WHO (2002a) has pointed out that the unregulated commercialization of traditional therapies may end up making them too expensive for many people living in low-income countries, who depend on traditional medicine as their primary source of healthcare. In response, the WHO adopted a "Global Strategy on Traditional and Alternative Medicine" in 2002 to assist countries with regulating traditional and alternative therapies. The WHO's embrace of CAM, and particularly traditional Chinese medicine, has continued since then, particularly under the leadership of Margaret Chan—a major proponent of traditional Chinese medicine—from 2006 to 2017. In 2018, the WHO sparked controversy by stating that it would recognize traditional Chinese medicine in its global medical compendium, devoting a chapter to Chinese medicine (Cyranoski 2018). Proponents argue that this will recognize the value of a much-used medical approach and stimulate research into the efficacy of alternative therapies. Opponents argue that the inclusion of traditional Chinese therapies, many of which currently have little grounding in evidence-based medicine, is potentially dangerous.

HEALTHCARE PROVISION AND ACCESS

Healthcare can be viewed as a type of economy with its own supply and demand. Increasing the supply of healthcare resources to meet demand is a basic means through which a society can improve the well-being of its members. Healthcare *provision*—where and how healthcare is provided—is, therefore, a significant topic with important policy implications. Although one would expect that there should be a clear correlation between healthy people and places with ample healthcare provision, in practice, this relationship is far from universal (Kearns and Joseph 1993). In analyzing this apparent paradox, health geographers are interested in the factors that mediate the use of available resources. This includes the study of *utilization*—where and how people seek healthcare—and *accessibility*—who has access to existing services.

Healthcare provision

By analyzing the relationship between health outcomes and healthcare indicators, such as number of doctors per capita or healthcare spending, it is possible to assess the role that healthcare plays in supporting health. For instance, Day et al. (2008) used cluster analysis to group countries by life expectancy. They then analyzed variation among these groups, revealing that differences were associated with factors such as healthcare spending, vaccination coverage, and number of hospital beds.

As Figure 10.3 illustrates, international inequalities in healthcare spending are significant. These inequalities have led to global transfers in healthcare resources. Perhaps most significantly, qualified medical staff are migrating in search of better working and living conditions. This migration has led to critical losses of qualified staff from many low-income countries and from rural areas—places that cannot afford to lose skilled professionals (Box 10.1). Lack of healthcare workers has been identified as a significant barrier to provision of effective healthcare. Although scholarly opinions are mixed about the precise relationship between density of healthcare personnel and health outcomes, most research suggests that population health suffers if a country lacks a critical threshold of healthcare workers. In one cross-country study, for instance, density of medical personnel was significant in accounting for differences in maternal mortality rates, infant mortality rates, and under-age-5 mortality rates (Anand and Barnighausen 2004).

Stark differences in the cost of care have also led to **medical tourism**, whereby patients from affluent countries seek care abroad, usually with the goal of finding quality care at a lower cost. The actual size of the medical tourism market is difficult to estimate, but there is little doubt that it is growing rapidly (Buzinde and Yarnal 2012). Some middle-income countries such as Brazil and India have become well known for offering procedures such as cosmetic and cardiovascular surgery at relatively low cost. In recent years, patients have also begun to travel internationally to avoid long waiting lists or receive treatments that are unavailable or even illegal in their home country. Parents are even seeking

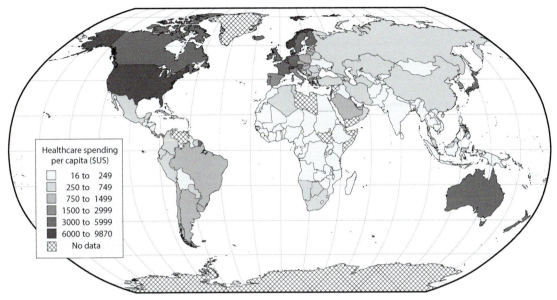

Figure 10.3 Healthcare spending per capita, 2016

This map shows the average annual per capita value of healthcare expenses, including both private and government expenditures. Expenses are given in 2016 US dollars.

Box 10.1

MIGRATION OF HEALTHCARE PERSONNEL

Economic migration is common in the modern globalzed world, even though most countries impose tight immigration restrictions to regulate population flows. Many countries offer special visas for people trained in occupations for which there is a domestic skills shortage, and healthcare workers often fall within these categories. There has been an effort to recruit healthcare workers for some years, particularly in affluent countries where aging populations not only increase the demand for healthcare, but also reduce the supply of workers entering the workforce. Undervaluation of public sector work, especially lower-level positions such as entry-level nursing and personal care assistants, has exacerbated this shortage of healthcare workers. Health and care industry organizations have lobbied for immigration restrictions to be relaxed to allow these jobs to be filled more easily or, as critics would argue, to fill them at lower wages. However, a lack of legal immigration channels for less-skilled workers in many countries leaves many immigrants in lower-level caring positions working in a "gray economy" on the margins of the legal economy (Eckenwiler 2014).

As early as the 1970s, significant numbers of doctors and nurses began to migrate from low- to high-income countries. In 1972, about 6 percent of physicians and 4 percent of nurses worldwide worked outside their countries of origin, mostly in the US, UK, and Canada (Bach 2004). Flows of workers often reflected colonial and linguistic ties, with India and Pakistan sending many health workers to the UK, for example. The migration of health workers has grown significantly over the past forty

continued

Box 10.1 *continued*

years and now healthcare migrations are occurring at "unprecedented rates" (Eckenwiler 2014), with the number of migrant doctors and nurses working in the affluent OECD countries increasing by 60 percent over the past decade (WHO 2019) (Table 10.1). Increasingly it is nurses rather than doctors who are moving (Brush and Sochalski 2007), and women are contributing a growing proportion of healthcare migrants (Eckenwiler 2014). Migration has also become less closely linked to cultural ties and more significantly to economic demands (Bach 2004). The ultimate destination for many workers is Europe, North America, or the Gulf States, with the US acting as the major recipient, but many migrants make **step migrations**, moving from a poor country to a middle-income and then to an affluent country. Additionally, many countries now act as both recipients and senders of health care personnel. For example, Canada has supplied nurses to the US for many years, while simultaneously receiving them from South Africa, which recruits from Cuba (Brush and Sochalski 2007).

Table 10.1 Doctors and nurses trained abroad in select OECD* countries, 2016

Country	Doctors		Nurses	
	Total	Percent	Total	Percent
Canada	23,569	24.6	31,356	7.9
France	24,420	11.2	19,405	2.9
Ireland	9,123	41.6	(no data)	(no data)
Israel	16,319	58.1	4,615	9.0
UK	45,732	28.4	105,811	15.2
US	215,630	25.0	(no data)	(no data)

*The OECD is the Organization for Economic Cooperation and Development, whose members comprise largely affluent Western countries.

Source: Data from OECD (2019)

Many authors have described the "push" and "pull" factors that stimulate these population movements (e.g., Okeke 2013; Dovlo 2007; WHO 2006). These factors explain not only cross-continental movements, but also the flow of trained personnel from rural to urban areas and among neighboring countries. One study of health workers' reasons for migrating from sub-Saharan Africa emphasized the "push" of poor promotion prospects, low pay, heavy workloads, inadequate living conditions, and high levels of violence and crime. By contrast, receiving countries were perceived to offer prospects for better remuneration and promotion, the opportunity to gain experience, and a generally safer environment (Awases, Gbary, and Chotora 2004). Another study concluded that even small declines in per capita GDP in sub-Saharan African countries led to increased physician emigrations to the US and UK (Okeke 2013). Recently, great emphasis has also been placed on the "pull" of deliberate recruitment of healthcare workers by government and private agencies in the affluent world (WHO 2006; Bach 2004).

The social and economic implications of these worker migrations are often profound. On the positive side, the influx of trained workers to wealthy countries has helped maintain their healthcare systems at relatively low cost. The migrants themselves may also find the better conditions and career opportunities they seek. In the early 2000s, a Filipino nurse working overseas for one year could earn the equivalent of twenty years' salary in the Philippines, for example (Brush and Sochalski 2007). The sending country may also benefit from skills brought back with returning workers, as well as remittances (the money that migrants send back to family). Indeed, some countries (most notably the Philippines but also others such as India) have set up institutional structures to facilitate the training of nurses, with the explicit goal of using remittances to develop their economies (Eckenwiler 2014).

Although countries with carefully controlled emigration policies or surplus personnel can reap financial benefits from health worker migrations, there are often significant negative implications of health worker migrations. Healthcare workers frequently do not return home and so the sending country never recoups any benefits from many migrants' experiences abroad. Additionally, many sending countries have critical shortages of healthcare workers and cannot afford to lose any staff, leading to facility and program closures and increased workload and stress for remaining workers. In the early 2000s, 23 percent of doctors and 5 percent of nurses and midwives trained in sub-Saharan Africa were working in OECD countries (WHO 2006, 99). Considering the cost of training medical staff, this effectively represents a huge subsidy paid by the low-income world to affluent healthcare systems (Pittman, Aiken, and Buchan 2007). Overall, the WHO estimates that the basic healthcare systems of 57 countries are affected by human resources shortages (Nair and Webster 2013).

Some bilateral agreements have emerged to regulate migration flows in ways that benefit both sending and receiving countries (Bach 2004). For example, South Africa and the UK signed a bilateral agreement in 2003 focused on giving international migrants time-limited placements (WHO 2006). In 2010, the WHO "Global Code of Practice on the International Recruitment of Health Personnel" was adopted by the 63rd World Health Assembly (Siyam, WHO, and Poz 2014). The Code promotes voluntary principles and practices for ethical international recruitment of health personnel. In particular, it discourages active recruitment of health personnel from low-income countries that face critical healthcare worker shortages. Another goal is to encourage the return of health migrants, so that their skills can be used to strengthen the home country's healthcare system. Despite these initiatives, health personnel continue to migrate in record numbers.

opportunities overseas to give birth in search of both superior medical care, but also often the opportunity to obtain citizenship for their child in a more affluent country—so-called **birth tourism** (Box 10.2).

Box 10.2

BIRTH TOURISM

Birth tourism refers to mothers traveling overseas to give birth. While some women engage in this practice to seek higher quality of care for themselves and their baby, many also aim to gain citizenship for their child in a country that offers **birthright citizenship** (automatic citizenship for those born in that country). Many countries have traditionally offered citizenship to anyone born there, but increasing mobility of populations has led to the widespread revocation of birthright citizenship laws over the past few decades. Countries that continue to have birthright citizenship are largely in the Americas, with Canada and the US remaining among the few affluent countries that still do, although there are debates in both countries over whether to end the practice.

The issue of birth tourism recently came to a head in the US, where an investigation in early 2019 revealed that affluent Chinese couples were paying as much as $100,000 to companies for birth-tourism packages. For many Chinese parents participating in the practice, this high cost is acceptable if it means that their son or daughter will be granted US citizenship, which is often viewed as a way to help children get back to the US for a prestigious US university degree. In addition, birth tourism companies advertised the high quality of US healthcare and easy access to epidural pain relief (Pak 2019).

continued

> **Box 10.2** *continued*
>
> The companies encouraged women to fly to the US during their pregnancy and then provided tips on navigating immigration and, in some cases, even provided advice on how to pay indigent rates (low rates for the poor) at hospitals (Jordan 2019). Critics of the practice claim that Chinese mothers are taking advantage of the legal loophole of birthright citizenship to gain access to opportunities and facilities intended for local populations, and note the burden that is placed on the US health system by those claiming indigency. Supporters point out that Chinese parents often pay high rates for private clinics and that some US clinics are fully exploiting birth tourism as a market opportunity (Pak 2019). Experts believe that thousands of Chinese women give birth in the US every year, with reports of increasing numbers of Russians, Middle Easterners, and Nigerians also coming to the US to give birth (Jordan 2019; Pak 2019).
>
> It is not just the US that is affected by birth tourism. Kaiser (2018) notes that Chinese women have been drawn to Hong Kong in search of similar advantages and generating similar criticisms. When Hong Kong became a Special Administrative Region of China in 1997, it gained some territorial autonomy for the following fifty years. Women in neighboring provinces in mainland China began coming to Hong Kong to give birth to avoid some of the limitations of China's "one-child" policy and to receive "right of abode" in Hong Kong. Hong Kong's world-class health facilities were an additional draw (*ibid.*). Given that Hong Kong has had a very low birth rate for a significant period of time and has been trying to encourage native residents to have larger families, it would seem that additional births would have been welcomed, particularly given the ethnic similarities between host and migrant populations. Over time, however, xenophobic attitudes towards mainland Chinese were expressed through concerns over birth tourism, with public opinion turning away from offering right of abode to children of mainland Chinese, and increased fees and restrictions being instituted on women coming to Hong Kong to give birth (Kaiser 2018).

Medical tourism has been encouraged by improved standards of medical care in the Global South, where many facilities now provide world-class care with international accreditation. Declining costs of international travel have also contributed—some medical facilities have even partnered with airlines to provide medical tourism packages that assist patients with the costs of transportation (Buzinde and Yarnal 2012). Although many low-income countries see medical tourism as an opportunity to increase their GDP and even improve medical facilities, scholars have pointed out that it is important to analyze the potential implications of this phenomenon, particularly if healthcare becomes less affordable for local residents.

Another global phenomenon is the growing international market in pharmaceuticals that has encouraged patients to look beyond national boundaries for cheaper drugs. Pharmacies in tourist areas in the Global South have advertised the availability of cheap drugs to international visitors for some years (Figure 10.4), but the trend has really taken off as pharmaceuticals became available on the Internet, making it relatively easy to purchase even prescription drugs from abroad. Where government-run healthcare systems can leverage economies of scale to negotiate lower drug prices, costs may also be considerably lower than in market-run healthcare systems. Such a framework prompts many US consumers to buy drugs at considerably lower cost in Canada, for instance. A downside of this international trade is that patients can order potentially dangerous drugs without a prescription from unscrupulous online retailers. Online pharmacies may also sell unapproved or counterfeit medication. Some rogue pharmacies prominently display a Canadian flag on their website, using Canada's reputation as a source of reliable but cheaper drugs to convince US consumers of the legitimacy of

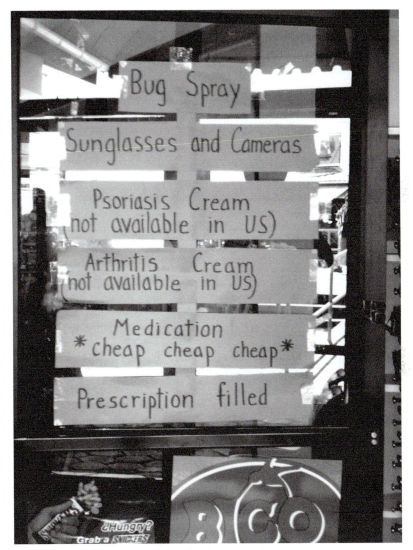

Figure 10.4 Pharmaceuticals for sale to tourists, Belize City, Belize, 2006

This pharmacy is located at the docking point for cruise ships and serves affluent tourists arriving from the US and other high-income countries.

their business, even though most have no connection to Canada (FDA 2018).

Although very poor countries tend to see steady improvements in health as health spending on basic interventions such as antibiotics and vaccinations increases, beyond a certain point the relationship between *spending* on healthcare and population health indicators becomes non-linear. Indeed, there may be little or no relationship between public spending on health and health outcomes such as infant mortality rates, once factors such as female literacy rate and income equality are accounted for (Schell et al. 2007). How healthcare is provided, who has access to it, and the efficiency of healthcare facilities mediate the relationship between healthcare resources and improvements in health.

In this context, *how* money is spent is vital. Spending on low-cost basic health interventions across the population is commonly cited as a way to efficiently achieve health improvements, particularly in low-income contexts. For example, diarrhea—a primary cause of infant mortality worldwide—can often be treated with oral rehydration therapy (ORT), whereby afflicted children are given a rehydration solution comprising salt, sugar, and water. This solution can be made easily at home if parents are provided with basic training on how to prepare it. Alternatively, small packets of oral rehydration salts, costing mere pennies, can be distributed and mixed directly with water. ORT has been credited with a major role in lowering infant mortality in many parts of the Global South. The dissemination of information about ORT is a powerful example of a low-cost but high-impact health practice. Implementing even this basic intervention has nonetheless proved challenging; despite its low cost and simplicity, communicating the technique to thousands of remote rural communities has taken years to achieve.

Bangladesh's experience with disseminating this sort of basic health information is a model in this respect. In the 1980s, a Bangladeshi non-governmental organization, the Bangladesh Rural Advancement Committee (BRAC), trained more than twelve million Bangladeshi mothers in how to prepare oral-rehydration solution using ingredients found in the home. Various agencies subsequently reinforced this initiative by distributing prepackaged oral-rehydration salts. An extensive follow-up assessment of the efficacy of this campaign in the mid-1990s found that ORT was being used to treat 60 percent of all diarrheal episodes. Furthermore, more than 70 percent of mothers could prepare an effective oral rehydration solution, even though many of them would have been too young to have heard the information first-hand during the original campaign, suggesting that knowledge was being passed among generations (Chowdhury et al. 1997). Key to this success was extensive community involvement in the project, with emphasis on community members rather than outsiders disseminating health information. The

community-participation model of BRAC continues to be used in Bangladesh to meet a variety of health and rural development needs.

This emphasis on community participation is at the heart of genuine **primary healthcare** campaigns. A strategy to promote primary healthcare, or "health for all," was adopted by the WHO in 1978 in the Declaration of Alma-Ata (Crooks and Andrews 2009) and has since been repeatedly endorsed by international agencies. The main tenets of primary care are: 1) use of appropriate technology, 2) rejection of medical elitism, and 3) promotion of health to enable social development (Cueto 2004). Primary healthcare campaigns should also encourage "community and individual self-reliance and participation, an emphasis on prevention, and a multisectoral approach" (WHO 2002a, 109). Although often associated with low-income countries, in the original declaration primary healthcare was designed to be applied in both high- and low-income contexts. Common aspects of primary health campaigns in low-income countries include sanitation, nutrition, health education, maternal and child healthcare, family planning, treatment of infectious disease, and provision of essential drugs. In high-income regions, primary healthcare debates often revolve around shortages of general practice doctors and nurses and healthcare restructuring. Geographers' interest in community participation, access, and equity place them in a good position to contribute to the under-researched topic of primary healthcare (Crooks and Andrews 2009).

While the original goal of primary healthcare was to promote health for all people in ways that encourage community self-reliance, most primary health campaigns have focused on meeting basic needs with less attention to community participation. The term "**selective primary healthcare**" refers to primary health campaigns that meet only some of the goals of the original declaration. Selective primary healthcare often translates to low-cost technical interventions that address disease problems in the low-income world (Cueto 2004), with many primary health campaigns falling short of being "delivered by, for, and in communities" (Crooks and Andrews 2009, 5). Indeed, many high-profile primary health campaigns have been led by large international non-governmental organizations (NGOs), as typified by UNICEF's GOBI campaign (Crooks and Andrews

2009). Adopted in 1982, the GOBI strategy advocated four main child health initiatives: **G**rowth monitoring, **O**ral rehydration therapy, **B**reastfeeding, and childhood **I**mmunization. The program was later expanded to include **F**amily planning, **F**ood supplementation, and **F**emale literacy (GOBI-FFF) (WHO 2002b). Through focusing on providing equitable access to basic healthcare, primary healthcare campaigns have arguably led to greater improvements in global health outcomes than any other healthcare intervention. Much scholarly work suggests that returning to the ideal of community participation could strengthen these programs even further.

As healthcare moves beyond low-cost, high-impact basic interventions such as antibiotics and immunization,

the relationship between health spending and health outcomes quickly weakens. Healthcare moves beyond interventions for basic survival towards therapies that improve quality of life as well as more expensive therapies that may be life-saving but only for a minority of the populace. In these cases, healthcare spending has far less impact on common health indicators such as infant mortality rates and life expectancy. Other factors, such as equality of access to healthcare, also play a powerful role in mediating the relationship between health spending and outcomes. The scatterplot graph in Figure 10.5 illustrates clearly how, for affluent countries, there is relatively little association between health spending and child mortality, used as a **proxy** for health

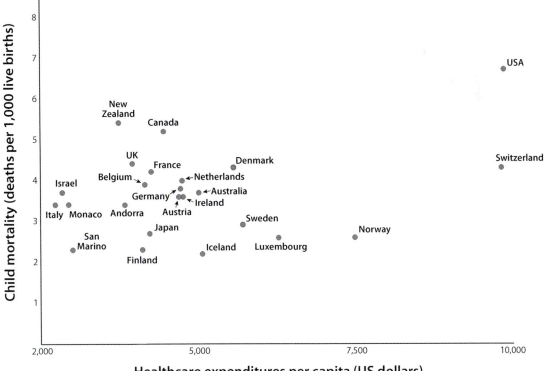

Figure 10.5 Health spending and child mortality in affluent countries, 2016

This scatterplot graph shows average per capita healthcare expenditure compared with the under-age-5 mortality rate, used as a general indicator of health, for high-income countries. Notice that healthcare expenditure does not bear much relation to health outcomes among these affluent countries.

outcomes. Indeed, the two countries spending by far the most per capita on healthcare, Switzerland and the US, have relatively poor child mortality statistics among affluent countries. Iceland, by contrast, spends only half as much as Switzerland and the US do on healthcare and yet has lower child mortality. In affluent countries, factors such as equality of healthcare access, provision of social programs, and even levels of inequality in society are often better predictors of a country's overall health than per capita expenditure on healthcare. Notably, the US spends more per capita on healthcare than any other country in the world and yet ranks below much of the industrialized world in terms of its infant mortality rate and life expectancy. Many commentators have argued that inequalities in access to care associated with the market-oriented nature of US healthcare as well as broader societal inequalities in the US underlie this finding. Examining different formalized healthcare systems can reveal strengths and weaknesses in healthcare delivery systems that can help explain findings such as this.

Formalized healthcare systems

Most governments have mechanisms to provide or regulate healthcare for their populations. The structure of healthcare systems, their efficiency of operation, and their accessibility can influence the degree to which healthcare spending translates into improved health. These factors are deeply embedded in cultural and political contexts.

While each of the world's healthcare systems is unique, most countries follow one of four primary models for delivering healthcare (Reid 2010). These four systems are distinguished largely by who manages the healthcare system and who pays for it. Although it may seem fairest to pay for healthcare on a per-use basis—with individual patients paying a market price for each service they use—advocates of **universal healthcare** argue that the people most in need of healthcare are often least able to pay for it. Indeed, as healthcare procedures become increasingly expensive and the range of known therapies expands, treatment for many conditions is now beyond the means of most individuals. Many also argue that healthcare is a fundamental human right, and so nobody should be denied access because of an inability

to pay. These arguments support the idea that healthcare provision should be organized to share the costs across a population, with healthy people effectively subsidizing treatment for the sick. For this reason, many countries mandate that all citizens must pay into the healthcare system, either through taxes or insurance programs, regardless of their health status. However, there are important differences in the extent to which societies are willing to enforce such cost-sharing mechanisms as well as differences in how they are implemented.

At one extreme, the government serves as the central organizing force behind healthcare, coordinating both the financing and management of healthcare. Healthcare is funded through taxes, with individual patients paying nothing or very little for care. This approach follows a model set by social reformer William Beveridge, who designed the UK's National Health Service (NHS) (Reid 2010). This type of healthcare system is sometimes called "**socialized medicine**," although this term can have negative connotations in the US where socialism has a tarnished reputation. Most countries that employ this model, including New Zealand, Sweden, and the UK, allow private medical practices to exist alongside government-run facilities. Some governments are also moving towards paying private providers to supply some of the care they offer.

In another group of countries, government works in an integrated way with private healthcare providers. The "Bismarck Model" was developed in nineteenth-century Germany under the Prussian Chancellor Otto von Bismarck. In this system, the government mandates a non-profit insurance system paid for by employers through employee paycheck deductions, but doctors and hospitals are privately owned and managed. Government-run insurance systems often operate alongside private insurers as a safety net for the unemployed or others falling through the cracks. France, Belgium, the Netherlands, Japan, and Switzerland operate healthcare systems modeled on this approach (Reid 2010).

The national health insurance model is similar to the Bismarck Model, except that health insurance is centralized through a non-profit, government-run insurer. This **single-payer system** leverages economies of scale, enabling a single insurer to negotiate aggressively with providers of health goods and services such as drug companies in order to keep healthcare costs down (Reid 2010). As noted earlier, Canada has been so successful in lowering drug costs through negotiating with pharmaceutical companies that US patients regularly travel to Canada to purchase drugs. Taiwan and South Korea have also adopted this model.

The "out-of-pocket" model is the healthcare equivalent of a free market. In this system, medical care is provided to those who can pay on a per-use basis. Many countries in the Global South utilize this model because they lack the government resources to manage or maintain a complex national healthcare system (Reid 2010). An obvious disadvantage is that it leaves the poorest members of society without healthcare. Healthcare resources also tend to become concentrated in urban areas where they are most profitable, leaving rural areas underserved.

The challenge of providing an efficient and equitable healthcare system can be daunting, even in high-income countries. The ethical, political, and social issues tied to healthcare delivery are perhaps best illustrated by analyzing healthcare in the US, where healthcare provision has traditionally been predicated on the idea that healthcare should be provided by the free market, leading to significant inequalities and repeated demands for reform. Most healthcare in the US is privately funded via employer-based insurance companies, although government-run programs provide healthcare at free or reduced cost to some vulnerable populations, specifically the very poor, disabled, war veterans, and elderly. Proponents of this free-market model of healthcare argue that enforcing healthcare provision through government subsidies impinges upon personal freedom and encourages the growth of inefficient government systems. Critics point to the large un- and underinsured population in the US and note that the country has poorer health statistics than would be expected for its affluence as evidence that this free-market system is not working to the benefit of the general population. Box 10.3 discusses contemporary efforts to reform the US healthcare system in response to these criticisms.

As lifespans increase, the proportion of the population aged over 60 is growing faster than any other age

Box 10.3

HEALTHCARE REFORM IN THE US

The US is one of the only countries in the Global North that does not attempt to provide universal healthcare to its citizens. Instead, most people in the US rely on private insurance companies via policies provided by their employer or bought through an individual healthcare market. Although US politicians have made efforts to extend healthcare to vulnerable populations over the past fifty years—notably *Medicaid* (for the poor and those with disabilities), *Medicare* (for the elderly), and veterans' health programs—many people fall through the cracks. People who are underemployed may be ineligible for employer-based programs and yet considered too affluent for government assistance. The self-employed may be unable to bear the high costs of premiums on the private insurance market. Those who lose or move jobs may find themselves temporarily without insurance. As a result, medical expenses are a common cause of bankruptcy in the US.

Although several US government administrations have explored offering universal healthcare, none succeeded in implementing significant reform until the Obama administration (2008–2016). After significant political wrangling, a healthcare reform bill was passed into law by a slim margin in March 2010. The Affordable Care Act (ACA, or "Obamacare") attempted to address the US's healthcare crisis in several ways. First, insurance companies were prohibited from refusing or dropping coverage for people with pre-existing conditions—so-called *guaranteed issue*—and companies were prevented from raising the cost of premiums for people with poor health. In recognition of the burden this placed on insurance companies, all individuals were required to participate in the health

continued

Box 10.3 *continued*

insurance system to ensure a large payment pool. Without that requirement, healthy individuals could avoid buying insurance until they became ill, leaving too few healthy people to pay premiums to support the cost of care. The so-called *individual mandate* was therefore introduced, requiring that everyone have health insurance coverage or pay a penalty. Other facets of the bill extended Medicaid to people with higher income levels than had been eligible previously, allowed young adults to stay on their parents' insurance plans until age 26, and provided tax breaks for small businesses to assist with the cost of providing health coverage for employees.

The reforms were successful in decreasing the number of uninsured in the US, particularly through the use of Medicaid. After passage of the bill, the number of uninsured individuals dropped to a historic low (Garfield and Orgera 2019). However, observers noted that many people found it challenging to find affordable insurance premiums on the individual market. Additionally, many people who failed to buy insurance were never fined but were instead given economic hardship exemptions, calling into question the extent to which the individual mandate was meaningfully implemented. Furthermore, although part of the intent of the bill was to curtail rapidly escalating healthcare costs, rising medical costs were not effectively addressed.

"Obamacare" was politically contentious when it was introduced, with many members of the conservative Republican party concerned that increased government oversight of healthcare is an inappropriate expansion of government power and an unwanted burden on taxpayers. The individual mandate was a particular source of controversy and a rallying cry for groups who believe that the bill represents a curtailment of individual liberties. Repeal of "Obamacare" became a key policy platform for many Republicans during the 2016 election cycle. The subsequent Republican administration under Donald Trump made a vocal priority of dismantling the program. To date, the Trump administration has removed the individual mandate as a legal requirement, defunded parts of the program, and limited access to healthcare exchanges (where people can buy health insurance policies) by reducing enrollment periods and cutting funding for enrollment navigators and advertising. Nonetheless, lacking any viable alternative plan, and with many newly insured people keen to maintain their insurance coverage, the ACA has rolled on. Although the uninsured population increased slightly in 2017 (Garfield and Orgera 2019), other indicators suggest that the ACA continues to operate moderately effectively, with reports of an increase in the number of insurance companies participating in the government insurance exchanges and a slight decrease in the cost of premiums by 2019. Whether the ACA will persist beyond the Trump presidency currently remains to be seen.

cohort in regions around the world. This demographic change means that the cost of providing healthcare is quickly rising, leaving large **welfare states** in an especially challenging position. Japan's projected population pyramid for 2020 provides a stark illustration of an aging population, but even low-income countries such as India are beginning to see aging population structures as birth rates decline (Figure 10.6).

In many affluent countries, a large proportion of the population is now living long enough to be in retirement for as long as they were workers. In combination with low birth rates, this development means that the proportion of workers in the population is rapidly diminishing. The relationship between dependent and working populations is known as the **dependency ratio**. The dependency ratio is usually estimated by age cohorts, based on the assumption that people under age 15 and over age 65 are dependent on those between the ages of 15 and 65, who are assumed to form the economically active population. Although the economic strain of having a

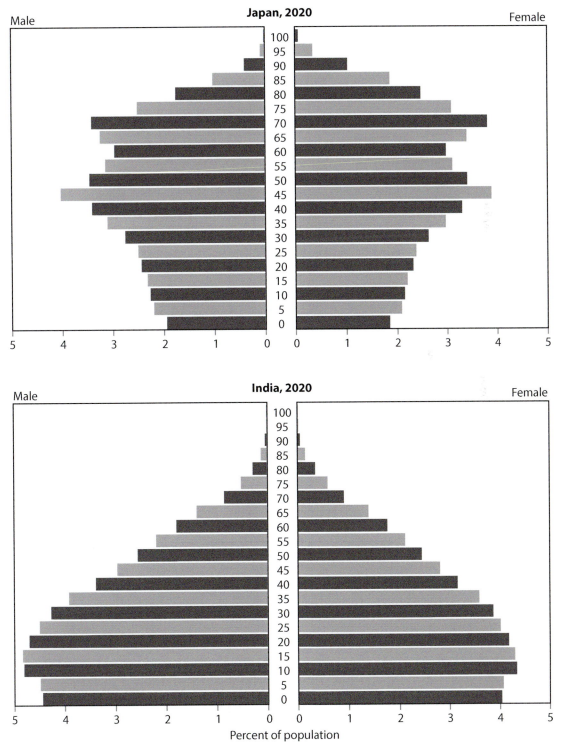

Figure 10.6 Projected population pyramids for Japan and India for 2020

Source: United Nations Population Division (2019)

high dependency ratio may be associated with either a very young or very old population, the burden of high healthcare costs associated with old age means that older populations face particularly grave economic challenges. The number of people living to *very* old age is also significant because people older than 85 years require significantly more care than the 65-to-85 age group, some of whom may also still be economically active (Figure 10.7). Indeed, one hopeful aspect of the aging of populations is that many people in their sixties and seventies are living healthier, more productive lives than

ever before, and so efforts to increase retirement age—although politically unpopular—might help alleviate the problem. Many governments are now grappling with these demographic issues as they face the prospect of having to scale back their welfare programs to cope with rapidly aging populations.

$$\text{Dependency ratio} = \frac{\text{Number of people under 15 and over 65 in a specified population}}{\text{Number of people aged 15 to 65 in a specified population}} \times 100$$

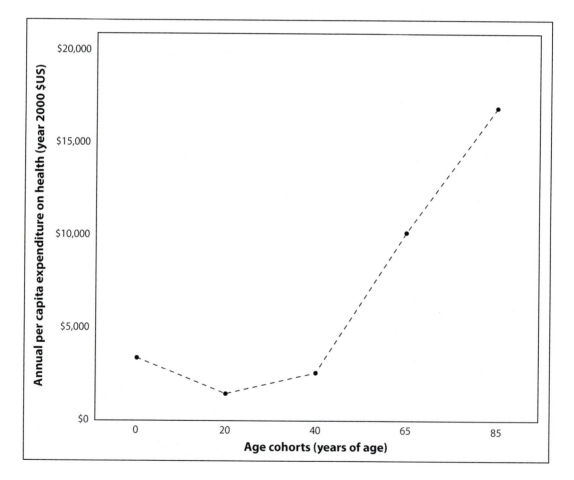

Figure 10.7 Annual per capita healthcare spending by age cohort in the US, 2000
Source: Data from Alemayehu and Warner (2004)

Another response to escalating medical costs is the restructuring of healthcare systems towards more market-oriented, neo-liberal models with the goal of improving efficiency (Lawson 2007), as has begun to occur in New Zealand, Canada, and the UK. These "restructuring" initiatives often involve shutting small and rural hospitals and clinics to centralize care in more efficient larger institutions. Additionally, facilities for vulnerable populations such as the mentally ill have been closed in favor of "care in the community," on the grounds that old, institutionalized models of care are outdated. Although justified by the logic of efficiency and new ideas of effective care, many view this kind of restructuring as a way for government to transfer costs of care from the state to individuals, and this trend has ultimately further diminished access for already marginalized populations (Lawson 2007).

Increasingly, many socialized medicine systems now charge fees for some services, such as dental care, eye care, or prescription drugs, to balance the books. Some governments are also attempting to more rigorously evaluate new treatments before they are widely adopted, as exemplified by the role of the National Institute for Health and Care Excellence (NICE) in the UK, which provides independent advice to the UK's National Health Service. Part of the organization's responsibility is to review medical therapies in order to recommend which ones provide sufficient benefits at reasonable cost.

Healthcare utilization and access

Inequalities in *access* to care mean that some groups do not benefit fully from existing services. Communities with the greatest health needs often have the poorest access to healthcare services, a phenomenon sometimes referred to as the "inverse care law" (Tudor Hart 1971). The study of healthcare accessibility and utilization often focuses on **healthcare barriers**—the geographic, economic, social, and cultural factors that prevent individuals from receiving effective healthcare. These sorts of barriers mean that there is an important distinction between *potential* access—how a population could use health services, and *realized* access—how a population actually uses health services (see Aday and Andersen 1974).

Geographers have long been interested in modeling access to healthcare using methods such as **Central Place**

Theory and **location-allocation modeling**. A major advantage of these methods is that reliable data are relatively easy to find and straightforward to interpret (Hanlon 2009). These models assume that a primary physical barrier to healthcare is *distance*. For poverty-stricken individuals, even the price of bus fare may prove a significant barrier to getting to a healthcare facility. In a study of healthcare utilization in rural West Kenya, for example, the authors found a significant **distance-decay effect** in the way that residents use healthcare clinics, starting at distances of as little as 1 km (Feikin et al. 2009). Another study, conducted in rural Mali, showed that lack of transportation was a significant barrier for pregnant women trying to maintain regular prenatal care and obtain professional assistance during delivery (Gage 2007). Healthcare facilities typically become centralized in urban areas if market pressures drive their location, and so many governments make a concerted effort to redirect health services towards rural areas. Costa Rica, for example, has built a network of rural clinic outposts to provide rural populations with better access to healthcare (Figure 10.8).

While distance is generally not as important a barrier in affluent countries, physical distance (sometimes referred to as "**spatial access**") can be important in terms of direct and indirect travel costs, the need for extra time off work for healthcare visits, and even the challenges of terrain and seasonal weather conditions (Panelli, Gallagher, and Kearns 2006). The impact of spatial access is especially important for rural communities, where low population densities mean that healthcare facilities are more dispersed and specialist healthcare may mean hours of travel. In one study in the US state of North Carolina, for instance, whether a resident held a driver's license was found to influence frequency of healthcare visits and health outcomes (Arcury, Gesler et al. 2005; Arcury, Preisser et al. 2005). This poorer access to healthcare services is combined with generally poorer health in rural communities—an example of Hart's "inverse care law" (Schuurman 2009). The impact of distance to healthcare in rural areas is also compounded by socioeconomic vulnerability (Wilson et al. 2009; Schuurman 2009). In many cases, pre-existing problems with healthcare provision are exacerbated by efforts to restructure health services to rationalize costs, frequently involving the loss of small, "inefficient" rural providers (Schuurman 2009).

Figure 10.8 A rural clinic outpost in Costa Rica

This loss of rural health facilities may also be significant beyond its direct impact on rural health, as health workers often play important social and economic roles within rural communities, which are lost when health services are consolidated (Farmer et al. 2003).

An additional problem for remote or rural areas is the recruitment and retention of health professionals (Wilson et al. 2009; Skinner and Rosenberg 2006). City jobs are often more attractive to healthcare personnel because of higher salaries, better infrastructure and facilities, and other trappings of urban life. Governments around the world have sought to draw health professionals to rural areas to reduce the rural-urban divide in healthcare provision, but often with little success. Because people raised in rural areas are most likely to remain there, preferential recruitment of rural applicants for healthcare training may provide one solution. Providing medical and nursing students with exposure to rural healthcare

settings may also encourage healthcare workers to take up rural posts (Wilson et al. 2009). In some countries, such as Costa Rica, governments require medical students to commit to a period of work in an underserved area. However, these policies often lead to high turnover of rural staff, as workers leave for higher-paid, more attractive city jobs once their term is completed. Furthermore, it may leave rural clinics operating primarily with newly qualified, inexperienced staff (*ibid.*).

Economic barriers to health are pervasive in most healthcare systems, particularly in places that do not provide **universal coverage**. Millions of people in low-income regions may go their entire lives without receiving any formal healthcare because they cannot afford it. In wealthier countries, where the cost of healthcare can be very high, the key economic barrier is high health insurance premiums or other charges associated with care. Additional barriers to healthcare, such as a lack of paid leave for medical appointments, childcare responsibilities, and transportation costs, also affect poor individuals the most. Even in places such as Western Europe where there is a strong tradition of ensuring equitable access to healthcare, people have been found to avoid seeking treatment because of potential expenses (Mackenbach and Bakker 2002).

Cultural factors can also serve as barriers to healthcare. Language is often a problem for immigrant groups, who may be intimidated by speaking an unfamiliar language or simply unable to communicate with healthcare staff (Whitehead and Dahlgren 2006; Documét and Sharma 2004; Jang et al. 2016). Older household members may have to rely on children or grandchildren to translate for them, posing significant potential for embarrassment or misunderstanding. The cultural appropriateness of care may also be significant. For some cultural groups, the gender of a physician is important. Even the way that doctors interact with patients may influence the acceptability of healthcare for some groups.

It is often only through qualitative analysis that we can explore factors such as individual care-seeking behaviors. For example, in a focus group study of immigrants in Ontario, Canada, participants reported that they felt rushed by doctors, and vocalized concerns about formulaic approaches to medicine where doctors followed strict protocols to avoid litigation (Asanin and Wilson 2008). Feelings of shyness or embarrassment may also be particularly significant for some ethnic groups. In a study investigating barriers to cervical screening in Auckland, New Zealand, for instance, concerns over exposing the body were found to be a significant barrier, particularly among Maori and South Pacific women, "who were more likely to see the bodily domains involved as sacred" (Lovell, Kearns, and Friesen 2007, 148). Differences in fundamental understandings of health and disease can also act as a barrier to effective treatment. Fadiman (1998) explored cultural issues related to the treatment of epilepsy for a Hmong child's family and her doctors in the US. Repeated misunderstandings arose between the child's family and her doctors, related to different ideas about the cause of the condition and divergent expectations of treatment. The resulting confusion resulted in the child being taken into state care at one point and several mistakes in her treatment regime.

Immigrant communities are active shapers of their healthcare experiences, taking a variety of steps to navigate the barriers that they encounter. For instance, Jang (2016) describes how Korean Americans use a variety of coping strategies, including visiting co-ethnic doctors, seeking out traditional Korean medicine, and taking trips to Korea for care. Despite these actions, many communities find themselves marginalized from mainstream medical care, and so much of the literature on healthcare access recognizes the challenges of social marginalization.

GEOGRAPHIES OF CARE

Some critical geographers have taken up the topic of care more generally as a process integral to human society and through which geographic inequalities become evident. Care, from this perspective, is viewed as an inherently political, economic, and cultural concern, bringing into question the roles played by gender, race, and class as drivers of who cares for whom (Tronto 2005). As Lawson (2007, 7) states: "Caring for person-to-person relations involves understanding how difference is socially constructed, and so a critical ethic of care must be coupled with analysis of the structures and institutions that reproduce exclusion, oppression, environmental degradation, and the like."

This work has roots in feminist considerations of who performs caring roles, and how society views the responsibility of providing care (Thien and Hanlon 2009). Feminist geographers have pointed out that the greatest burden of caring is usually shouldered by women, as well as immigrants and ethnic minorities, and that care work is often marginalized and devalued (Williams and Crooks 2008; Tronto 2005) (Figure 10.9). The space of home as a site of care-giving has received particular attention (e.g., Crooks and Andrews 2009; Brown 2003; Williams 2002; Milligan 2000), often focusing on the dichotomy between the therapeutic nature of home for those cared for and the strain placed on family members who serve as caregivers.

Another focus of geographies of care is a social justice perspective on healthcare access, inequality, and political and organizational reform (Parr 2003). In

Figure 10.9 Mother and child, Peru

A critical feminist examination of care encourages us to think about who shoulders the greatest burden of care in society. Child-care and eldercare have traditionally fallen to women, with women often essentialized as nurturing and therefore ideally suited to caring roles. This results in women undertaking a large burden of caring labor in many households.

particular, healthcare restructuring has affected where care is provided and who provides care. One concern is that neo-liberal economic policies have pushed healthcare provision towards the privatization of care (Williams and Crooks 2008; Lawson 2007), leading to a greater domestic care burden (Williams 2002). While deinstitutionalization and the associated notion of providing "care in the community" for vulnerable populations may move people out of institutions that do not effectively promote health, this restructuring presupposes that community sources of care are available. In many cases this has not been the case, leaving vulnerable individuals homeless or not receiving the regular care and counseling that they need (Lee, Wolch, and Walsh 1998). In response, policymakers have begun to accept that this model has not worked as intended, and some countries have begun to return to an institutional model (see Moon 2000; Wolch, Nelson, and Rubalcaba 1988).

THERAPEUTIC LANDSCAPES

As researchers have broadened their understanding of influences on health to include humanistic approaches that consider individual experiences, new potential sources of healing have come to light. A primary focus for health geographers within this expanding field has been how individuals' **sense of place** influences their health. In Chapter 7, we considered the importance of place with respect to how landscapes of deprivation can have *detrimental* impacts on health. Here we consider the growing literature that explores **therapeutic landscapes**, places that are believed to be *beneficial* to health.

"A therapeutic landscape arises when physical and built environments, social conditions and human perceptions combine to produce an atmosphere which is conducive to healing" (Gesler 1996, 96). Intriguingly, modern healthcare facilities are rarely considered in popular conceptions of therapeutic landscapes. Instead, research on therapeutic landscapes has focused on places with longstanding folk reputations for healing, such as spas and sacred sites (Figure 10.10). For instance, Bath, England, has a reputation for healing that stems from ideas about the therapeutic properties of mineral springs

(Gesler 1998); Lourdes in France is renowned for healing because of its significance as a place of religious pilgrimage and miraculous events (Gesler 1996). Mystical, spiritual, or cultural beliefs can generate and reinforce the reputation of sites as healing places (Williams 2010). It is not only the places themselves that are important in explaining their healing properties, but also how people have constructed these places in personal and public imaginations. Geores (1998), for instance, describes how the founders of the town of Hot Springs, South Dakota, made a conscious decision to construct the identity of their town as a site of healing.

Researchers have also considered the importance of landscapes whose therapeutic nature is not explicitly articulated. Much work has considered *natural* landscapes as therapeutic, often advocating the valuable role that wilderness experiences can play in healing. In Japan, the practice of "forest bathing," or Shinrin-yoku, was developed in the 1980s to exploit the therapeutic benefits of spending time in nature. The practice has now become popular in other countries, with proponents advocating that Western medicine look at its potential benefits to health (Hansen, Jones, and Tocchini 2017). Other studies have looked at the significance of natural areas to health more broadly. For instance, Lea (2008) describes the role of the natural environment in generating a recuperative experience at a massage retreat in Spain, Palka (1999) discusses the therapeutic nature of landscapes in Denali National Park, Finlay et al. (2015) suggest that natural areas promote well-being among older adults, and Bell et al. (2015) look at the therapeutic nature of coasts.

Scholars have also extended the concept of therapeutic landscapes to encompass the therapeutic nature of everyday spaces, such as living and working environments, even public libraries (Brewster 2014). The notion of "home" as a therapeutic landscape has received particular attention (Mowl, Pain, and Talbot 2000; Dyck 1998). For instance, Nagib and Williams (2018) explore whether modifications to the home can generate a place of healing for children with autism. Coyle (2004) discusses how women suffering from environmental illness—a condition in which sufferers report negative health consequences from everyday exposures to chemicals, even at supposedly safe levels—often

Figure 10.10 A therapeutic garden near Tokyo, Japan

This therapeutic garden provides hot spring baths for its customers. Hot springs and baths have a long tradition as places of healing, which continues to this day. In some countries, doctors may prescribe a period at a thermal bath as a holistic treatment for ailments, which may be covered by health insurance.

construct "home" as a safe haven from the disease. This involves not only removing chemical threats from their home spaces, but also constructing the home as a space of trust and confidence in order to buffer the emotional strain of living with a condition that elicits derision from others.

Some researchers have argued that the health implications of therapeutic landscapes may derive as much from the social interactions they foster as from the physical landscape itself. For instance, Milligan et al. (2004) considered the role of community gardens in improving quality of life and well-being for the elderly in Northern England, and concluded that the inclusionary, mutually supporting spaces of these gardens help to combat social isolation and develop social networks. The therapeutic qualities of landscapes are thus often analyzed in terms of interactions among their environmental, social, and symbolic aspects (Conradson 2005). In this way, we should not assume that there is some inherent therapeutic experience associated with any particular landscape. Instead, it is important to consider how personal interactions with landscapes generate unique "therapeutic landscape experiences" (Conradson 2005, 346).

CONCLUSION

Healthcare sits at a critical intersection of political, economic, and cultural geographies. Political and economic structures frame how care is provided, how it is funded, and even who is eligible to receive treatment. Cultural preferences influence factors such as which conditions are deemed to require treatment, how treatment is administered, and who is considered competent to provide treatment.

Much geographic research on healthcare focuses on equity, considering patterns in provision of and access to healthcare. Many of the methods and techniques of geography are well suited to such analyses, particularly the ability to combine spatial and social approaches. We extend this effort to combine approaches by concluding with a chapter that considers ways to effectively combine ecological, spatial, and social approaches.

DISCUSSION QUESTIONS

1   Do people have a right to healthcare? What procedures should we consider a privilege? Think about categories such as: primary healthcare, elective procedures, cosmetic procedures, curative versus life-prolonging procedures, procedures that improve quality of life, etc.

2   Is a biomedical understanding of the mechanism through which a particular therapy works required for it to be a valid healthcare method? If not, how can we effectively regulate alternative and complementary therapies that do not fit the biomedical mold?

3   Is the flow of healthcare professionals around the world according to market demands an overall positive phenomenon, or should we strengthen regulations to control this flow? How else might we address shortages of healthcare personnel?

4   What system is used to provide healthcare in the country where you live? What are some of the pros and cons of this system? What are some of the current controversies surrounding this system?

5   What barriers to healthcare can you identify in your local community? How might healthcare access be improved?

6   What types of landscapes do you perceive as therapeutic? How do your ideas of therapeutic landscapes differ from those of a classmate? Why do you think your ideas differ?

SUGGESTED READING

Eckenwiler, L. 2014. "Care Worker Migration, Global Health Equity, and Ethical Place-Making." *Women's Studies International Forum* 47: 212–22.

Finlay, J., T. Franke, H. McKay, and Sims-Gould. 2015. "Therapeutic Landscapes and Wellbeing in Later Life: Impacts of Blue and Green Spaces for Older Adults." *Health & Place* 34: 97–106.

Gage, A. J. 2007. "Barriers to the Utilization of Maternal Health Care in Rural Mali." *Social Science & Medicine* 65: 1666–82.

Gesler, W. 1996. "Lourdes: Healing in a Place of Pilgrimage." *Health and Place* 2: 95–105.

Lawson, V. 2007. "Geographies of Care and Responsibility." *Annals of the Association of American Geographers* 97 (1): 1–11.

Reid, T. R. 2010. *The Healing of America: A Global Quest for Better, Cheaper, and Fairer Health Care.* New York: Penguin Press.

VIDEO RESOURCES

"Sick Around America." Frontline, PBS. More information at: www.pbs.org/wgbh/pages/frontline/sickaroundamerica/.

"Sick Around the World." Frontline, PBS. More information at: www.pbs.org/wgbh/pages/frontline/sickaroundtheworld/.

REFERENCES

Aday, L. A., and R. Andersen. 1974. "A Framework for the Study of Access to Medical Care." *Health Services Research* 9 (3): 208–20.

Alemayehu, Berhanu, and Kenneth E. Warner. 2004. "The Lifetime Distribution of Health Care Costs." *Health Services Research* 39 (3): 627–42. doi:10.1111/j.1475-6773.2004.00248.x.

Anand, S., and T. Barnighausen. 2004. "Human Resources and Health Outcomes: Cross-Country Econometric Study." *Lancet* 364 (9445): 1603–9. doi:10.1016/s0140-6736(04)17313-3.

Arcury, T. A., Wilbert M. Gesler, John S. Preisser, Jill Sherman, et al. 2005. "The Effects of Geography and Spatial Behavior on Health Care Utilization Among the Residents of a Rural Region." *Health Services Research* 40 (1): 135–55. doi:10.1111/j.1475-6773.2005.00346.x.

Arcury, T. A., J. S. Preisser, W. M. Gesler, and J. M. Powers. 2005. "Access to Transportation and Health Care Utilization in a Rural Region." *Journal of Rural Health* 21 (1): 31–38.

Asanin, J., and K. Wilson. 2008. "'I Spent Nine Years Looking for a Doctor': Exploring Access to Health Care Among Immigrants in Mississauga, Ontario, Canada." *Social Science & Medicine* 66 (6): 1271–83. doi:10.1016/j.socscimed.2007.11.043.

Awases, M., A. Gbary, and R. Chotora. 2004. "Migration of Health Professionals in Six Countries: A Synthesis Report." Accessed May 14, 2019. https://www.afro.who.int/sites/default/files/2017-06/hrh%20migration_en.pdf.

Bach, Stephen. 2004. "Migration Patterns of Physicians and Nurses: Still the Same Story?" *Bulletin of the World Health Organization* 82 (8): 624–25.

Bell, Sarah L., Cassandra Phoenix, Rebecca Lovell, and Benedict W. Wheeler. 2015. "Seeking Everyday Wellbeing: The Coast as a Therapeutic Landscape." *Social Science & Medicine* 142: 56–67. doi:10.1016/j.socscimed.2015.08.011.

Brewster, Liz. 2014. "The Public Library as Therapeutic Landscape: A Qualitative Case Study." *Health & Place* 26: 94–99. doi:10.1016/j.healthplace.2013.12.015.

British House of Commons. 2010. "Science and Technology Committee—Fourth Report. Evidence Check 2: Homeopathy." Accessed August 22, 2019. https://publications.parliament.uk/pa/cm200910/cmselect/cmsctech/45/4502.htm.

Brown, Michael. 2003. "Hospice and the Spatial Paradoxes of Terminal Care." *Environment and Planning A: Economy and Space* 35 (5): 833–51. doi:10.1068/a35121.

Brush, B. L., and J. Sochalski. 2007. "International Nurse Migration: Lessons from the Philippines." *Policy, Politics, & Nursing Practice* 8 (1): 37–46. doi:10.1177/1527154407301393.

Buzinde, C. N., and C. Yarnal. 2012. "Therapeutic Landscapes and Postcolonial Theory: A Theoretical Approach to Medical Tourism." *Social Science & Medicine* 74 (5): 783–87. doi:10.1016/j.socscimed.2011.11.016.

Chowdhury, A. Mushtaque R., Fazlul Karim, S. K. Sarkar, Richard A. Cash, and Abbas Bhuiya. 1997. "The Status of ORT in Bangladesh: How Widely Is It Used?" *Health Policy and Planning* 12 (1): 58–66. doi:10.1093/heapol/12.1.58.

Conradson, David. 2005. "Landscape, Care and the Relational Self: Therapeutic Encounters in Rural England." *Health & Place* 11 (4): 337–48. doi:10.1016/j.healthplace.2005.02.004.

[CORDIS] Community Research and Development Information Service. 2012. "Complementary Medicine Popular Across Europe." Accessed July 20, 2019. https://cordis.europa.eu/news/rcn/35388/en.

Coyle, Fiona. 2004. "'Safe Space' as Counter-Space: Women, Environmental Illness and 'Corporeal Chaos'." *The Canadian Geographer/Le Géographe canadien* 48 (1): 62–75. doi:10.1111/j.1085-9489.2004.01e08.x.

Crooks, V. A., and G. J. Andrews. 2009. *Primary Health Care: People, Practice, Place.* Ashgate.

Cueto, Marcos. 2004. "The Origins of Primary Health Care and Selective Primary Health Care." *American Journal of Public Health* 94 (11): 1864–74. doi:10.2105/ajph.94.11.1864.

Cyranoski, D. 2018. "Why Chinese Medicine Is Heading for Clinics Around the World." *Nature* 561 (7724): 448–50. doi:10.1038/d41586-018-06782-7.

Day, P., J. Pearce, and D. Dorling. 2008. "Twelve Worlds: A Geo-Demographic Comparison of Global Inequalities in Mortality." *Journal of Epidemiology and Community Health* 62 (11): 1002–10. doi:10.1136/jech.2007.067702.

Documét, Patricia I., and Ravi K. Sharma. 2004. "Latinos' Health Care Access: Financial and Cultural Barriers." *Journal of Immigrant Health* 6 (1): 5–13. doi:10.1023/B:JOIH.0000014638.87569.2e.

Dovlo, D. 2007. "Migration of Nurses from Sub-Saharan Africa: A Review of Issues and Challenges." *Health Services Research* 42 (3 Pt 2): 1373–88. doi:10.1111/j.1475-6773.2007.00712.x.

Dyck, I. 1998. "Women with Disabilities and Everyday Geographies." In *Putting Health into Place: Landscape, Identity, and Well-being*, edited by R. A. Kearns and W. M. Gesler. Syracuse University Press.

Eckenwiler, Lisa. 2014. "Care Worker Migration, Global Health Equity, and Ethical Place-Making." *Women's Studies International Forum* 47: 213–22. doi:10.1016/j.wsif.2014.04.003.

Ernst, E. 2015. "There Is No Scientific Case for Homeopathy: The Debate Is Over." *The Guardian*. Accessed August 24, 2019. www.theguardian.com/commentisfree/2015/mar/12/no-scientific-case-homeopathy-remedies-pharmacists-placebos.

Esmail, N. 2017. "Complementary and Alternative Medicine: Use and Public Attitudes 1997, 2006, and 2016." Frasier Institute. Accessed July 20, 2019. www.fraserinstitute.org/studies/complementary-and-alternative-medicine-use-and-public-attitudes-1997-2006-and-2016.

Fadiman, A. 1998. *The Spirit Catches You and You Fall Down: A Hmong Child, Her American Doctors, and the Collision of Two Cultures*. Farrar, Straus and Giroux.

Farmer, J., W. Lauder, H. Richards, and S. Sharkey. 2003. "Dr John Has Gone: Assessing Health Professionals' Contribution to Remote Rural Community Sustainability in the UK." *Social Science & Medicine* 57 (4): 673–86.

Feikin, D. R., L. M. Nguyen, K. Adazu, M. Ombok, et al. 2009. "The Impact of Distance of Residence from a Peripheral Health Facility on Pediatric Health Utilisation in Rural Western Kenya." *Tropical Medicine & International Health* 14 (1): 54–61. doi:10.1111/j.1365-3156.2008.02193.x.

Finlay, Jessica, Thea Franke, Heather McKay, and Joanie Sims-Gould. 2015. "Therapeutic Landscapes and Wellbeing in Later Life: Impacts of Blue and Green Spaces for Older Adults." *Health & Place* 34: 97–106. doi:10.1016/j.healthplace.2015.05.001.

Gage, A. J. 2007. "Barriers to the Utilization of Maternal Health Care in Rural Mali." *Social Science & Medicine* 65 (8): 1666–82. doi:10.1016/j.socscimed.2007.06.001.

Garfield, R., and K. Orgera. 2019. "The Uninsured and the ACA: A Primer—Key Facts About Health Insurance and the Uninsured Amidst Changes to the Affordable Care Act." Accessed July 23, 2019. www.kff.org/uninsured/report/the-uninsured-and-the-aca-a-primer-key-facts-about-health-insurance-and-the-uninsured-amidst-changes-to-the-affordable-care-act/.

Geores, M. 1998. "Surviving on Metaphor: How 'Health = Hot Springs' Created and Sustained a Town." In *Putting Health into Place: Landscape, Identity, and Well-being*, edited by R. A. Kearns and W. M. Gesler. Syracuse University Press.

Gesler, W. 1996. "Lourdes: Healing in a Place of Pilgrimage." *Health & Place* 2 (2): 95–105. doi:10.1016/1353-8292(96)00004-4.

Gesler, W. 1998. "Bath's Reputation as a Healing Place." In *Putting Health into Place: Landscape, Identity, and Well-being*, edited by R. A. Kearns and W. M. Gesler. Syracuse University Press.

Grady, D. 1998. "To Aid Doctors, A.M.A. Journal Devotes Entire Issue to Alternative Medicine." *New York Times*, November 11.

Hanlon, N. 2009. "Access and Utilization Reconsidered: Towards a Broader Understanding of the Spatial Ordering of Primary Healthcare." In *Primary Health Care: People, Practice, Place*, edited by V. A. Crooks and G. J. Andrews. Ashgate.

Hansen, Margaret M., Reo Jones, and Kirsten Tocchini. 2017. "Shinrin-Yoku (Forest Bathing) and Nature Therapy: A State-of-the-Art Review." *International Journal of Environmental Research and Public Health* 14 (8): 851. doi:10.3390/ijerph14080851.

Hao, Q., Z. Lu, B. R. Dong, C. Q. Huang, and T. Wu. 2011. "Probiotics for Preventing Acute Upper Respiratory Tract Infections." *The Cochrane Database of Systematic Reviews* (9): Cd006895. doi:10.1002/14651858. CD006895.pub2.

Harvard Medical School. 2004. "Why the FDA Banned Ephedra." Accessed July 20, 2019. www.health.harvard. edu/staying-healthy/ephedra-ban.

Helman, C. G. 2007. *Culture, Health and Illness.* 5th ed. Taylor & Francis.

Jang, Yuri, Nan Sook Park, David A. Chiriboga, Hyunwoo Yoon, et al. 2016. "Risk Factors for Social Isolation in Older Korean Americans." *Journal of Aging and Health* 28 (1): 3–18. doi:10.1177/0898264315584578.

Jordan, M. 2019. "3 Arrested in Crackdown on Multimillion-Dollar 'Birth Tourism' Businesses." *New York Times*, January 31. www.nytimes.com/2019/01/31/ us/anchor-baby-birth-tourism.html.

Kaiser, R. 2018. "Birth and Biopolitics." In *Reproductive Geographies*, edited by M. R. England, M. Fannin, and H. Hazen, 161–83. Routledge.

Kearns, Robin A., and Alun E. Joseph. 1993. "Space in Its Place: Developing the Link in Medical Geography." *Social Science & Medicine* 37 (6): 711–17. doi:10.1016/0277-9536(93)90364-A.

Kemppainen, Laura M., Teemu T. Kemppainen, Jutta A. Reippainen, S. T. Salmenniemi, et al. 2018. "Use of Complementary and Alternative Medicine in Europe: Health-Related and Sociodemographic Determinants." *Scandinavian Journal of Public Health* 46 (4): 448–55. doi:10.1177/1403494817733869.

Lassi, Z. S., A. Moin, and Z. A. Bhutta. 2016. "Zinc Supplementation for the Prevention of Pneumonia in Children Aged 2 Months to 59 Months." *The Cochrane Database of Systematic Reviews* 12: Cd005978. doi:10.1002/14651858.CD005978.pub3.

Lawson, Victoria. 2007. "Geographies of Care and Responsibility." *Annals of the Association of American Geographers* 97 (1): 1–11. doi:10.1111/j.1467-8306.2007.00520.x.

Lea, J. 2008. "Retreating to Nature: Rethinking 'Therapeutic Landscapes'." *Area* 40: 90–98.

Lee, A., and L. T. Fan. 2009. "Stimulation of the Wrist Acupuncture Point P6 for Preventing Postoperative Nausea and Vomiting." *The Cochrane Database of Systematic Reviews* (2): Cd003281. doi:10.1002/14651858. CD003281.pub3.

Lee, J., J. Wolch, and J. Walsh. 1998. "Homeless Health and Service Needs." In *Putting Health into Place: Landscape, Identity, and Well-being*, edited by R. A. Kearns and W. M. Gesler. Syracuse University Press.

Lovell, S., R. A. Kearns, and W. Friesen. 2007. "Sociocultural Barriers to Cervical Screening in South Auckland, New Zealand." *Social Science & Medicine* 65 (1): 138–50. doi:10.1016/j.socscimed.2007.02.042.

Mackenbach, J. P., and M. Bakker. 2002. *Reducing Inequalities in Health: A European Perspective.* Routledge.

Magner, L. N. 1992. *A History of Medicine.* Taylor & Francis.

Meyer, S., L. Gortner, A. Larsen, G. Kutschke, S. Gottschling, S. Graber, and N. Schroeder. 2013. "Complementary and Alternative Medicine in Paediatrics: A Systematic Overview/Synthesis of Cochrane Collaboration Reviews." *Swiss Medical Weekly* 143: w13794. doi:10.4414/smw.2013.13794.

Milligan, Christine. 2000. "'Bearing the Burden': Towards a Restructured Geography of Caring." *Area* 32 (1): 49–58. doi:10.1111/j.1475-4762.2000.tb00114.x.

Milligan, Christine, Anthony Gatrell, and Amanda Bingley. 2004. "'Cultivating Health': Therapeutic Landscapes and Older People in Northern England." *Social Science & Medicine* 58 (9): 1781–93. doi:10.1016/ S0277-9536(03)00397-6.

Moon, G. 2000. "Risk and Protection: The Discourse of Confinement in Contemporary Mental Health Policy." *Health and Place* 6 (3): 239–50.

Mowl, Graham, Rachel Pain, and Carol Talbot. 2000. "The Ageing Body and the Homespace." *Area* 32 (2): 189–97.

Nagib, Wasan, and Allison Williams. 2018. "Creating 'Therapeutic Landscapes' at Home: The Experiences of Families of Children with Autism." *Health & Place* 52: 46–54. doi:10.1016/j.healthplace.2018.05.001.

Nahin, R. L., P. M. Barnes, and B. J. Stussman. 2016. "Expenditures on Complementary Health Approaches: United States, 2012." *National Health Statistics Reports* (95): 1–11.

Nair, M., and P. Webster. 2013. "Health Professionals' Migration in Emerging Market Economies: Patterns, Causes and Possible Solutions." *Journal of Public Health* 35 (1): 157–63. doi:10.1093/pubmed/fds087.

National Center for Complementary and Integrative Health. 2016. "Ephedra." Accessed July 20, 2019. https://nccih.nih.gov/health/ephedra.

[OECD] Organisation for Economic Cooperation and Development. 2019. "OECD Data." Accessed September 1, 2019. https://data.oecd.org/.

Okeke, E. N. 2013. "Brain Drain: Do Economic Conditions 'Push' Doctors out of Developing Countries?" *Social Science & Medicine* 98: 169–78. doi:10.1016/j.socscimed.2013.09.010.

Pak, J. 2019. "Providing Medical Care to 'Birth Tourists' from China." Accessed July 23, 2019. www.marketplace.org/2019/03/13/world/providing-medical-care-birth-tourists-china/.

Palka, E. 1999. "Accessible Wilderness as a Therapeutic Landscape: Experiencing the Nature of Denali National Park, Alaska." In *Therapeutic Landscapes: The Dynamic Between Place and Wellness*, edited by A. Williams. New York: University of Press of America.

Panelli, Ruth, Lou Gallagher, and Robin Kearns. 2006. "Access to Rural Health Services: Research as Community Action and Policy Critique." *Social Science & Medicine* 62 (5): 1103–14. doi:10.1016/j.socscimed.2005.07.018.

Parr, Hester. 2003. "Medical Geography: Care and Caring." *Progress in Human Geography* 27 (2): 212–21. doi:10.1191/0309132503ph423pr.

Pittman, Patricia, Linda H. Aiken, and James Buchan. 2007. "International Migration of Nurses: Introduction." *Health Services Research* 42 (3 Pt 2): 1275–80. doi:10.1111/j.1475-6773.2007.00713.x.

Reid, T. R. 2010. *The Healing of America: A Global Quest for Better, Cheaper, and Fairer Health Care*. Penguin Publishing Group.

Schell, C. O., M. Reilly, H. Rosling, S. Peterson, and A. M. Ekstrom. 2007. "Socioeconomic Determinants of Infant Mortality: A Worldwide Study of 152 Low-, Middle-, and High-Income Countries." *Scandinavian Journal of Public Health* 35 (3): 288–97. doi:10.1080/14034940600979171.

Schuurman, N. 2009. "The Effects of Population Density, Physical Distance, and Socio-Economic Vulnerability on Access to Primary Health Care in Rural and Remote British Colombia, Canada." In *Primary Health Care: People, Practice, Place*, edited by V. A. Crooks and G. J. Andrews. Farnham, Surrey: Ashgate.

Siyam, A., WHO, and M. R. D. Poz. 2014. *Migration of Health Workers: WHO Code of Practice and the Global Economic Crisis*. Geneva: World Health Organization.

Skinner, Mark W., and Mark W. Rosenberg. 2006. "Managing Competition in the Countryside: Non-Profit and For-Profit Perceptions of Long-Term Care in Rural Ontario." *Social Science & Medicine* 63 (11): 2864–76. doi:10.1016/j.socscimed.2006.07.028.

Tabish, S. A. 2008. "Complementary and Alternative Healthcare: Is It Evidence-Based?" *International Journal of Health Sciences* 2 (1): V–IX.

Thien, D., and N. Hanlon. 2009. "Unfolding Dialogues About Gender, Care and 'The North': An Introduction." *Gender, Place & Culture* 8.

Tronto, J. 2005. "The Value of Care." Accessed July 20, 2011. https://bostonreview.net/archives/BR27.1/tronto.html.

Tudor Hart, Julian. 1971. "The Inverse Care Law." *The Lancet* 297 (7696): 405–12. doi:10.1016/S0140-6736(71)92410-X.

United Nations Population Division. 2019. *World Population Prospects: The 2019 Revision*. Geneva: United Nations Population Division.

US Food and Drug Administration (FDA). 2018. "How to Buy Medicines Safely from an Online Pharmacy." Accessed August 24, 2019. www.fda.gov/consumers/consumer-updates/how-buy-medicines-safely-online-pharmacy.

Whitehead, M., and Göran Dahlgren. 2006. *Levelling Up (part 1): A Discussion Paper on Concepts and Principles for Tackling Social Inequities in Health*, Edited by World Health Organization. Copenhagen, Denmark: WHO, Regional Office for Europe.

[WHO] World Health Organization. 1946. *Constitution of the World Health Organization*. New York: WHO Interim Commission.

[WHO] World Health Organization. 2002a. *WHO Traditional Medicine Strategy 2002–2005*. Geneva: World Health Organization.

[WHO] World Health Organization. 2002b. *The World Health Report 2002: Reducing Risks, Promoting Healthy Life*. Geneva: World Health Organization.

[WHO] World Health Organization. 2006. *World Health Report 2006 (The): Working Together for Health*. Geneva: World Health Organization.

[WHO] World Health Organization. 2019. "Health Workforce—Migration." Accessed July 23, 2019. www.who.int/hrh/migration/en/.

Williams, Allison. 2002. "Changing Geographies of Care: Employing the Concept of Therapeutic Landscapes as a Framework in Examining Home Space." *Social Science & Medicine* 55 (1): 141–54. doi:10.1016/S0277-9536(01)00209-X.

Williams, Allison. 2010. "Spiritual Therapeutic Landscapes and Healing: A Case Study of St. Anne de Beaupre, Quebec, Canada." *Social Science & Medicine* 70 (10): 1633–40. doi:10.1016/j.socscimed.2010.01.012.

Williams, Allison, and Valorie A. Crooks. 2008. "Introduction: Space, Place and the Geographies of Women's Caregiving Work." *Gender, Place & Culture* 15 (3): 243–47. doi:10.1080/09663690801996254.

Wilson, N. W., I. D. Couper, E. De Vries, S. Reid, et al. 2009. "A Critical Review of Interventions to Redress the Inequitable Distribution of Healthcare Professionals to Rural and Remote Areas." *Rural Remote Health* 9 (2): 1060.

Wolch, J., C. Nelson, and A. Rubalcaba. 1988. "To Backwards? Prospects for the Reinstitutionalization of the Mentally Disabled." In *Location and Stigma: Contemporary Perspectives on Mental Health and Mental Health Care*, edited by Christopher J. Smith and J. A. Giggs. Unwin Hyman.

11

INTEGRATING APPROACHES TO THE STUDY OF THE GEOGRAPHY OF HEALTH

POLICYMAKING FROM GEOGRAPHIC PERSPECTIVES

At the core of the geography of health is an interest in the ways that space and place influence health. In this book, we first discussed how humans interact with their environment in ways that promote or hinder health—an ecological approach. Second, we examined how the social environment, including economic, cultural, and political facets, drive health. Throughout the book we have also considered explicitly spatial approaches, which explore how the spatial arrangement of organisms, societies, and environments influence health outcomes.

Although these approaches offer important perspectives on health, a combination of approaches ultimately offers the optimal way to explore critical questions and formulate policy. While it is useful to understand the biological role of mosquitos in the transmission of malaria, and ecological and spatial approaches may help us to identify where people are most vulnerable to contracting the disease, these approaches do little to help us understand why some people do not use bed nets even when they are aware of their effectiveness at reducing transmission of the disease. To address this question, we may need to delve into questions of economic resources, political will for helping marginalized populations, and the influence of individuals' life-worlds on

their behavior. Conversely, studying beliefs and practices related to malaria may contribute little to preventing the disease if we do not have a solid grasp of the disease's ecology. Large-scale disease control programs must, therefore, integrate information and strategies from multiple perspectives. The *Roll Back Malaria* partnership, for instance, is integrating research on vaccine development, efforts to improve vector control, and public education in its effort to develop successful interventions for malaria (RBM Partnership to End Malaria 2019).

Although synthesizing diverse approaches can offer a rich understanding of human health, it is seldom an easy or straightforward task. Theoretical divides remain within academia and health organizations, particularly between scholars who focus on **positivist** approaches to problem-solving and those who emphasize post-positivist approaches and diverse forms of knowledge. Both groups make considerable contributions, however, and so we urge students to consider the effectiveness and usefulness of a diversity of approaches. This is not to suggest that every research project should use multiple methods or combine all theoretical approaches. Instead, we gain the most complete understanding of health through the synthesis of findings from a variety of different studies that approach similar questions in different ways.

In this chapter, we provide examples of efforts to combine methods and approaches, using two topics where the intersection of ecological, social, and spatial approaches is easily demonstrated. First, we look at disease **eradication** and control using vaccination, considering smallpox and polio eradication efforts in particular. Second, we look at the global obesity crisis to show how ecological and social perspectives can inform effective interventions for a chronic disease.

GLOBAL INFECTIOUS DISEASE CAMPAIGNS

Effectively tackling infectious disease requires a thorough understanding of the ecology of disease and a familiarity with the social context in which infection occurs. Before the twentieth century, limited understanding of the biology of microorganisms and the mechanisms of disease cycles hindered control efforts, although numerous attempts were made to control disease. For instance, a common response to outbreaks of bubonic plague in fourteenth-century Europe was to isolate infected individuals and abandon densely populated areas, even though the causative agent and life cycle of plague remained a mystery until the late 1800s. In this early era of formal medicine, behavioral adaptations such as this were often the only recourse for combating disease.

The notion of a formal **quarantine** also emerged in response to the plague. In the fourteenth century, the port of Dubrovnik, Croatia, enforced a period of isolation on boats that arrived at the port to verify that people on board did not develop plague symptoms. As a major trading hub, Dubrovnik could not afford to close its port during repeated outbreaks of infectious disease, and so in 1377 the city council decreed that newcomers would be subject to a thirty-day, and subsequently forty-day, period of isolation. The term "quarantine" derives from the Latin "*quaranta*," meaning forty. Quarantining efforts were subsequently made with other diseases, including dysentery, leprosy, smallpox, syphilis, and yellow fever.

As our understanding of the **etiology** of disease has improved, so too has our ability to design effective campaigns to control disease. With the advent of modern biomedical advances, the challenges are now often social as much as technical. Insufficient political will, lack of funding, and poor comprehension of cultural issues are common stumbling blocks that impede public health campaigns. Although there have been great strides in reducing the incidence or range of diseases in some regions, coordinating efforts at the global scale has proved especially difficult. To date, smallpox remains the only human disease that has been eradicated, although the prevalence of polio and dracunculiasis (guinea worm) have been reduced to the point where eradication is now possible.

A key geographic question for disease control is the scale at which the problem should be approached. Should limited health care dollars be spent on large-scale disease eradication programs or on smaller-scale efforts to control or eliminate disease in local areas? Eradication can yield significant economic gains, not only from the reduction of healthcare costs to treat the illness, but also from the diminished need for vaccination, surveillance, and prevention programs (Barrett 2004). The incentive for countries to participate in ambitious eradication programs, and particularly for high-income countries to help fund them, is also often stronger than for localized control programs. Not only does the excitement generated by eradication appeal to funders, but also the investment offers significant returns to both global and local communities, as every country will ultimately benefit from eradication when the need for vaccination programs is obviated (*ibid.*). The likelihood of success is uncertain, however, making eradication a major gamble. An entire campaign can be threatened if a single country refuses to participate or if conflict prevents health workers from gaining access to an infected community. Indeed, the difficulties experienced in eradication attempts have led some observers to argue that the struggle required for eradication may be simply too great. Donald Henderson, one of the leaders of the smallpox eradication program, for example, has suggested that the "siren song of eradication" may lead public health officials to underestimate the challenges involved and that the goal of eradication may be more "evangelical" than attainable (McNeil and Dugger 2006).

Despite these concerns, eradication efforts continue. Current eradication campaigns focus on polio and dracunculiasis, although there are also hopes that regional elimination (or even possibly eradication) may be possible also for some other **neglected tropical diseases**. In theory, all infectious diseases should be eradicable given the right tools (Dowdle 1998); in practice, there are distinct technical, biological, and social factors that make some diseases easier targets than others as we illustrate in the cases of polio and smallpox. Ultimately, which diseases get attention is often more a political or economic decision than a medical one, with many diseases neglected for economic or cultural reasons. The term "**orphan disease**" refers to diseases that have been neglected because so few people suffer from them, or the populations that suffer from them are so poor that developing therapies makes little economic sense. Recently, legislation and incentive programs have been created to try to encourage research on some of these diseases (Box 11.1).

Box 11.1

ORPHAN DISEASES AND NEGLECTED TROPICAL DISEASES

Economic structures are influential in driving research agendas and funding priorities, with significant impacts in terms of how and if health problems are treated. The term "orphan disease" refers to a disease that has been ignored by the pharmaceutical industry as a target for research because it afflicts too few people for a drug to be profitable or because it is prevalent in places where people are unable to pay for drug therapies. Cystic fibrosis and Tourette's syndrome are relatively well-known examples of orphan diseases; others such as Job syndrome and Jumping Frenchmen of Maine remain largely unknown (FDA 2018). Although each disease may only affect a small number of people, the US National Institutes of Health estimate that as many as 25 million people in the US alone may have an orphan disease (*ibid.*). In recognition of this problem, rare-disease patient advocacy organizations have lobbied for government intervention in the market to encourage pharmaceutical companies to invest in researching rare diseases. The US, Japan, Australia, and the European Union, among others, have adopted legislation to encourage investment in searching for therapies for rare conditions.

The US "Orphan Drug Act" of 1983 attempted to tackle the problem by offering manufacturers of drugs for orphan diseases seven years of exclusive marketing of the product after approval by the Food and Drug Administration (FDA). The act also provides tax incentives for clinical trials, and the FDA oversees grants for research on orphan diseases. The disease must affect fewer than 200,000 people to be classified as an "orphan disease" in the US, although conditions that afflict more people may qualify if products developed to treat them are unlikely to be profitable. According to the FDA (2009), the act has brought more than two hundred products to market, compared with fewer than ten in the previous decade. In Europe, similar steps were introduced with the adoption of the "Regulation on Orphan Medicinal Products" in 2000 by the European Parliament. Incentives include market exclusivity for the product for the first ten years and lower fees for getting the product authorized for the market. By 2011, sixty orphan drugs had received marketing authorization (Westermark et al. 2011).

Despite these efforts in the Global North, many argue that there is still a need for funding and political will to tackle diseases of the Global South that do not offer significant potential profit owing to the poverty of the populations affected. In particular, a group of neglected tropical diseases (NTDs), including dracunculiasis, lymphatic filariasis, and leishmaniasis, has been identified that collectively afflict more than a billion people (CDC 2018). Many of these diseases, including nine diseases caused by parasitic worms, leprosy, and trachoma,

continued

Box 11.1 *continued*

have known, powerful, and low-cost preventative or curative interventions. For example, filtering water through cheesecloth can prevent dracunculiasis, while insecticide-treated bed nets can break lymphatic filariasis' cycle of transmission (Sachs 2007). In 2019, the WHO added snakebite as a target for intervention, with the goal of halving the number of deaths and cases of disability from snakebite over the next twelve years (WHO 2019g). Basic public health and sanitation could effectively eradicate several of these diseases, but most of those affected live in urban slums, remote rural regions, or conflict zones, complicating efforts to serve them. In 2011, 11 percent of the global population did not have access to safe water sources and more than one third still lacked access to "improved sanitation" (WHO 2013), leading

political ecologists to argue that political structures that influence investment decisions are key underlying causes of mortality.

The WHO (2019e) argues that public-private partnerships could provide solutions to addressing neglected tropical diseases. We discuss the donation of doses of the drug Mectizan for treating river blindness by Merck & Co., Inc. in Box 11.2, but at least twelve major drug donation programs are now in place (Crompton et al. 2010), with Merck & Co., GlaxoSmithKline, Johnson & Johnson, Pfizer, Novartis, and Sanofi-Pasteur all donating medicines and other contributions towards fighting NTDs (Sachs 2007). Non-governmental organizations have played an important role in lobbying and fundraising for disease control programs for these neglected diseases.

Onchocerciasis (river blindness) illustrates many of the challenges of tackling **neglected tropical diseases**. The parasitic worm that causes river blindness, transmitted through the bite of the black fly, persisted in many parts of the low-income world as the afflicted populations were simply unable to afford treatment, creating little economic incentive to fund research on the disease. The chance discovery that a veterinary drug could treat

river blindness finally created an effective treatment in the early 1990s, but the inability of those affected to pay for the treatment raised questions about how the drug could be administered to those who needed it. In this case, the ecology of the disease was well known, but even after an effective treatment had been devised, political and economic structures remained significant barriers (Box 11.2).

Box 11.2

RIVER BLINDNESS AND MERCK

River blindness (onchocerciasis) is caused by a worm that is contracted through the bite of the black fly. Symptoms include skin rash, itching, and weight loss. If the worms enter the eye, infection can lead to blindness, making the disease a leading cause of blindness in affected communities. Today, more than 99 percent of cases occur in sub-Saharan Africa, although the disease also occurs in Yemen, Venezuela, and Brazil (WHO

2019c). Several control programs have been attempted since the 1960s, coordinated by international organizations such as the WHO. Mapping the distribution of factors such as prevalence of the disease, prevalence of blindness, and vegetation coverage played a significant part in informing these early efforts, offering an example of the powerful role that spatial methods can play in health studies (see Hunter 2010).

Between 1974 and 2002, the Onchocerciasis Control Program brought the disease under control in West Africa, through application of insecticides from helicopters (WHO 2019c). Unfortunately, the fast-flowing streams in which black flies live rapidly render insecticide ineffective as it is diluted and washed away (Hosmer 2006). It was the development of a drug for deworming animals, Ivermectin (sold under the brand name Mectizan), that subsequently led to the development of an effective therapy in humans and a viable disease control program. In 1995, the African Programme for Onchocerciasis Control (APOC) initiated a strategy to deliver the drug to affected communities using community volunteers (WHO 2019a). The drug requires no refrigeration and can be administered by minimally trained personnel, making it ideal for distribution in this way. By 2007, the APOC had achieved a 73 percent reduction in infection prevalence, and was delivering doses of the drug to 54.6 million people annually (WHO 2009b). The APOC program was phased out in 2015 as the focus shifted towards elimination of the disease (WHO 2019c).

The success of the program was facilitated by the availability of free doses of Mectizan, donated by the pharmaceutical corporation Merck & Co. The story of how Mectizan came to be donated by Merck illustrates how public health programs hinge on political and economic decisions. In the mid-1970s, researchers at Merck discovered a drug that was effective against a relatively harmless parasite afflicting horses. Although the drug appeared to have little profit potential in human populations, one of the researchers leveraged a scientific freedom policy at Merck to pursue its potential for human use. This policy enabled him to spend US$500,000 of company funds and up to a year of his time to pursue a drug concept before analysis of the commercial viability of the drug had been performed (Hosmer 2006). Within a few months, he had demonstrated that the drug was effective at killing the microfilariae that cause river blindness in human tissues but had not yet determined whether it could be safely used as a clinical treatment.

Merck was then faced with the difficult decision of whether to pursue the drug through expensive human trials, with full knowledge that there was unlikely to be a profitable market for the drug at the end of the process owing to the extreme poverty of afflicted populations. During the testing and certifying of the drug for human use, Merck approached the US Agency for International Development (USAID) and the World Health Organization (WHO) to seek support for developing Ivermectin, but without success (Hosmer 2006). Nonetheless, Merck completed the trials, which indicated that the drug was effective in treating river blindness. The company had absorbed US$100 million in costs to this point. Once the effectiveness of the drug had been established, Merck's board of directors voted to spend an additional US$100 million per year to assist with distribution of the drug (*ibid.*). In October 1987, Merck initiated a donation program of an unlimited supply of the drug to anyone who needed it for as long as was required (Colatrella 2008). Investors and financial analysts expressed concern that the company was neglecting its responsibility to shareholders (Hosmer 2006), although many others praised the company for its actions.

The Mectizan Donation Program is the longest-running public-private partnership of its kind (Colatrella 2008). Since its inception in 1987, Merck has enabled more than 530 million treatments to be administered, reaching more than 68 million people in 33 endemic countries. In parts of Latin America and Africa, transmission of the disease has now all but ceased. In 2017, Merck expanded its donation program to provide Mectizan to treat lymphatic filariasis (Merck 2017). The community volunteer structure developed to deliver Mectizan is now being employed to address other public health problems, such as delivering bed nets against malaria, immunization campaigns, and cataract identification (Colatrella 2008).

The importance of understanding social context in concert with biomedical and ecological concerns is also illustrated by the global initiative to eradicate malaria in the 1950s. The development of an inexpensive insecticide (DDT), which showed considerable promise in controlling malaria vectors, stimulated global malaria eradication efforts. However, a lack of experience with eradication programs and failure to complete large-scale pilot studies meant that potentially significant challenges were overlooked (Henderson 1998). Initiated in 1955, the WHO malaria eradication campaign consisted of spraying the interiors of houses with insecticides (DDT in particular), administering anti-malarial drug therapy, and conducting disease surveillance. During its 15-year history, the malaria eradication campaign absorbed more than one third of WHO expenditures (*ibid.*). The program achieved considerable success in eradicating malaria from temperate regions and areas with seasonal transmission. Additionally, some tropical countries, including India and Sri Lanka, achieved substantial reductions in the prevalence of the disease.

Eventually, however, a combination of ecological and social factors caused the program to collapse. From an ecological perspective, efforts to tackle malaria were thwarted by increasing resistance to DDT among mosquito populations. Socio-politically, war, population movements, and waning donor support contributed to the collapse of the program, while a rigid, highly centralized operational structure provided little place-specific cultural flexibility (see Henderson 1998). By the late 1960s, the program's focus had switched from *eradication* of malaria to *control*, further eroding interest and support for the program. Today, malaria is once again a significant public health problem, especially in sub-Saharan Africa. Eradication is no longer considered a viable short-term goal, although it remains a long-term target.

With improved understanding of the social and ecological complexity of eradication programs, the International Task Force for Disease Eradication was convened in 1993 to review the potential for eradicating a list of more than eighty infectious diseases within the next generation. Six diseases were considered eradicable; an additional four diseases were subsequently identified by the WHO as targets for eradication in 1997 (Dowdle 1998) (Table 11.1). That year, the Dahlem Workshop on the Eradication of Infectious Diseases convened to discuss possible public health interventions for infectious disease, including control, **elimination** (removing a disease from a region), and **eradication** (removing a disease globally). Workshop participants concluded that, for eradication to be feasible, several prerequisites must be in place: eradication must be biologically and technically possible, it must yield a benefit that exceeds the cost, and it must have considerable political support (Barrett 2004; Aylward et al. 2000). Biological and technical feasibility hinge upon: 1) the availability of a suitable intervention to interrupt transmission of the disease, 2) the existence of effective diagnostic tools to identify levels of infection that could lead to transmission, and 3) humans being the central component to the life cycle, with no other vertebrate or environmental reservoirs (Dowdle 1998). The importance of political will and economic logic should not be overlooked, however, as Aylward et al. (2000, 1515) state, "Of all the lessons learned in the past 85 years, none is more important than the recognition that societal and political considerations ultimately determine the success of a disease eradication effort."

Eradication campaigns bring power structures to the surface because they require a considerable degree of centralized administration from public health establishments. Furthermore, vaccination campaigns often require careful surveillance to ensure compliance, which is endorsed or implemented by privileged groups, sometimes prompting resistance to campaigns from minority groups. As Greenough (1995, 633) states, "Public health measures derive their power from the police powers of the state, and people do not lightly offer themselves (or their immune systems) to government." Significantly, vaccination campaigns target healthy individuals, extending the reach of public health beyond the sick. This leaves disease control campaigns vulnerable to accusations of coercion as campaign workers must convince healthy individuals to participate in a program that may offer little immediate personal benefit. Ideally, public health education campaigns can encourage communities to participate willingly (Figure 11.1), but this has not always proved sufficient when a program goal is ambitious as in the

Table 11.1 Top ten candidates for disease eradication

Rank	Disease	Estimated cases (date)	Transmission	Breaking cycle
1	Polio (virus)	1936 (2005)	Sewage-contaminated water	Oral vaccine (drops)
2	Guinea worm (worm)	12,000 (2005)	Infected drinking water	Filter or treat pond water; dig wells deeper to avoid contamination
3	Lymphatic filariasis (worm)	120 million (1996)	Mosquitos	De-worming pills; patients must be treated annually for six years
4	Measles (virus)	30 million annually; 500,000 deaths	Airborne droplet	Vaccine (injection)
5	Blinding trachoma (bacteria)	84 million affected; two million blind	Spread by flies	Antibiotics, access to clean water, covered latrines; surgery in late stages
6	Onchocerciasis (river blindness) (worm)	18 million affected; 500,000 blind	Spread by bite of black flies	Insecticides for flies; de-worming pills for patients
7	Hepatitis B (virus)	350 million carriers	By blood or body fluids	Three vaccine doses
8	Leprosy (bacteria)	2.8 million affected; 1-2 million disabled	Transmission by extended physical contact	Antibiotic triple therapy daily for a year
9	Neonatal tetanus (bacteria)	200,000 deaths per year; 95% death rate	Umbilical cord cut with dirty blade	Clean delivery practices; vaccine for mother and baby
10	Iodine deficiency	740 million	Goiters in adults; brain damage in fetus	Iodized salt

Source: New York Times (2006)

case of eradication campaigns. Charges of coercion have been leveled at various control campaigns, including both smallpox and polio efforts.

Smallpox eradication

The eradication of smallpox remains a high point in global public health efforts. In 1979, two years after the last known naturally transmitted case, the WHO officially declared the disease eradicated (Figure 11.2). By 1986, routine vaccination had ceased in all countries (WHO 2010a). Not only did smallpox eradication remove a significant source of human suffering, but it is also estimated that the economic benefits have been enormous. The smallpox eradication campaign from 1967 to 1979 probably cost around US$200–300 million but may have resulted in annual savings of

US$2,500 million. Money was saved by removing the need for vaccination programs and treating complications, as well as elimination of the indirect costs of death, disability, and loss of earnings (Wickett 2002, 69). Another estimate suggests that the economic benefit of eradicating smallpox may be as high as US$450 saved for every dollar spent (Barrett 2004, 684).

Sadly, over the past thirty years, we have not managed to repeat this success with any other human disease. The ecology and social context of smallpox made it a relatively "easy" target for eradication in many ways, calling into question whether we can repeat this success with any other disease.

Written descriptions of smallpox date as far back as AD 400 (CDC 2016). The disease may have emerged from an animal poxvirus as early as 10,000 BC (Wickett 2002). The disease is primarily spread by airborne

Figure 11.1 Public health education campaign encouraging vaccination, Copacabana, Bolivia, 2005
Credit: Heike Alberts

transmission but can also be spread via *fomites* such as infected bedding, although with much lower transmissibility. The disease had an approximately 30 percent mortality rate, and no effective treatment has ever been developed (WHO 2019f). Most survivors were left with disfiguring scars, often with profound social implications. In particular, the identity of "smallpox victim" left many people unmarriageable in their communities. The disease was also a significant cause of blindness among survivors. As recently as the eighteenth century, smallpox killed 10 percent of children born in Sweden and France. Even after an effective vaccine became available in the early 1950s, approximately fifty million cases continued to occur each year (*ibid.*).

When the WHO's global eradication program was launched in 1967, there was still no effective treatment for smallpox, which continued to threaten 60 percent of the global population (WHO 2019f). An effective vaccination was available, however, following a long history of experimentation with smallpox inoculation. As early as the late 1700s, Edward Jenner and others realized that humans who had been infected with the closely related but milder cowpox pathogen developed some resistance to smallpox infection and that inoculation with cowpox could protect people against smallpox.

A variety of factors related to the ecology of smallpox made it a good target for eradication. The disease has no known animal reservoir or vector, and so eradication

Figure 11.2 Smallpox eradication commemoration poster
Source: WHO (2010b)

efforts could target the human population without risk of reinfection from other sources. The disease's highly distinctive symptoms made it easy for lay people to identify infected individuals. The disease is not infectious during its incubation period, and so the only time that people were at significant risk of contracting the disease was in the presence of another person with the rash (WHO 2019f). Interrupting the spread of disease could therefore be achieved by isolating patients with the obvious rash. Vaccination is at least partially effective for up to four days after exposure to the virus, meaning that people in close contact with patients could benefit from vaccination even if they had already been infected. Containment of outbreaks of the disease could, therefore,

break the transmission cycle relatively easily (*ibid.*). The vaccination procedure itself was also straightforward. A freeze-dried, easily administered vaccine was developed in the 1940s that did not require refrigeration, allowing the vaccine to be transported easily. The vaccine was administered via a short needle into a scratch on the skin, requiring limited equipment or training for vaccinators. Once vaccinated, most people are protected for about ten years, making repeat vaccinations unnecessary during the campaign.

In many ways, the social aspects of the eradication campaign were more complex than its biological and ecological facets. In 1958, the USSR proposed a global eradication program to the World Health Assembly, but genuine progress did not begin until the World Health Assembly intensified its efforts in 1967 by providing a special budget to fund the effort. At the start of the campaign, an estimated ten to 15 million cases occurred in 31 countries, with one billion people living in countries where the disease was endemic (CDC 2002).

Fortunately, the smallpox campaign was able to draw from past experiences. The Pan American Sanitary Organization (now the Pan American Health Organization, PAHO) had already experimented with a smallpox elimination campaign in the Western hemisphere. Although ultimately unsuccessful, it informed the global smallpox eradication campaign. The campaign also drew from lessons learned in the previous, unsuccessful malaria campaign; in particular, it recognized the value of using pre-existing health service structures and resources, rather than an autonomous body, to administer the campaign. In this way, the program gained considerable community support. The smallpox campaign was also designed to proceed in a more flexible manner than the malaria campaign, taking account of local conditions (Henderson 1998). For instance, campaign workers in parts of India monitored shrines to a smallpox goddess on the understanding that smallpox sufferers were often brought to the goddess to entreat her for a favorable outcome.

From the start, the smallpox program gave keen attention to the spatial components of its implementation. Four key focus regions were identified through mapping the distribution of the disease: sub-Saharan Africa, the Indian subcontinent extending north into Nepal and Afghanistan, Indonesia, and Brazil (CDC 2002). The campaign then used a two-pronged strategy for administering the vaccine. The primary goal was mass vaccination of at least 80 percent of afflicted populations to achieve **herd immunity** to protect the remaining population. It was soon realized, however, that reaching even this target would be impossible in places with challenging terrain, dispersed populations, and poor infrastructure. Fortunately, an understanding of the disease's diffusion patterns suggested another strategy: **ring vaccination**. Ring vaccination concentrates on identifying cases of the disease and then vaccinating anyone with whom they have had contact, plus anyone with whom their friends or family members could have come

into contact (Figure 11.3). The basic idea is to form a protective buffer of vaccinated individuals around every infected individual. Owing to the specific ecology of smallpox, with its clear symptoms and four-day window for vaccinating people exposed to the disease, ring vaccination proved effective.

Previous experience with mass vaccination campaigns had emphasized the importance of active cooperation of community members. In many cases, public health workers relied on local contacts such as teachers and missionaries—people who possessed an in-depth understanding of local cultural practices and attitudes—for guidance on program implementation. Campaigners nonetheless met opposition. In Ethiopia, for instance, armed guards were sometimes used to enforce containment strategies (Aylward et al. 2000). In South Asia in the early 1970s, there were reports that some physicians working for the WHO intimidated local officials and coerced individuals into being vaccinated (Greenough 1995).

By the end of 1975, smallpox persisted only in parts of the Horn of Africa, where civil war, poor infrastructure, and famine hindered the campaign. After a concerted ring vaccination effort in the spring and summer of 1977, the world's last naturally acquired case of smallpox was recorded in Somalia in October 1977. Today, the virus remains in several high-security facilities, and some countries continue to keep stockpiles of the vaccine, although the WHO (2019f) argues that the relatively high incidence of adverse side effects from the smallpox vaccine make its use unwarranted unless there is genuine fear of exposure. There is some concern over the potential threat of bioterrorism from the continued existence of the virus, and policymakers continue to debate whether to destroy the remaining stocks of the virus.

Changes in social and political context have altered the playing field for eradication efforts since the smallpox success. As populations become increasingly mobile, it becomes more difficult to contain disease outbreaks. Additionally, the global spread of HIV has left millions of people immuno-suppressed and unable to be vaccinated with many traditional vaccines. The WHO Director General's representative for polio eradication has noted that it was fortuitous that the smallpox campaign

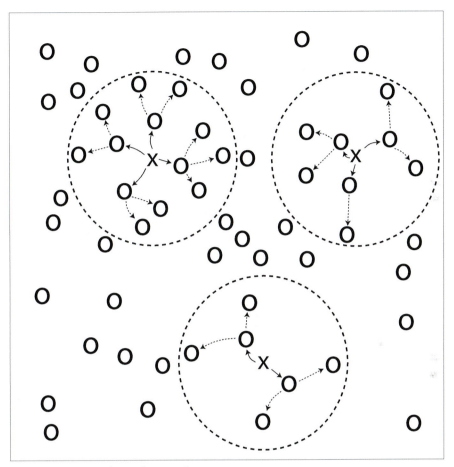

Figure 11.3 Diagram illustrating the mechanism of ring vaccination

In this diagram, an "x" represents an infected individual and the "o"s represent other community members. A dotted line between two individuals indicates contact between them. Each circle on the diagram encompasses all the people who would be vaccinated in a ring vaccination campaign, comprising the infected person's contacts and contacts of those contacts.

took place before widespread HIV infection (McNeil 2006). There has been an effort to develop a smallpox vaccine that is safe for HIV-infected populations in case a renewed vaccination campaign is ever needed (WHO 2019f). Additionally, many populations have become skeptical of large-scale, top-down projects that curtail individual freedoms, and an anti-vaccination movement is now flourishing (Box 11.3). Had rumors about the potentially serious side effects of the smallpox vaccine spread via social media in the 1970s, it would have

posed a significant challenge to the campaign. Nonetheless, the lure of eradication remains, and many public health officials are keen to repeat the success of smallpox eradication. Key current eradication targets are polio and dracunculiasis (Box 11.4).

Polio eradication

Poliomyelitis (polio) has been a target for eradication since the Global Polio Eradication Initiative (GPEI) was

Box 11.3

VACCINE HESITANCY

After the provision of clean water, vaccination may have led to the greatest reduction in the burden of infectious disease in human history. Vaccinations prevent two to three million deaths per year, with the potential to increase this by another 1.5 million if vaccination coverage were improved (WHO 2019h). Vaccination is one of the most cost-effective ways of avoiding disease, and vaccination has traditionally been a key pillar of primary health campaigns. Globally, the direct savings achieved from vaccines could be on the order of tens of billions of US dollars (Ehreth 2003). In the Global North, vaccination campaigns had made outbreaks of many traditional childhood diseases rare by the late twentieth century, and rates have dropped dramatically in many low-income countries over the past fifty years.

Despite these achievements, a vociferous anti-vaccination ("anti-vax") lobby has emerged, especially in the West, where the greatest benefits of vaccination have been realized. Indeed, the fact that Western parents are now unfamiliar with many infectious diseases may have contributed to complacency about their return. Concerns with vaccination typically center on fears that vaccines are unsafe or unnatural, or overtax a child's developing immune system, especially if several immunizations are received at once. In the 1990s, these concerns were fueled by the proposal that the measles-mumps-rubella (MMR) combined vaccine was linked with the onset of autism. Although the idea was only a hypothesis and had very limited supporting evidence, it was widely publicized and led to significant drops in the uptake of the MMR vaccine in many countries. Subsequently, numerous scientific studies have concluded that there is no causal link, and the doctor who proposed the link has been discredited, but the idea that the MMR could cause autism continues to circulate within the "anti-vax" community. Concerns over certain constituents of vaccines, particularly mercury and aluminum, also remain common, with misinformation circulating about both what is in vaccines and the risk that many ingredients pose.

The "anti-vax" movement has now become a cultural phenomenon in its own right, with the Internet often credited with uniting people with concerns over vaccination and generating a culture of **vaccine hesitancy** that has spread widely. The democratization of information on the Internet means that blogs from individuals without medical training can compete with public health information generated from years of data and research (Forster 2017). Indeed, Internet searches are likely to generate more anti-vaccination than pro-vaccination results, leaving many parents confused over which sources to trust (McClure, Cataldi, and O'Leary 2017). The highly controversial nature of the vaccine debate leads some to defend vaccine rejection to the point where it becomes deeply entrenched — and even a marker of belonging within certain communities. Research suggests that, once views become ingrained in this way, simply providing more information to parents may not be effective in changing beliefs. Instead, new information must be tailored to meet the parents' framework of understanding (Dubé and MacDonald 2016). Considerable research has therefore sought to understand vaccine-hesitant parents' perspectives in order to develop culturally appropriate interventions.

One major challenge is that parents tend to focus on the needs of their own child, while the public health community is focused on issues at the population scale. While there is clearly considerable individual benefit from the protection offered by many vaccines, there is also some risk. Some people will have a reaction to vaccines — typically these are mild such as a rash, but in rare cases reactions can be serious — although both the WHO and many independent reports have shown that modern vaccines are safer than therapeutic medicines (Andre et al. 2008). Nonetheless, public health policymakers must weigh these risks with the potential benefits to the community, often estimated via the number of children who would be likely to die from the disease in an unvaccinated population.

For a vaccination program to proceed in an evidence-based public health program, the risk to individuals must be justified by the population benefits. This relation may be time- and place-specific. For instance, we could argue that administering the smallpox vaccine to communities with a genuine risk of smallpox in the 1960s and '70s was justified, but that a smallpox vaccination campaign today would not be, as the potential side effects of the vaccine outweigh the near impossibility of infection. Vaccines currently in use have low risks of side effects and target diseases that continue to pose a genuine public health threat, providing a favorable cost-benefit balance.

In some cases, individual benefits of vaccination are limited, however, complicating cost-benefit analyses. For instance, rubella is usually only a mild disease in children, providing little incentive to vaccinate children for their own personal benefit. If a developing fetus is exposed to rubella, however, it can cause congenital rubella syndrome—a serious condition that can render the child blind and deaf, and with significant developmental difficulties. Rubella vaccination campaigns are therefore premised on the idea of limiting the likelihood of pregnant women being exposed, rather than concerns for the children actually being vaccinated. In other cases, **herd immunity** may be leveraged to protect individuals who are unable to be vaccinated because of compromised immune systems or other underlying health problems. In such cases, vaccinated individuals bear the risk in pursuit of the common good of maintaining sufficient levels of vaccination coverage to protect vulnerable populations.

Recent measles outbreaks that have occurred in response to declining vaccination rates demonstrate the delicate balance inherent in efforts to achieve herd immunity. Owing to the ease with which measles can spread, very high vaccination rates are required to prevent circulation of the disease. Estimates suggest that 95 percent of a community must be vaccinated to prevent outbreaks of the disease, but many communities now have vaccination rates well below this. In 2015, a multistate outbreak developed after just one infectious individual at Disneyland infected others, leading to 147 cases by the end of the outbreak (CDC 2019). Other recent outbreaks have occurred within communities with a history of low vaccination rates. The US experienced 17 measles outbreaks in 2017, mostly in the New York area where several outbreaks focused on unvaccinated members of Orthodox Jewish communities.

In Europe, measles cases hit a record high in 2018 with 41,000 cases in the first six months of the year (Baynes 2019). Worldwide, the WHO estimates that measles cases increased four-fold in the first three months of 2019 compared with the same period in 2018. In Africa, cases rose seven-fold (BBC 2019c). In many poor countries, vaccination rates have stalled below the 95 percent needed to prevent outbreaks owing to disruptions in funding and rising skepticism of top-down public health campaigns, while in rich countries misinformation and broader vaccine hesitancy have fueled renewed outbreaks. The WHO declared vaccine hesitancy to be one of the top threats to public health in 2019 (WHO 2019h).

In response, public health authorities have begun practicing increasingly punitive measures to encourage vaccination. Australia has amended child tax benefit rules to withhold some of the benefit for children not meeting immunization requirements and noted that vaccine objection would no longer be considered a legitimate exemption from vaccination programs (Australian Government 2018). In March 2019, the state of New York developed a policy to administer fines to parents of unvaccinated children after a "state of emergency" was declared related to measles outbreaks. Two months later, Italy banned unvaccinated children from schools (BBC 2019a, 2019b).

Box 11.4

DRACUNCULIASIS ERADICATION

One target for eradication, little known outside endemic regions, is dracunculiasis (guinea worm). Originally found across parts of Asia and Africa, the disease has been eradicated from all but a few pockets of sub-Saharan Africa. Despite the WHO (2009a) declaring that the disease is "extremely easy to combat and should no longer be prevalent," eradication has proved more challenging than expected.

As early as the end of the nineteenth century, the life cycle of the dracunculiasis pathogen was well understood, and preventative measures had been developed and practiced. The disease is caused by the parasitic worm *Dracunculus medinensis*, which lives in water fleas (*Cyclops*) (Hopkins et al. 2005). Once a person drinks contaminated water, the fleas are rapidly dissolved by stomach acids, releasing the parasites into the human host where they breed. About a year after infection, the female worms emerge from the human body, typically through the feet or lower leg, and release their eggs into water. At the point of emergence of the female worm, a painful blister develops on the host's skin. To soothe the pain, people are driven to soak the affected limb in water, stimulating the adult worm to release her eggs and re-contaminating the water supply (WHO 2009a).

Although rarely fatal, the disease is debilitating, incapacitating people for an average of two to three months during the worm's emergence (Hopkins et al. 2005). Additionally, migration of the worm around the body through subcutaneous tissues can cause severe pain (WHO 2009a). The disease can lead to a vicious cycle in which impoverished people are prevented from working by infection, drawing them further into poverty. In some parts of Africa, the disease is referred to as "empty granary" because of its tendency to occur at harvest time, preventing people from working at one of the most critical times of the agricultural year (McNeil 2006).

Individual infections last only one year, but people never develop immunity and there is currently no vaccine or drug treatment for the disease (Hopkins et al. 2005). Preventive measures must therefore rely on behavioral changes. Infection can be prevented if drinking water is filtered or treated to remove the water fleas before consumption. Alternatively, the disease cycle can be broken if infected people do not put affected limbs in open water when the worm emerges. Both measures require education and community cooperation but are low-tech and cost-effective. Fortunately, diagnosis is straightforward and simple, the water fleas that transmit the disease have limited mobility, and the disease has a limited geographic range.

By the 1980s, several countries had eliminated the disease, and there was political will to repeat the success elsewhere (WHO 2009a). The CDC initiated a dracunculiasis eradication campaign in 1980. Since 1986, the Carter Foundation, in association with the WHO, UNICEF, and the CDC, joined in the effort (Hopkins et al. 2005). At that time, there were still an estimated three million cases of the disease per year across twenty countries in Africa and Asia. Gaining and maintaining political support has been critical to the success of the program, but the campaign also relies on thousands of volunteers and supervisory health staff in affected countries (Hopkins et al. 2005). By 2008, cases of the disease had been reduced by 99 percent, and the WHO had declared 168 of the world's countries as free from dracunculiasis (Hopkins et al. 2005), bringing the disease to the verge of eradication (Carter Foundation 2008).

It was initially believed that the disease could be eradicated within ten years, but progress has been slower than expected. Civil unrest was an important barrier to the program's success in Ghana and Sudan, which together reported 90 percent of all cases in 2004 (Hopkins et al. 2005). In other cases, progress has been slowed trying to encourage the cooperation of marginalized nomadic populations, including the Black Tuareg of the Burkina Faso-Mali-Niger border region and the Konkomba people of Ghana and Togo (Hopkins et al. 2005). The campaign also met resistance from communities who were unwilling to use pesticides, particularly in sacred waters (McNeil 2006). Nonetheless, by 2007 there were fewer than ten thousand cases in just six African countries. By the end of 2018, only 28 cases of the disease were reported, primarily in Chad and South Sudan, although one case unexpectedly appeared in Angola, illustrating the critical need for continued surveillance (WHO 2019b). To date, only US$350 million has been spent on the campaign, making it an incredibly cost-effective program (by contrast, US$9.5 billion has been spent to date on polio eradication) (The Lancet Infectious Diseases 2016).

Unfortunately, the project recently experienced a setback. In 2018, guinea worm infections were reported in dogs in Chad, revealing that there was an animal reservoir of the disease, which significantly complicates eradication efforts. Canine infections have since been observed in Chad, Ethiopia, and Mali, with evidence that cats and a baboon might also be infected in Ethiopia; by the end of 2018, more than one thousand infected animals had been reported (WHO 2019b). One possible route of exposure for village dogs may be through consumption of fish harboring the larvae (The Lancet Infectious Diseases 2016). Cases in animals caused considerable confusion as they seemed to appear just as the disease was on the brink of eradication: had the disease really just mutated in such a way as to suddenly be able to infect other animals at this critical moment in the process of eradication? A more likely explanation is that increasing attention to the disease and rewards for the reporting of infections stimulated local people to report animal infections that had occurred for a long time but out of sight of the research team (Galán-Puchades 2016), reinforcing the critical need for local expertise in public health campaigns. These new cases have ushered in a new era of dracunculiasis management in which animal surveillance must now be incorporated into efforts.

initiated in 1988, spearheaded by the WHO, UNICEF, and Rotary International, with assistance from the Gates Foundation and the CDC. Policymakers hoped that the disease would be eradicated by the year 2000, but eradication remains out of reach so far, although the disease has been eliminated from many places. A variety of ecological and social aspects of the disease make it a more difficult target for eradication than smallpox.

Polio is a highly infectious viral disease. Symptoms include fever, fatigue, headache, and vomiting (WHO 2019d), making it difficult for a layperson to distinguish polio from many other infections. In addition, many cases are asymptomatic or show only mild symptoms. For a minority of infected individuals, the disease can be very serious, however. Infection leads to irreversible paralysis in approximately one in two hundred cases; among those paralyzed, 5 to 10 percent die when breathing muscles become immobilized (*ibid.*).

The disease is spread primarily via the fecal-oral route and is commonly contracted from water supplies. Contaminated water, sewage, and soil can all act as reservoirs of the virus (Fine and Carneiro 1999). The virus is also found at low concentrations in oral secretions, enabling it to be transmitted directly between people through contact with saliva or airborne droplets. Environmental factors may influence the degree to which each route of transmission is favored. In conditions of poor sanitation, the fecal-oral route is more significant; where standards of hygiene are high, the respiratory route may be more important. The ratio of these two types of transmission is therefore closely correlated with a community's socioeconomic status (*ibid.*).

Although there is no cure for polio, two effective vaccines were developed in the 1950s: an injectable inactivated vaccine and a live oral polio vaccine. The oral vaccine is often preferred because it is cheap to produce and easy to administer (Figure 11.4). The live viruses in the oral vaccine can also be transmitted from vaccinated individuals to their contacts, potentially resulting in the immunization of unvaccinated people (Fine and Carneiro 1999). On the other hand, the presence of live viruses maintains a source of the pathogen in the population and, in rare cases, can cause a vaccinated individual to develop symptoms of the disease, even to the point of paralysis—a condition known as vaccine-associated paralytic poliomyelitis (VAPP). The likelihood of paralysis occurring from the oral vaccine is many times higher among people who have compromised immune systems, such as people living with HIV/AIDS, and has led to

questions over the ethics of using the live vaccine (Paul and Dawson 2005). Although the inactivated vaccine avoids the risk of VAPP, it is five times more expensive to produce than the oral vaccine and requires injection by a trained healthcare worker using sterile equipment. Many industrialized countries now preferentially offer the safer inactivated vaccine, but the oral polio vaccine is still used during outbreaks (GPEI 2019a).

Recognizing the potential to control polio using these vaccines, several countries introduced national elimination campaigns in the mid-twentieth century. In Cuba, two annual vaccination campaigns succeeded in interrupting endemic transmission of the disease by

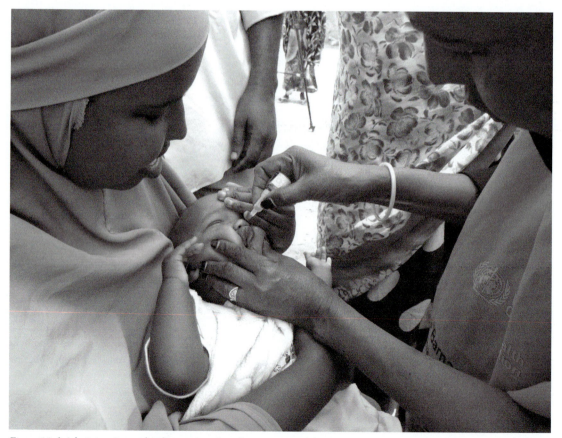

Figure 11.4 Administering oral polio vaccine, Somalia
Source: World Health Organization Photo Gallery/WHO 2010c
Credit: UNICEF/Morooka

1962; the US was declared polio-free by the end of the 1970s (Rodríguez-Cruz 1984). The basic practices of the Cuban campaign were subsequently assumed by PAHO in a campaign that eliminated the disease from the Americas by the early 1990s. Worldwide, the disease continued to exact a significant toll, however. In 1988, when the Global Polio Eradication Initiative started, 125 countries still had endemic polio or notable epidemic activity (Aylward 2006, 403), and more than 350,000 children were estimated to have been paralyzed by the virus in that year alone (Hull et al. 1994).

The GPEI campaign focused on the four approaches that had proved successful in the Americas: 1) high infant immunization coverage, 2) additional child immunizations for children under age 5 during supplementary campaigns, 3) surveillance for wild poliovirus and laboratory investigation of paralysis cases in children, and 4) targeted "mop-up" campaigns when isolated pockets of transmission were identified (WHO 2019d). By 2002, polio transmission had been interrupted worldwide in all but six countries. This success was a result of strong political and economic commitment from governments and non-governmental organizations, an enormous investment in surveillance, and deployment of large numbers of public health workers (Aylward 2006).

Between 2003 and 2005 the campaign began to falter, however, as place-specific challenges became apparent in the remaining pockets of endemic transmission. Despite an active campaign, high vaccination rates, and few operational problems, transmission was sustained in several areas with high population density and suboptimal hygiene, such as Cairo, Egypt, and Mumbai, India. This led to concerns that the use of the oral vaccine might be ineffective in some conditions and eventually resulted in the introduction of the injectable vaccine. In Pakistan and Afghanistan, vaccination rates were also generally high, but the infection persisted in remote, sparsely populated areas where problems of inaccessibility, conflict, and cultural opposition to the program slowed progress. Substantial cross-border movements also repeatedly recirculated the virus to susceptible populations. In Nigeria and Niger, the final two countries with persistent endemic infection, the key problem was low vaccination rates, which remained under 50 percent in some regions (Aylward 2006, 407). In response,

the GPEI implemented an aggressive supplementary immunization program in 2003. Unfortunately, the heightened profile of this more assertive campaign led to concerns among local populations about its motives (ibid.), generating rumors that the vaccine was contaminated and that the campaign was intended to sterilize young girls (Heymann and Aylward 2004). These concerns prompted some leaders in northern Nigerian states to suspend vaccination efforts. The year-long controversy resulted in the diffusion of Nigerian polioviruses across parts of Africa and the Arabian Peninsula and as far as Indonesia, leading to the reinfection of 18 previously polio-free countries (Aylward 2006, 407). Responses to these problems focused on providing additional financial and human resources and strengthening community support for the campaign through collaboration with community and religious leaders. All told, between 2003 and 2005, no fewer than 25 countries that had been polio free were reinfected by imported cases (WHO 2019d).

By 2010, polio was limited again to just four countries: Nigeria, India, Pakistan, and Afghanistan (WHO 2019d). In the thirty years of the eradication campaign, the total number of cases has fallen by 99 percent, and the campaign has probably prevented more than five million cases of paralysis (ibid.). The achievements of the campaign are fragile, however, as the remaining reservoirs of the disease can quickly lead to outbreaks of the disease in the absence of effective surveillance and response. Furthermore, support for polio eradication has waned in many quarters, owing to fatigue over repeated polio vaccination efforts (Aylward 2006). The situation is now being monitored so closely that the GPEI reports individual cases of polio. By 2018, only 32 cases of wild poliovirus were reported, in Afghanistan and Pakistan, although 104 cases of vaccine-derived poliovirus infection also occurred, mostly in Nigeria, Papua New Guinea, the Democratic Republic of the Congo, and Somalia (GPEI 2019b).

The ecology of the poliovirus has been a huge challenge to the success of the campaign. The persistence of the virus in the natural environment or within human populations that were recently vaccinated with the live virus means that reservoirs of reinfection will remain, potentially for several months after the cessation of the vaccination campaign. Although few people are likely to be long-term carriers of the poliovirus, it is possible that immuno-suppressed individuals could excrete the virus for months or even years after infection. Predictions about the persistence of the virus remain speculative, but Fine and Carnerio (1999, 1016) argue that "given the variety, heterogeneity, interconnectedness, and sheer number of human populations, the possibility cannot be excluded that OPV [oral polio] viruses could succeed in persisting for several years, somewhere, in one or another population network." Should a source of reinfection persist for several years, it is possible that a new epidemic could be initiated when a fresh cohort of unvaccinated individuals is born (*ibid.*). Vaccine-derived polioviruses are known to have caused six polio outbreaks between 2000 and 2005 and could lead to hundreds of thousands of cases again per year (Aylward 2006, 409). These concerns suggest that a rapid, globally coordinated cessation of vaccination may be needed once eradication has been achieved (*ibid.*). Steps in this direction were taken in 2015–16 with the introduction of the inactivated vaccine to a number of countries of the Global South to replace the live vaccine (GPEI 2015). Today, the GPEI is pursuing its "Polio Endgame Strategy," establishing a regional hub to oversee efforts to control the remaining wild polioviruses circulating in Afghanistan and Pakistan, improving surveillance techniques, and providing support for countries to bolster their ongoing vaccination programs (GPEI 2015).

As the cases of smallpox and polio eradication efforts demonstrate, an understanding of the ecology of a disease is critical but insufficient to stemming its spread. The battle for eradication is often won or lost depending on human factors such as availability of funding, support for the goals of the program, and conflict. The fact that polio has persisted in Afghanistan and Pakistan has more to do with fragile political and economic situations than any inherent ecological issues. As our next case study shows, chronic disease can similarly be tied to social and ecological conditions.

THE OBESITY EPIDEMIC

The cause of obesity may appear to be obvious at first sight: more calories consumed than expended in activity leads to the accumulation of fat reserves and potentially, obesity. The underlying causes of this imbalance of "energy-in and energy-out" are incredibly complex, however, with researchers struggling to disentangle the effects of genetics, culture, and built and social environments.

Limited access to material resources can diminish health by precluding individuals from nourishing themselves sufficiently. Undernutrition and malnourishment are, therefore, typically problems of poverty, where certain groups lack the resources to purchase or grow sufficient food, particularly protein-rich foods, such as meat and dairy products (Figure 11.5). Many primary health programs use height and weight measurements as an important indicator of child health.

Although malnutrition persists in many regions, many countries now experience higher rates of overnutrition than undernutrition. Overnutrition includes the categories "overweight" and "obesity." Overweight and obesity are estimated by **body mass index (BMI)**, which is calculated by dividing an individual's weight in kilograms by height in meters squared. The CDC defines "overweight" and "obesity" as "labels for ranges of weight that are greater than what is generally considered healthy for a given height" (CDC 2017), although the actual risk of being overweight is still contested because BMI provides only a rough estimate of an individual's disease risk. Being obese, by contrast, has powerful negative impacts on health. The WHO reports that high BMI is a major risk factor for heart disease, diabetes, and musculo-skeletal disorders (WHO 2018b). Obesity-related health problems that were once only observed in adults are now also being reported in children, increasing the risk of other health problems later in life (Singh et al. 2008).

The WHO estimates that globally 1.9 billion adults (39 percent of the population) were overweight in 2016, of whom more than 650 million (13 percent) were obese. In addition, more than 380 million children and adolescents were overweight or obese (WHO 2018b). The worldwide prevalence of obesity almost tripled between 1975 and 2016 (WHO 2018b), a trend now often called the "obesity pandemic" (Figure 11.6).

Although the proximate cause of overweight and obesity is consumption of too many calories, the root causes are complex. In low-income countries, obesity rates tend

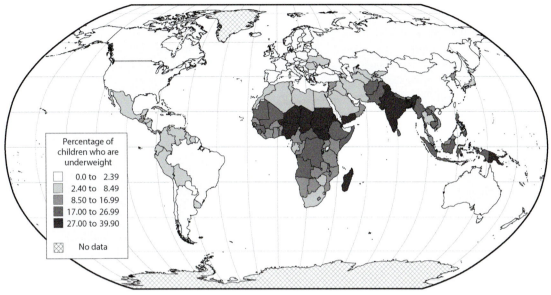

Figure 11.5 Percentage of children under age 5 who are underweight, by country

Source: Data from WHO (2018a)

Children are considered underweight if they are under two standard deviations of the weight considered to be healthy for their height. Data are from the latest available year.

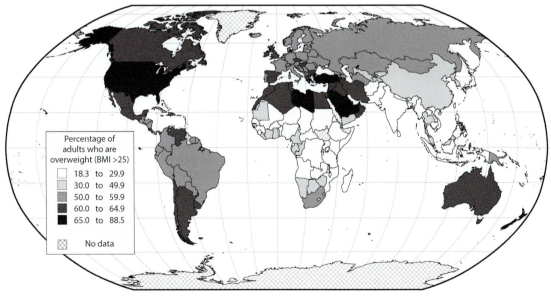

Figure 11.6 Percentage of adults who are overweight by country, 2016

Source: Data from WHO (2018a)

Adults are considered overweight if they have a BMI greater than 25.

to be highest among affluent urbanites who have the greatest access to an abundance of foods, particularly processed foods. Poorer rural communities, by contrast, often remain dependent on traditional agricultural produce, and may be underweight (Figure 11.7). To this point, the relationship between affluence and food is straightforward. As a country's affluence increases, however, overweight and obesity tend to increase more rapidly among lower-income populations, and obesity is focused in poorer segments of the population in many wealthy countries. In the US, for example, children who live below the poverty level are roughly twice as likely to be obese as children living above the poverty line (Freedman et al. 2007). Some middle-income countries, too, have seen a recent shift in obesity prevalence from the rich to the poor, as Monteiro, Conde, and Popkin

(2004) document in Brazil. At this point, the relationship between obesity and socioeconomic status becomes complicated, going beyond the ability to purchase food. Research suggests that the social environment plays an important role in mediating income and obesity, through influencing factors such as how much and what people choose to eat. Teasing apart the place-specific connections between social and economic environments and overconsumption of food is a critical challenge for health geographers and is the focus of the remainder of this section.

Geographic research on obesity

Geographic approaches to obesity have focused on the influence of physical and social environments on obesity

Figure 11.7 Food market in the Peruvian Andes

This street market displays foods traditionally available in the Peruvian Andes. Owing to limited incomes and seasonal fluctuations in the availability of produce, rural communities such as this have traditionally suffered more from under- than overnutrition.

rates, and particularly the emergence of **obesogenic** societies "characterized by environments that promote increased food intake, non-healthful foods, and physical inactivity" (CDC 2009). The impact of environment on obesity can be analyzed at both a macro scale, including factors such as the role of the food industry, and at the micro scale, considering aspects such as neighborhood socioeconomic environments, the built environment, and even the school environment (Kim and Kawachi 2010). Low-income populations are vulnerable because they are most likely to live in obesogenic environments; some research suggest that the poor may also be more sensitive to the effects of their environment (Merchant et al. 2007; Swinburn et al. 2004).

One way in which the economic environment is important is through its influence on food choices (see Hoek and MacLean 2010). As modern agricultural techniques have made food cheaper, calories are readily available for most people in affluent countries. However, the relative cost of different types of foods tends to encourage the consumption of energy-dense foods—particularly foods high in fats and carbohydrates—as they often provide the lowest-cost way to feed a family (Drewnowski 2004). Cooking meals from scratch, using fresh ingredients such as fish, fruits, and vegetables, is frequently not only more expensive but also more time-consuming for people with limited financial resources, who are often also working long hours (Figure 11.8). Government

Figure 11.8 Fresh fish market, Venice, Italy
Credit: Melissa Gould

Preparing meals from fresh, healthy, raw ingredients such as fish, fruits, and vegetables is beneficial for health. Research suggests that people of low socioeconomic status may struggle to find the time and money to make this investment in health.

agricultural subsidies have also been implicated in making processed, high-energy foods relatively cheap compared with more nutritious whole foods. In the US, for example, long-standing government support for corn production has led to such an abundance of corn that large quantities are converted into cheap animal fodder and sweeteners. This, in turn, has reduced the cost of high-energy meat, dairy, and sweetened foods (Pollan 2006). Improved efficiency in food processing has also made many processed foods cheaper than whole-food alternatives. These developments, combined with the attractive taste and convenience of pre-packaged foods, tends to bias food choices towards processed foods (Drewnowski 2004).

Qualitative research lends support to these ideas. For instance, in interviews with low-income consumers, Wiig-Dammann and Smith (2009) concluded that price is a very significant factor in food choice, with many interviewees reporting that they believed that improving the healthfulness of their families' diets would be unaffordable. As Drewnowski (2004, 161) concludes, "The ability to adopt healthier diets may have less to do with psychosocial factors, self-efficacy, or readiness to change than with household economic resources and purchasing power." Paradoxically, therefore, uncertain access to food, or **food insecurity**, may actually be linked to obesity (Drewnowski 2004; Wiig Dammann and Smith 2009).

Characteristics of an individual's neighborhood can also affect food choice and availability, although research has produced variable results on this topic (see Thornton and Kavanaugh 2010). Some research suggests that low-income neighborhoods have limited access to supermarkets and grocery stores that offer healthy foods such as fruits and vegetables. Such neighborhoods are often referred to as "food deserts" (Mead 2008). Obstacles such as a lack of available land, demanding regulations, depopulating neighborhoods, and urban crime discourage supermarkets from trading in inner-city neighborhoods, leading to race-based as well as socioeconomic inequalities in food access (Crowe, Lacey, and Columbus 2018). In one study, researchers used GIS to map the location of supermarkets in London, Ontario, and concluded that residents of low-income inner-city neighborhoods have the poorest access to supermarkets

as a result of supermarket chains moving to the suburbs (Larsen and Gilliland 2008). Other studies have found an over-abundance of outlets providing "unhealthy" options, such as fast food or alcohol, in poorer or minority neighborhoods (e.g., Block, Scribner, and DeSalvo 2004; Reidpath et al. 2002).

Not all research supports these ideas, however (Cummins and Macintyre 2002), and there may be critical differences around the world. For example, Pearce et al. (2008) found that access to supermarkets was actually better in deprived neighborhoods in New Zealand. In another study, Pearson et al. (2005) concluded that deprivation and a lack of healthy food outlets were not associated with fruit and vegetable consumption in a region of England, and that socio-cultural attitudes to food were more significant. Despite these mixed results, the concept of "food deserts" has become popular in the media and is used to inform policy (Cummins and Macintyre 2002).

The relationship between poverty and obesity may also be influenced by opportunities for exercise (Panter and Jones 2008; Giles-Corti and Donovan 2002). Research suggests that people with lower income and education are less likely to engage in physical activity (Brownson et al. 2001). There are a variety of possible reasons for this, but of particular relevance to our argument about obesogenic environments is the idea that facets of the neighborhood (**area effects**) could discourage exercise in areas of deprivation. For instance, poor air quality, heavy traffic, and a lack of parks may limit opportunities for exercise in inner cities. By contrast, a variety of positive neighborhood characteristics—such as the availability of green spaces (Takano, Nakamura, and Watanabe 2002), the presence of public open spaces (Giles-Corti and Donovan 2002), the walkability of streets (Panter and Jones 2008; Leslie et al. 2007), and mixed land use design—have been associated with increased physical activity (Brownson et al. 2001). We must be wary of making assumptions about cause and effect from these studies, however. For instance, one study of eight European countries found that physical activity and not being overweight were positively correlated with high levels of greenery and low levels of graffiti and litter (Ellaway, Macintyre, and Kearns 2001). Although it is tempting to conclude that pleasant surroundings encourage people to exercise, it is also possible that aesthetically pleasing environments simply attract wealthier residents who are more likely to exercise for other reasons. Wealthy individuals are commonly reported to have more leisure time available for exercise, for example. Even here, evidence is contradictory, however, as some research suggests that people living in deprived neighborhoods may actually be more likely to walk, perhaps because of limited transportation options (e.g., Santana et al. 2009). Such contradictory findings once again emphasize the place-specific and complex relationship between obesity and environment.

The social environment may also influence exercise patterns (Ming, Browning, and Cagney 2007). For instance, fear of crime may make individuals less likely to engage in outdoor physical activities. By contrast, residents of neighborhoods with high levels of **collective efficacy** may have a greater sense of safety, resulting in more time spent outdoors, and better access to recreational facilities. Results are again mixed, however, suggesting that effects may be small or place-specific.

A variety of spatial studies have also improved our understanding of the obesity epidemic. At a global scale, organizations like the WHO closely monitor obesity rates by country, revealing how rapidly the problem has spread from the Global North to middle-income and even low-income countries. At national scales, mapping reveals significant regional differences in obesity, enabling us to generate hypotheses related to how cultural, economic, or other factors might drive these patterns. At local scales, we can explore how different communities have more or less access to health-promoting facilities like grocery stores, parks, and gyms, as well as the clustering of health-damaging outlets such as fast food chains and liquor stores. At even finer scales, such as the scale of an individual playground, spatial methods can further elucidate connections between health and environment. For instance, different playground designs may influence children's physical activity (Anthamatten et al. 2014).

Viewing obesity as an environmental problem encourages approaches to the obesity epidemic that move beyond individuals' role in making "healthy choices" towards an interest in structural changes that nudge entire societies in healthier directions. Structural solutions to the obesity epidemic include removing government subsidies on unhealthy foods (such as US support for overproduction of corn) or adding taxes to unhealthy foods (such as the UK's "sugar tax," which targets the soft drinks industry). Others have suggested increasing support for farmers' markets and school gardening projects as ways to increase access and exposure to healthy foods (Langellotto and Gupta 2012; Freedman et al. 2013; Leung et al. 2011).

If impoverished communities are indeed at a disadvantage in maintaining healthy eating and exercise regimes as a result of their social and economic surroundings, it is critical that policy addresses questions of equity in tackling obesity, however. These arguments require us to step back even further from food itself to examine the significance of poverty and disempowerment as risk factors for obesity. Lee (2013), for instance, notes that researchers and policymakers have struggled to make sense of the complexity of issues that connect poverty to obesity, and argues that alleviating poverty and improving access to education directly may end up having a more significant impact on obesity than any number of targeted anti-obesity interventions. As she notes: "Junk food tax revenues might be better spent on scholarships for low-income families than on explicit obesity-prevention strategies. . . . A college degree is likely to be a more reliable obesity prevention tool than any amount of school gardens, farmers' markets, or new grocery stores" (Lee 2013). Such approaches do not require that we abandon efforts to improve food environments but encourage us to take a multi-pronged approach where we complement existing targeted anti-obesity interventions with an honest engagement with the role of inequality in health outcomes.

CONCLUSION

Global campaigns to control disease provide an excellent illustration of ways in which geographic approaches can be applied to health problems. The successful eradication of smallpox could not have happened without serious study of both space and place, using the techniques from health geography highlighted in this book. Similarly, we are unlikely to make much progress in reducing obesity if we do not engage with both the physical and social environments.

As we strive to cope with unprecedented and rapid changes in varied facets of the planet's geography, an appreciation for and understanding of space and place to health is more critical than ever before. Climate change, environmental degradation, population growth, and the rapid flow of people, goods, and information that have accompanied globalization have significant impacts on human health in nuanced and complex ways. While in-depth study of the individual topics covered in each of the chapters in this book is a critical first step, a study of health is only complete if it strives to understand the ways in which these topics are inextricably linked. If there is a single message we wish to convey from this book, it is that health has many facets: it is biomedical, ecological, social, cultural, and political, and it is deeply linked to location. Geography, through the integral role it gives to space and place, explicitly strives to combine this variety of perspectives to achieve a holistic and complete understanding of health.

DISCUSSION QUESTIONS

1 Considering the ecological, spatial, and social aspects of polio and dracunculiasis, which do you consider to have a greater likelihood of eradication? Taking into account both social and ecological factors, discuss whether you would support control, elimination, or eradication of another disease you know about.

2 To what degree should the rights of the individual be protected in eradication campaigns? Should individuals have the right to decline vaccination for themselves or their children? Why, or why not, or in what circumstances?

3 What are the pros and cons of a biomedical approach to health?

4 What aspects of the social and economic environment could be considered obesogenic in your community? Are different sub-populations within your community (e.g., minority groups, children, the poor) affected in different ways or to different degrees?

SUGGESTED READING

Aylward, B., K. A. Hennessey, N. Zagaria, J. M. Olivé, and S. Cochi. 2000. "When Is a Disease Eradicable? 100 Years of Lessons Learned." *American Journal of Public Health* 90 (10): 1515–20.

Aylward, R. B. 2006. "Eradicating Polio: Today's Challenges and Tomorrow's Legacy." *Annals of Tropical Medicine and Parasitology* 100 (5–6): 401–13.

Crowe, J., Lacey, C., and Columbus, Y. 2018. "Barriers to Food Security and Community Stress in an Urban Food Desert." *Urban Science* 2 (2): 46. www.mdpi.com/2413-8851/2/2/46/htm.

Larsen, K., and Gilliland, J. 2008. "Mapping the Evolution of 'Food Deserts' in a Canadian City: Supermarket Accessibility in London, Ontario, 1961–2005." *International Journal of Health Geographics* 7 (1): 16.

Ritchie, H., and Vanderslott, S. 2019. "How Many People Support Vaccination Across the World?" *OurWorldInData.org*. https://ourworldindata.org/support-for-vaccination.

VIDEO RESOURCES

PBS "Rx for Survival: Disease Warriors." More information at: www.pbs.org/wgbh/rxforsurvival/.

REFERENCES

Andre, Francis E., Robert Booy, Hans L. Bock, John Clemens, et al. 2008. "Vaccination Greatly Reduces Disease, Disability, Death and Inequity Worldwide." *Bulletin of the World Health Organization* 86: 140–46.

Anthamatten, P., E. Fiene, E. Kutchman, M. Mainar, et al. 2014. "A Microgeographic Analysis of Physical Activity Behavior Within Elementary School Grounds." *American Journal of Health Promotion* 28 (6): 403–12. doi:10.4278/ajhp.121116-QUAN-566.

Australian Government, Department of Social Services. 2018. "Immunisation and Health Check Requirements for Family Tax Benefit." Accessed July 23, 2019. www.

dss.gov.au/our-responsibilities/families-and-children/benefits-payments/strengthening-immunisation-for-young-children.

Aylward, R. B. 2006. "Eradicating Polio: Today's Challenges and Tomorrow's Legacy." *Annals of Tropical Medicine and Parasitology* 100 (5–6): 401–13. doi:10.1179/136485906x97354.

Aylward, R. B., K. A. Hennessey, N. Zagaria, J. M. Olivé, and S. Cochi. 2000. "When Is a Disease Eradicable? 100 Years of Lessons Learned." *American Journal of Public Health* 90 (10): 1515–20. doi:10.2105/ajph.90.10.1515.

Barrett, Scott. 2004. "Eradication Versus Control: The Economics of Global Infectious Disease Policies." *Bulletin of the World Health Organization* 82 (9): 683–88.

Baynes, C. 2019. "Anti-Vaccine Movement 'a Top Threat to Global Health in 2019' Says WHO." *Independent*. Accessed July 23, 2019. www.independent.co.uk/news/health/anti-vaccine-movement-world-health-organisation-who-diseases-mmr-antivax-a8732621.html.

BBC. 2019a. "Italy Bans Unvaccinated Children from School." Accessed July 23, 2019. www.bbc.com/news/world-europe-47536981.

BBC. 2019b. "New York Measles Emergency Declared in Brooklyn." Accessed July 24, 2019. www.bbc.com/news/world-us-canada-47872952.

BBC. 2019c. "Measles Cases Quadruple Globally in 2019, Says UN." Accessed July 23, 2019. www.bbc.com/news/health-47940710.

Block, J. P., R. A. Scribner, and K. B. DeSalvo. 2004. "Fast Food, Race/Ethnicity, and Income: A Geographic Analysis." *American Journal of Preventive Medicine* 27 (3): 211–17. doi:10.1016/j.amepre.2004.06.007.

Brownson, R. C., E. A. Baker, R. A. Housemann, L. K. Brennan, and S. J. Bacak. 2001. "Environmental and Policy Determinants of Physical Activity in the United States." *American Journal of Public Health* 91 (12): 1995–2003. doi:10.2105/ajph.91.12.1995.

Carter Foundation. 2008. "Guinea Worm Eradication Program." Accessed January 1, 2009. www.cartercenter.org/health/guinea_worm/index.html.

[CDC] Centers for Disease Control. 2002. "Smallpox: 30th Anniversary of Global Eradication." Accessed July 22, 2011. www.cdc.gov/ncidod/EID/vol8no1/cover.htm.

[CDC] Centers for Disease Control. 2009. "Defining Overweight and Obesity." Accessed December 23, 2009. www.cdc.gov/obesity/defining.html.

[CDC] Centers for Disease Control and Prevention. 2016. "History of Smallpox." Accessed July 20, 2019. www.cdc.gov/smallpox/history/history.html.

[CDC] Centers for Disease Control and Prevention. 2017. "Defining Adult Overweight and Obesity." Accessed July 21, 2019. www.cdc.gov/obesity/adult/defining.html.

[CDC] Centers for Disease Control and Prevention. 2018. "Neglected Tropical Disease." Accessed July 23, 2019. www.cdc.gov/globalhealth/ntd/index.html.

[CDC] Centers for Disease Control and Prevention. 2019. "Measles Cases and Outbreaks." Accessed July 23, 2019. www.cdc.gov/measles/cases-outbreaks.html.

Colatrella, B. 2008. "The Mectizan Donation Program: 20 Years of Successful Collaboration—A Retrospective." *Annals of Tropical Medicine and Parasitology* 102 (Suppl 1): 7–11. doi:10.1179/136485908x337418.

Crompton, D. W. T., World Health Organization, Department of Control of Neglected Tropical Diseases, D. Daumerie, World Health Organization, P. Peters, and L. Savioli. 2010. *Working to Overcome the Global Impact of Neglected Tropical Diseases: First WHO Report on Neglected Tropical Diseases*. Geneva: World Health Organization.

Crowe, Jessica, Constance Lacy, and Yolanda Columbus. 2018. "Barriers to Food Security and Community Stress in an Urban Food Desert." *Urban Science* 2 (2). doi:10.3390/urbansci2020046.

Cummins, Steven, and Sally Macintyre. 2002. "A Systematic Study of an Urban Foodscape: The Price and Availability of Food in Greater Glasgow." *Urban Studies* 39 (11): 2115–30. doi:10.1080/0042098022000011399.

Dowdle, W. 1998. "The Principles of Disease Elimination and Eradication." *Bulletin of the World Health Organization* 76 (Suppl 2): 22–25.

Drewnowski, A. 2004. "Obesity and the Food Environment: Dietary Energy Density and Diet Costs." *American Journal of Preventive Medicine* 27 (Suppl 3): 154–62. doi:10.1016/j.amepre.2004.06.011.

Dubé, Eve, and Noni E. MacDonald. 2016. "Addressing Vaccine Hesitancy and Refusal in Canada." *Canadian Medical Association Journal = journal de l'Association medicale canadienne* 188 (1): E17–E18. doi:10.1503/cmaj.150707.

Ehreth, Jenifer. 2003. "The Global Value of Vaccination." *Vaccine* 21 (7): 596–600. doi:10.1016/S0264-410X(02)00623-0.

Ellaway, Anne, Sally Macintyre, and Ade Kearns. 2001. "Perceptions of Place and Health in Socially Contrasting Neighbourhoods." *Urban Studies* 38 (12): 2299–316. doi:10.1080/00420980120087171.

[FDA] Food and Drug Administration. 2018. "Orphan Products: Hope for People with Rare Diseases." Accessed July 23, 2019. www.fda.gov/drugs/drug-information-consumers/orphan-products-hope-people-rare-diseases.

Fine, Paul E. M., and Ilona A. M. Carneiro. 1999. "Transmissibility and Persistence of Oral Polio Vaccine Viruses: Implications for the Global Poliomyelitis Eradication Initiative." *American Journal of Epidemiology* 150 (10): 1001–21. doi:10.1093/oxfordjournals.aje.a009924.

[FDA] Food and Drug Administration. 2009. "Developing Products for Rare Diseases and Conditions." Accessed 25 July 2009. www.fda.gov/ForIndustry/DevelopingProductsforRareDiseasesConditions/default.htm.

Forster, K. 2017. "Revealed: How Dangerous Fake News Conquered Facebook." Last modified January 7, 2017. Accessed August 22, 2019. www.independent.co.uk/life-style/health-and-families/health-news/fake-news-health-facebook-cruel-damaging-social-media-mike-adams-natural-health-ranger-conspiracy-a7498201.html.

Freedman, D. A., S. K. Choi, T. Hurley, E. Anadu, and J. R. Hebert. 2013. "A Farmers' Market at a Federally Qualified Health Center Improves Fruit and Vegetable Intake Among Low-Income Diabetics." *Preventive Medicine* 56 (5): 288–92. doi:10.1016/j.ypmed.2013.01.018.

Freedman, D. S., C. L. Ogden, K. M. Flegal, L. K. Khan, et al. 2007. "Childhood Overweight and Family Income." *Medscape General Medicine* 9 (2): 26.

Galán-Puchades, M. Teresa. 2016. "Dogs and Guinea Worm Eradication." *The Lancet Infectious Diseases* 16 (7): 770. doi:10.1016/S1473-3099(16)30080-9.

Giles-Corti, B., and R. J. Donovan. 2002. "The Relative Influence of Individual, Social and Physical Environment Determinants of Physical Activity." *Social Science & Medicine* 54 (12): 1793–812.

[GPEI] Global Polio Eradication Initiative. 2015. "Introducing the Inactivated Polio Vaccine: How Far Have We Come?" Accessed July 21, 2019. http://polioeradication.org/news-post/introducing-the-inactivated-polio-vaccine-how-far-have-we-come/.

[GPEI] Global Polio Eradication Initiative. 2019a. "Inactivated Poliovirus Vaccine." Accessed July 21, 2019. http://polioeradication.org/polio-today/polio-prevention/the-vaccines/ipv/.

[GPEI] Global Polio Eradication Initiative. 2019b. "Polio This Week as of 22 August 2019." Last modified August 22, 2019. Accessed August 22, 2019. http://polioeradication.org/polio-today/polio-now/this-week/.

Greenough, P. 1995. "Intimidation, Coercion and Resistance in the Final Stages of the South Asian Smallpox Eradication Campaign, 1973–1975." *Social Science & Medicine* 41 (5): 633–45.

Henderson, D. A. 1998. "Eradication: Lessons from the Past." *Bulletin of the World Health Organization* 76 (Suppl 2): 17–21.

Heymann, D. L., and R. B. Aylward. 2004. "Eradicating Polio." *New England Journal of Medicine* 351 (13): 1275–77. doi:10.1056/NEJMp048204.

Hoek, J., and R. MacLean. 2010. *Changing Food Environment and Obesity: An Overview*, edited by J. Pearce and K. Witten, *Geographies of Obesity: Environmental Understandings of the Obesity Epidemic*. Farnham, Surrey and Burlington, VT: Ashgate.

Hopkins, D. R., E. Ruiz-Tiben, P. Downs, P. C. Withers, Jr., and J. H. Maguire. 2005. "Dracunculiasis Eradication: The Final Inch." *The American Journal of Tropical Medicine and Hygiene* 73 (4): 669–75.

Hosmer, L. R. T. 2006. *The Ethics of Management*. McGraw-Hill Companies, Incorporated.

Hull, H. F., N. A. Ward, B. P. Hull, J. B. Milstien, and C. de Quadros. 1994. "Paralytic Poliomyelitis: Seasoned Strategies, Disappearing Disease." *Lancet* 343 (8909): 1331–37. doi:10.1016/s0140-6736(94)92472-4.

Hunter, John M. 2010. "River Blindness Revisited." *Geographical Review* 100 (4): 559–82.

Kim, D., and I. Kawachi. 2010. *Contextual Determinants of Obesity: An Overview*, edited by J. Pearce and K. Witten, *Geographies of Obesity: Environmental Understandings of the Obesity Epidemic*. Farnham, Surrey and Burlington, VT: Ashgate.

Langellotto, G. A., and A. Gupta. 2012. "Gardening Increases Vegetable Consumption in School-aged Children: A Meta-analytical Synthesis." *Horttechnology* 22 (4): 430–45.

Larsen, Kristian, and Jason Gilliland. 2008. "Mapping the Evolution of 'Food Deserts' in a Canadian City: Supermarket Accessibility in London, Ontario, 1961–2005." *International Journal of Health Geographics* 7 (1): 16. doi:10.1186/1476-072X-7-16.

Lee, H. 2013. "The Making of the Obesity Epidemic: How Food Activism Led Public Health Astray." *Breakthrough* 3 (Winter).

Leslie, E., N. Coffee, L. Frank, N. Owen, A. Bauman, and G. Hugo. 2007. "Walkability of Local Communities: Using Geographic Information Systems to Objectively Assess Relevant Environmental Attributes." *Health and Place* 13 (1): 111–22. doi:10.1016/j.healthplace.2005.11.001.

Leung, C. W., B. A. Laraia, M. Kelly, D. Nickleach, et al. 2011. "The Influence of Neighborhood Food Stores on Change in Young Girls' Body Mass Index." *American Journal of Preventive Medicine* 41 (1): 43–51. doi:10.1016/j.amepre.2011.03.013.

McClure, C. C., J. R. Cataldi, and S. T. O'Leary. 2017. "Vaccine Hesitancy: Where We Are and Where We Are Going." *Clinical Therapeutics* 39 (8): 1550–62. doi:10.1016/j.clinthera.2017.07.003.

McNeil, D. 2006. "Dose of Tenacity Wears Down a Horrific Disease." *New York Times*, March 26.

McNeil, D., and C. Dugger. 2006. "To Conquer, or Control? Disease Strategy Debated." Last modified March 20, 2006. www.nytimes.com/2006/03/20/world/to-conquer-or-control-disease-strategy-debated.html.

Mead, M. Nathaniel. 2008. "The Sprawl of Food Deserts." *Environmental Health Perspectives* 116 (8): A335. doi:10.1289/ehp.116-a335a.

Merchant, Anwar T., Mahshid Dehghan, Deanna Behnke-Cook, and Sonia S. Anand. 2007. "Diet, Physical Activity, and Adiposity in Children in Poor and Rich Neighbourhoods: A Cross-Sectional Comparison." *Nutrition Journal* 6: 1.

Merck. 2017. "Merck Commemorates 30 Years of MECTIZAN® Donation Program Progress." Accessed July 23. https://investors.merck.com/news/press-release-details/2017/Merck-Commemorates-30-Years-of-MECTIZAN-Donation-Program-Progress/default.aspx.

Ming, Wen, Christopher R. Browning, and Kathleen A. Cagney. 2007. "Neighbourhood Deprivation, Social Capital and Regular Exercise During Adulthood: A Multilevel Study in Chicago." *Urban Studies* 44 (13): 2651–71. doi:10.1080/00420980701558418.

Monteiro, C. A., W. L. Conde, and B. M. Popkin. 2004. "The Burden of Disease from Undernutrition and Overnutrition in Countries Undergoing Rapid Nutrition Transition: A View from Brazil." *American Journal of Public Health* 94 (3): 433–34. doi:10.2105/ajph.94.3.433.

New York Times. 2006. "Diseases on the Brink." Last modified March 20, 2006. Accessed January 31, 2011. www.nytimes.com/ref/health/2006_BRINK_SERIES.html.

Panter, J. R., and A. P. Jones. 2008. "Associations Between Physical Activity, Perceptions of the Neighbourhood Environment and Access to Facilities in an English City." *Social Science & Medicine* 67 (11): 1917–1923. doi:10.1016/j.socscimed.2008.09.001.

Paul, Y., and A. Dawson. 2005. "Some Ethical Issues Arising from Polio Eradication Programmes in India." *Bioethics* 19 (4): 393–406.

Pearce, Jamie, and Simon Kingham. 2008. "Environmental Inequalities in New Zealand: A National Study of Air Pollution and Environmental Justice." *Geoforum* 39 (2): 980–93. doi:10.1016/j.geoforum.2007.10.007.

Pearson, T., J. Russell, M. J. Campbell, and M. E. Barker. 2005. "Do 'Food Deserts' Influence Fruit and Vegetable Consumption? A Cross-Sectional Study." *Appetite* 45 (2): 195–97. doi:10.1016/j.appet.2005.04.003.

Pollan, Michael. 2006. *The Omnivore's Dilemma: A Natural History of Four Meals*. New York: Penguin Press.

RBM Partnership to End Malaria. 2019. "endmalaria.org." Accessed July 20, 2019. https://endmalaria.org/.

Reidpath, D. D., C. Burns, J. Garrard, M. Mahoney, and M. Townsend. 2002. "An Ecological Study of the Relationship Between Social and Environmental Determinants of Obesity." *Health and Place* 8 (2): 141–45.

Rodríguez-Cruz, Rodolfo. 1984. "Cuba: Mass Polio Vaccination Program, 1962–1982." *Reviews of Infectious Diseases* 6: S408–12.

Sachs, J. D. 2007. "The Neglected Tropical Diseases - For the Equivalent of a Few Days' Worth of Military Spending, Devastating Illnesses of the Global Poor could be Controlled Worldwide." *Scientific American* 296: 19B.

Santana, Paula, Rita Santos, and Helena Nogueira. 2009. "The Link Between Local Environment and Obesity: A Multilevel Analysis in the Lisbon Metropolitan Area, Portugal." *Social Science & Medicine* 68.

Singh, G. K., M. D. Kogan, P. C. Van Dyck, and M. Siahpush. 2008. "Racial/Ethnic, Socioeconomic, and Behavioral Determinants of Childhood and Adolescent Obesity in the United States: Analyzing Independent

and Joint Associations." *Annals of Epidemiology* 18 (9): 682–95. doi:10.1016/j.annepidem.2008.05.001.

Swinburn, B. A., I. Caterson, J. C. Seidell, and W. P. James. 2004. "Diet, Nutrition and the Prevention of Excess Weight Gain and Obesity." *Public Health Nutrition* 7 (1a): 123–46.

Takano, T., K. Nakamura, and M. Watanabe. 2002. "Urban Residential Environments and Senior Citizens' Longevity in Megacity Areas: The Importance of Walkable Green Spaces." *Journal of Epidemiology and Community Health* 56 (12): 913. doi:10.1136/jech.56.12.913.

The Lancet Infectious Diseases. 2016. "Guinea Worm Disease Nears Eradication." *The Lancet Infectious Diseases* 16 (2): 131. doi:10.1016/S1473-3099(16)00020-7.

Thornton, Lukar, and Anne Kavanaugh. 2010. "Understanding the Local Food Environment and Obesity." In *Geographies of Obesity: Environmental Understandings of the Obesity Epidemic*, edited by J. Pearce and K. Witten. Farnham, Surrey and Burlington, VT: Ashgate.

Westermark, K., B. B. Holm, M. Söderholm, J. Llinares-Garcia, et al. 2011. "European Regulation on Orphan Medicinal Products: 10 Years of Experience and Future Perspectives." *Nature Reviews Drug Discovery* 10 (5): 341–49. doi:10.1038/nrd3445.

[WHO] World Health Organization. 2009a. "About Dracunculiasis." Accessed January 1, 2019. www.who.int/dracunculiasis/disease/en/.

[WHO] World Health Organization. 2009b. "Achievements of Community-directed Treatment with Ivermectin." Accessed July 27, 2009. www.who.int/apoc/cdti/achievements/en/.

[WHO] World Health Organization. 2010a. Smallpox Factsheet. Accessed March 16, 2010. www.who.int/mediacentre/factsheets/smallpox/en/.

[WHO] World Health Organization. 2010b. "Archives of the Smallpox Eradication Programme." Accessed May 7, 2011. www.who.int/archives/fonds_collections/bytitle/fonds_6/en/index.html.

[WHO] World Health Organization. 2010c. "Somalia: Three Years Polio-Free." Accessed September 1, 2019. www.who.int/features/galleries/somalia_photo_gallery/en/.

[WHO] World Health Organization. 2013. "Progress on Sanitation and Drinking Water: 2013." Accessed July 23, 2019. www.who.int/water_sanitation_health/publications/2013/jmp_fast_facts.pdf?ua=1.

[WHO] World Health Organization. 2018a. *Global Health Observatory Data*. Geneva: World Health Organization.

[WHO] World Health Organization. 2018b. "Obesity and Overweight." Accessed July 21, 2019. www.who.int/news-room/fact-sheets/detail/obesity-and-overweight.

[WHO] World Health Organization. 2019a. "Achievements of Community-Directed Treatment with Ivermectin (CDTI)." Accessed July 23, 2019. www.who.int/apoc/cdti/achievements/en/.

[WHO] World Health Organization. 2019b. "Dracunculiasis Eradication Portal." Accessed September 1, 2019. www.who.int/dracunculiasis/portal/en/.

[WHO] World Health Organization. 2019c. "Onchocerciasis." Accessed July 23, 2019. www.who.int/news-room/fact-sheets/detail/onchocerciasis.

[WHO] World Health Organization. 2019d. "Poliomyelitis." Accessed July 21, 2019. www.who.int/news-room/fact-sheets/detail/poliomyelitis.

[WHO] World Health Organization. 2019e. "Public-Private Partnerships (PPPs)." Accessed July 23, 2019. www.who.int/intellectualproperty/topics/ppp/en/.

[WHO] World Health Organization. 2019f. "Smallpox." Accessed July 20, 2019. www.who.int/csr/disease/smallpox/en/.

[WHO] World Health Organization. 2019g. "Snakebite: WHO Targets 50% Reduction in Deaths and Disabilities." Accessed July 23, 2019. www.who.int/neglected_diseases/news/WHO-new-strategy-prevent-control-snakebite-envenoming/en/.

[WHO] World Health Organization. 2019h. "Ten Threats to Global Health in 2019." Accessed July 23, 2019. https://investors.merck.com/news/press-release-details/2017/Merck-Commemorates-30-Years-of-MECTIZAN-Donation-Program-Progress/default.aspx.

Wickett, John F. 2002. "The Final Inch: The Eradication of Smallpox and Beyond." *Social Scientist* 30 (5/6): 62–78. doi:10.2307/3518002.

Wiig Dammann, Kristen, and Chery Smith. 2009. "Factors Affecting Low-Income Women's Food Choices and the Perceived Impact of Dietary Intake and Socioeconomic Status on Their Health and Weight." *Journal of Nutrition Education and Behavior* 41 (4): 242–53. doi:10.1016/j.jneb.2008.07.003.

GLOSSARY

absolute count: the total number of a phenomenon without any further information about factors such as population size or age-structure (e.g., "23 hospitals")

acculturation: the process of adapting to a new culture, particularly the dominant culture

acute: severe, intense, but typically short-lived; as of a disease

aerial photography: a method for collecting data on the earth from photographs taken from aircraft

age adjustment (or age standardization): the statistical manipulation of data to remove the influence of age structure so that mortality or morbidity data from communities with different age structures can be compared

agency: the capacity of individuals to make their own decisions and control their own destiny; often used in contrast to "structure"

agent: an organism or substance that causes disease

age-specific mortality rates: mortality rates that refer to only a specified age range within the population (e.g., infant mortality rate); age-specific mortality rates allow populations with different age structures to be compared

aggregated data: data combined over time or space; often done to avoid the statistical problem of having very few data points for an analysis; for example, we could aggregate annual data on the occurrence of homicide in the UK over a decade, or include UK homicide data in an aggregated pool of data from all countries in the European Union

aging population: a population with an increasing proportion of people in older age cohorts

alternative medicine: healing practices that fall outside the biomedical approach

ambient: relating to the local environment (e.g., ambient air pollution, ambient temperatures)

animated map: a map animation that involves a series of maps to show change over time

Anthropocene: proposed name for the current geological epoch that reflects the dominance of human activities on natural systems

anthropozoonosis: a disease that commonly affects both animals and humans

area (or neighborhood) effects: the impact of an area on an individual or community's health and well-being

asymptomatic: lacking symptoms of disease despite infection

attribute: information about a spatial feature that is stored in a GIS, such as the name or population of a clinic

autosomal recessive condition: a genetic condition that develops when an individual inherits two copies of an abnormal gene

bands: specific ranges of the electromagnetic spectrum, as used in remote sensing

behavioral adaptation: adjusting behavior to cope with conditions

bimodal distribution: a bimodal distribution has two data peaks; for instance, deaths in a population may occur primarily in infants and the elderly, with few deaths in middle years

biocontrol: an organism that is used to reduce the population of a pest species

biomedical perspective: medical practice that focuses on empirical testing as a way to seek connections between particular causative agents and symptoms

biometeorology: the study of interactions between the earth's atmosphere and biosphere

birthright citizenship: system whereby any individual born in a country will automatically have citizenship in that country

birth tourism: traveling overseas to give birth to take advantage of better-quality healthcare and/or seek citizenship for the baby in a country that offers birthright citizenship

body mass index (BMI): ratio of weight to height, calculated by dividing an individual's weight in kilograms by height in meters squared

built environment: human-constructed aspects of one's surroundings

carcinogenic: having the potential to cause cancer; a **carcinogen** is a substance that is carcinogenic

carrier: an individual infected with a particular pathogen and able to transmit it to others, but who shows no symptoms of disease; can also refer to an individual who carries an unexpressed gene for a particular disease that could lead to the disease in the carrier's children

carrying capacity: the number of people (or other organisms) that can be supported by an ecological region

case-control study: an observational study that compares the exposures of individuals with a particular disease or outcome (the cases) with the outcomes of similar individuals without the disease or outcome (controls)

case fatality rate: the proportion of people with a particular disease who subsequently die from it

Central Place Theory: a theory developed by Walter Christaller to explain the location and size of human settlements as a function of their role in delivering goods and services

choropleth map: a thematic map that displays information by shading administrative units according to the value of the variable being displayed

chronic: persisting or recurring over a long period of time

clinical: a disease is said to be clinical when it is identifiable by its symptoms or clinical tests

cohort: a group of study subjects defined as a group, often by age (e.g., 10- to 14-year-olds)

cohort study: an observational study that compares the disease outcomes of an exposed and unexposed group

collective efficacy: the idea that individuals in a community believe that they can operate on behalf of the greater good and have the institutional capacity to act upon it

communicable: able to be passed from person to person

complementary medicine: often used interchangeably with the term "alternative medicine"; complementary medicine can also carry the additional implication that alternative healing approaches are being combined with mainstream techniques

confounding factor: a variable, unaccounted for in a particular statistical study, that is correlated with both the independent and dependent variables

congenital: present from birth, as of a disease or abnormality

contagious: transmitted *very easily* from person to person

continuous data: data with measurements at all points on the earth's surface, such as temperature or elevation

contour maps: maps that contain isolines, which connect points of equal value; contour maps are ideal for representing continuous data

control group: a group that is not exposed to the risk factor of interest but is otherwise similar to the treatment group

correlation: an association between two variables; if two variables increase or decrease together, they are positively correlated, if they move in opposite directions, they are negatively correlated

covariates: factors that may be related to the analysis in question (e.g., age, income, or race)

critical geography: a movement within geography that incorporates a wide variety of philosophical approaches that share an interest in opposing inequitable or repressive power relations

cross-sectional study: an observational study that considers the relationship between health outcomes and exposure for a "cross-section" of the population that is assumed to be representative of that population; provides a snapshot of a population's health at a particular moment in time

crude rates (or raw rates): rates that are calculated without statistical manipulation to adjust for factors such as differing age structures; crude rates account only for population size

cultural ecology: the study of how cultural practices influence people's interactions with their environment

culture: the beliefs and practices acquired from broader society that are lived by a particular group of people

data aggregation: the process of combining data (e.g., a morbidity rate for a county is an example of data aggregated from individuals); data can be combined over time or space

data smoothing: a technique used to remove aberrations in the data in order to highlight regional patterns; this cartographic technique can serve to remove some "statistical noise" to more clearly represent patterns

dead-end transmission: a situation in which a pathogen is unable to spread to new hosts, usually because the pathogen is not well adapted to that species of host

deaths of despair: deaths associated with drug abuse, suicide, and alcoholism; often interpreted as reflecting wider social problems

degenerative diseases: conditions that involve a slow breakdown of physiological function; often as significant for their debilitating effects as for their potential to cause death

demographic momentum: the potential for a population to continue to grow, despite declines in fertility, as a large, young population moves into its child-rearing years

Demographic Transition Model: model illustrating changes in births and deaths over time in human populations

density mapping: mapping or analyzing the concentration of a class of features over space

dependency ratio: the ratio between the economically dependent and economically active segments of a specified population

diffusion: the spread of something across space

direct transmission: transmission of a pathogen to a new host by physical contact or through the air

Disability Adjusted Life Years (DALY): a measure of disease burden that takes into account the number of years of life lost as the consequence of disease or disability

discrete data: phenomena that exist in some locations but not others (e.g., clinics or roads)

disease cluster: a grouping of disease cases or high incidence rates beyond what would normally be expected

disease ecology: an approach to disease that considers humans as one part of an inter-related ecological community, necessitating the study of pathogens, vectors, hosts, and the environments that support them

diseases of affluence: non-communicable diseases associated with poor diet and sedentary lifestyle, including obesity, heart disease, and cancer

diseases of childhood: diseases that typically affect only children because most adults are exposed in childhood and therefore carry resistance to the disease into adulthood

disease registry: systematic collection of data about patients with a particular disease

disease surveillance: data collected to monitor and model the incidence of disease

distance-decay effect: the declining likelihood of interactions between two places, as distance between the places increases

doctrine of specific etiology: the idea that one cause, such as a pathogen or toxin, is necessary and sufficient cause for a particular set of symptoms

dot density map: a map that uses points to visually represent the density of a phenomenon; points in these maps do not represent the actual location of the mapped features

double burden of disease: communities suffering significantly from both infectious diseases and diseases associated with increasing affluence such as obesity are sometimes said to suffer from a "double burden of disease"

ecological fallacy: the mistake of assuming that inferences about a group of people made from aggregated data will necessarily hold for individual members of that group

ecological niche: a living space in an ecosystem, including things such as available food and habitat; in an ecosystem, many species may compete for an ecological niche, and the fittest species will outcompete others to fill that niche

ecological study: a study that examines the health of populations using aggregated data

ecosystem simplification: the loss of species from an ecosystem; this reduction in biodiversity can allow "pest" species, including disease-causing organisms, to become dominant

[disease] elimination: removing a disease from a specific region

emerging infectious diseases: an infectious disease that has only recently been identified, or has recently shown a dramatic increase in incidence or geographic range

empiricism: a scientific approach based on the assumption that knowledge can only be gained through objective observations

endemic: prevalent in a particular area; an endemic disease is one that circulates constantly at low levels in a specified region

endocrine disruptor: a substance that can interfere with the normal hormonal functioning of the body

environmental determinism: a philosophical approach that suggests that humans are overwhelmingly influenced by their environment, particularly climate, with respect to characteristics such as behavior, physiology, and personality

environmental (in)justice: the idea that populations with less power (especially poor and minority groups) experience greater exposure to unhealthy environments; **environmental racism** links such inequalities specifically with race

environmental Kuznets curve: a model that suggests that environmental degradation initially increases as a population grows in affluence; at some point, the negative consequences of environmental destruction become significant and people's affluence has increased to the point where they are willing to invest in environmental measures, leading to an improvement in environmental conditions

environmental migrant: an individual forced to move to a new area because of environmental changes

epidemic: a sudden, large outbreak of a disease with far more cases than expected for a time and place

Epidemiologic Transition Model: model suggesting that deaths from infectious disease tend to decline over time as deaths from chronic and degenerative disease increase

epidemiology: the scientific methods and study of factors that affect the health of populations

epigenetics: the study of changes in the genome associated with environmental influences

epigenome: chemical compounds and proteins that attach to the genome and influence how genes are expressed

[disease] eradication: the complete, worldwide removal of a disease from human or animal populations

essentialism: the idea that there are certain characteristics that all members of a group of people naturally possess; for instance, females are inherently caring and nurturing

ethnicity: an ethnic group is a group of people identified as a community owing to shared cultural traits; although the term may be used in distinction from "race," which is often more explicitly associated with skin color or other biological variation; the terms "race" and "ethnicity" are often used interchangeably

etiology: cause of an illness or condition

eugenics: study of and policies related to selective breeding; focused on the "improvement" of the human gene pool

excess mortality or excess deaths: deaths beyond the number that we would expect under normal conditions

expansion (or contact) diffusion: the spread of a phenomenon by interactions among neighbors

experimental study: a study in which the exposures of different groups are manipulated by the researcher, and conclusions are drawn from the outcomes of these different groups

exposure: something that is believed to have the potential to cause disease; this can refer to substances such as lead or behaviors such as smoking

family planning: efforts to control fertility using contraceptive technologies and techniques

fate and transport study: studies that model the transport, transformation, and deposition of toxic substances using environmental data such as weather patterns

feature: a representation of a geographic phenomenon in a vector GIS

feature classes: types of features, such as points, lines, or areas

fecal-oral route: the transmission of pathogens through the consumption of fecal matter in contaminated food or water

fertility rate: average number of children born per woman over her lifetime

flow maps: maps that use line symbols to represent the flow of a phenomenon over space

focus group: a research approach that uses small groups of study participants to discuss a topic, as prompted by an interviewer

folk medicine: medical beliefs passed down through generations as general knowledge within a particular community

fomite: an inanimate object/substance that can harbor a pathogen

food insecurity: lacking a stable and guaranteed supply of sufficient food

GDP: gross domestic product; a measure of the market value of goods and services provided over a time period; it provides a good way to compare the relative affluence of different countries

gender: aspects of masculinity and femininity that are learned from cultural context; in contrast to "sex," which is a biological phenomenon, gender is socially constructed

gendered division of labor: the delegation of different tasks to men and women in a society, consistent with expectations that men and women have different abilities and preferences

genetic adaptation: the selection of genetic traits in a population over generations that offer advantages for living in local conditions

genome: the complete genetic information of a particular individual

geocoding: the process of converting textual information, such as street addresses or latitude and longitude coordinates, into spatially referenced features that can be displayed in a GIS

geogen: an inanimate object/substance that can cause disease

geographic scale: the area encompassed by a topic or study; a global study is considered to be a "large scale" study and a study of a neighborhood is considered to be a "small scale" study

georeferencing: defining the location of a feature in geographic space with a coordinate system

germ theory: the idea that microbes invade human bodies and cause alterations that result in disease

globalization: the economic integration of the world, facilitated by advances in logistics and communication

Global North: affluent countries, including Western Europe, the US, Canada, Australia, New Zealand, Israel, and Japan

global positioning system (GPS): a system of satellites that enables one to record the coordinates of a given location using a GPS receiver

Global South: countries marginalized by the global economy, including most countries of Africa, Latin America, and Asia

graduated symbols: symbols that are scaled according to data about the feature they represent

Green Revolution: an effort to increase crop production in the Global South by introducing modern industrial farming techniques (particularly high-yielding seed varieties, fertilizers, pesticides, irrigation, and mechanization), which began in the 1950s

healthcare barriers: the geographic, economic, social, and cultural reasons why individuals do not receive effective healthcare

health indicator: an objective measure used to estimate the health of an individual or group of people; infant mortality is a common health indicator

healthy immigrant phenomenon: the idea that recent immigrant groups may be healthier than host populations on first arriving in a new country

hemorrhagic: causing bleeding

herd immunity: a situation in which a sufficiently large proportion of a population is resistant to a disease that the likelihood of a susceptible individual meeting an infected individual becomes so low that disease transmission effectively ceases

hierarchical diffusion: the spread of a phenomenon through a spatial hierarchy, reaching large settlements before small ones

holistic medicine: an approach to health that considers physical, mental, spiritual, and emotional health as inter-related

horizontal transmission: the passing of a disease from one independent host to another

host: the human or animal in which a pathogen resides; technically, a **primary host** is where the organism completes the sexual stage of its life cycle, and an **intermediate host** is where the organism completes a larval stage

hotspot: a grouping of disease cases or high incidence rates beyond what would normally be expected; the term is usually reserved for non-contagious diseases

human ecology: interactions between humans and their environments

human microbiome: the microbe community of the human body, most of which live in the gut

humanistic approaches to health: focus on the individual, arguing that the everyday life experiences and goals of the individual are critical to constructing knowledge

hygiene hypothesis: the idea that exposure to microorganisms, particularly early in life, may help prevent the development of allergies and other immune conditions

identity: characteristics that distinguish a particular individual, as shaped by factors such as experience, personality, and position in society

immunologically naïve: lacking exposure to a particular pathogen

incidence rate: the number of *new* cases of a disease in a particular population over a specified time period per unit of that population (e.g., per 1,000 people)

in-depth interviews: interviews that focus on exploring a topic in detail, usually via open-ended questions

indigenous: native to a region

indirect transmission: the transmission of a pathogen from one host to another via a fomite or vector

infant mortality rate: the number of infants who die between birth and the age of 1 in a given year, usually reported per 1,000 live births

infectious: an infectious disease is one transmitted by a pathogen; in general usage the term often also implies that a disease is communicable

integrated medicine: the combination of modern medicine with traditional forms of healing

interactive map: a map that a user can alter by entering information

intermediate host: see host

interpolation: a method for producing estimates for locations where no data were collected, based on information collected at other locations

intersectionality: the idea that different aspects of societal oppression inter-relate in ways that reinforce discrimination

intersex: a situation in which an individual is born with a reproductive anomaly that makes it difficult to designate the baby as male or female

isoline: a line on a map that joins all points of equal value (e.g., contour lines)

knowledges: the plural term "knowledges" is used to explicitly acknowledge that different individuals or groups of people construct their understandings of the world in different ways

landscape epidemiology: the ways in which regions impart patterns to disease distributions via factors such as vegetation, geology, and climate

landscapes of deprivation: deprived areas, such as inner cities; the notion of a *landscape* of deprivation calls attention to the significance of the sense of place generated by these places

Latino Paradox: the health advantage observed among Latino groups in the US despite their relatively low average socioeconomic status

layer: a data layer is a collection of related information in GIS; a GIS for a city might include a roads layer, a city boundary layer, and a clinic location layer, for example

least cost path analysis: a form of network analysis that incorporates information about the transportation network to calculate the shortest distance and/or travel time from one point to another

LGBTQ+: acronym that collectively recognizes and promotes the needs of individuals who identify as lesbian, gay, bisexual, transgender, queer or questioning their sexuality

libertarianism: political approach that advocates for minimal state intervention

life-course perspective: the idea that health is affected by a lifetime of experiences rather than just recent events

life expectancy: the average years an infant born in a particular year is expected to live, assuming that current mortality rates continue to apply

life-world: the lived experiences of an individual, forming part of a philosophical approach that focuses on the uniqueness of individual experience

lithosphere: the crust and mantle of the Earth

location-allocation modeling: a method for modeling geographic access to health care that incorporates information about the serving capacity of healthcare facilities

map symbol: a graphical feature on a map that conveys meaning

marker of difference: a characteristic that distinguishes an individual from members of the surrounding community

maternal mortality rate: the number of women who die within 42 days of childbirth (regardless of the cause), among women who bore a child in that particular year; usually reported per 100,000 live births

medical pluralism: the coexistence of several different medical systems

medical tourism: traveling, usually internationally, to seek better, more accessible, or cheaper healthcare

megacity: a huge city; the term is often reserved for cities in the Global South that are not able to keep up with providing infrastructure to their rapidly growing populations

mental map: an imagined map that exists in a person's mind

meta-analysis: a study that reviews other recent studies on a topic, often using a statistical analysis to summarize the collective findings of the other studies

microcephaly: a condition in which a baby is born with an unusually small skull and brain

modernism: a philosophical movement that emerged in the West that emphasizes rationality, progress, and certainty

modifiable areal unit problem: the statistical effect that boundaries can have on results; the modifiable areal unit problem occurs when a result changes as a consequence of redrawn boundaries, even if the total area within those boundaries does not change

morbidity: poor health, disease

mortality rate: average number of deaths in a population per unit of that population (often per 1,000 or 100,000 people) over a specified time (typically one year)

neglected tropical diseases: a group of diseases that receive little attention from public health research and policy, owing to the lack of political and economic power of those affected

network analysis: analysis of a road or utility network; a network analysis can indicate the quickest route between two points, for example

non-communicable: not able to be passed from person to person

nosocomial infections: infections acquired in a hospital or hospital-like setting

obesogenic: characterized by factors that promote obesity

observational study: studies in which the researcher does not alter existing conditions; instead, the researcher relies on observing existing phenomena

orphan disease: diseases that have been largely ignored as targets for research because they afflict very few people or are concentrated in poor populations and so are unlikely to return much profit on investment

"Other": things that are marginalized or undervalued because they fall outside mainstream understandings of how things should be are sometimes referred to as the "Other" or being "othered"

outlier: a data point that is set apart from others in a data set

pandemic: a widespread epidemic, usually continent-wide or global in reach

parasite: an organism that lives off another to the detriment of the host

participatory research: a research approach that emphasizes the inclusion of members of the community being studied in the research process

pathogen: a living agent of disease

patriarchy: the structuring of society around men as the dominant figure

phenotype: genetically based, observable characteristics in an individual

physiological adaptation: changes to the physical body that allow an individual to better tolerate local conditions

place: *specific* geographic settings; spaces with meaning attached to them

placebo effect: the therapeutic benefits that can be generated when an individual believes that she is receiving treatment, even if that treatment has no objective curative effect

point distribution map: a thematic map that uses points to symbolize the precise location of specific geographic features

political ecology: a theoretical approach that focuses on the political nature of environmental issues

political economy: approaches that focus on the influence of political and economic structures

pollution: substances emitted into the environment by human activities, leading to detrimental environmental or health consequences; **primary pollutants** are caused directly by human activities; **secondary pollutants** are those that have undergone chemical changes after their release, converting them into new compounds

positivism: a philosophical approach that emphasizes measuring, counting, and observing physical phenomena as a way to generate knowledge—the basis of the scientific method; critics argue that not all important phenomena can be easily observed or measured in this way

postmodernism: a philosophical approach that rejects the apparent rationality of the modern era, urging us to be aware of and question the assumptions that we make, often unconsciously, from the dominant discourse of modernism; postmodernism emphasizes the importance of listening to many different voices within society

prevalence: the number of *existing* cases of a disease in a particular population per unit of that population (e.g., per 1,000 people)

preventative health: efforts to promote health and well-being by encouraging healthy behaviors *before* the onset of disease, as well as early detection of disease via screening programs

primary healthcare: healthcare designed explicitly with the goal of improving health for all, implemented by and for local communities, and providing the initial point of contact between patients and carers; sometimes distinguished from "**selective primary healthcare**," which refers to relatively basic health interventions such as vaccination and antibiotics that may be administered without true community participation

prion: an infectious particle composed primarily of protein and containing no genetic material

probiotics: live microorganisms, particularly bacteria and yeasts, introduced to the body to attempt to confer a health advantage, usually digestive health

pro-natalism: policy or practice of encouraging child-bearing

prophylaxis: a measure taken to prevent, rather than cure, disease

proximate causes of ill health: the immediate concerns that lead directly to illness, such as lack of food or exposure to an infectious agent

proxy: something that stands in for something else

public health: the study of health topics at the population scale, emphasizing health-promoting behaviors that will generate benefit for the broader population

Public Participation GIS (PPGIS): geographic information systems designed to empower community groups or grassroots organizations; frequently places an emphasis on preventative rather than curative approaches to health

qualitative approaches: techniques such as interviewing, focus groups, and participant observation that deepen our understanding of individual behavior, asking how and why people make the decisions they do

qualitative (or categorical) data: data that can be easily categorized based on characteristics; qualitative data are expressed by verbal descriptions

quantitative approaches: methods of inquiry based on counting, measuring, and other numerical techniques

quantitative data: data that can be counted or measured; quantitative data are expressed in numbers

quarantine: isolation of an infectious individual (or someone suspected of being infectious) from uninfected individuals

query: a GIS query enables the user to select particular features based on user-specified parameters (e.g., select all roads within a mile of a clinic)

race: regional groups of people, identified by differences in appearance such as skin color; the biological validity of the concept of race is highly contested

randomized controlled clinical trial: experimental study that randomly assigns people to treatment and control groups

rate: the number of cases per unit of the population (often per 1,000 or per 100,000 people); rates are used to remove the influence of population size on figures

reductionism: a philosophical approach that attempts to reduce complex systems to their parts, which can then be studied independently; critics argue that this can lead to oversimplification

reference map: a map designed to show the geographic distribution of general features

relocation diffusion: the introduction of a phenomenon to a location outside its current range

remotely sensed data: geographic data collected by measuring the amount of radiation reflected from the earth's surface

reporting bias: the idea that differences in reporting make us more aware of certain events than others; this may be significant temporally (recent phenomena are more likely to be reported than historic events) or spatially (e.g., affluent countries often have more resources for surveillance)

reservoir: a pool of pathogens that can act as a source of reinfection of human populations; reservoirs can be either other animal species or inanimate, often water or soil

resilience: the ability to cope with or rebound from challenges such as climate change or health insults

resolution: the smallest area for which a map shows unique data (e.g., the resolution of a choropleth map refers to the size of the administrative unit that is mapped—a map of provinces has a higher resolution than a map of countries)

ring vaccination: a vaccination approach in which all the contacts of an infected individual are vaccinated in order to stem the transmission of disease

risk analysis: a field of inquiry that assesses the relative risk associated with an exposure or policy

risk environment: the context in which risky behaviors occur; often used to emphasize the riskiness of places instead of behaviors

risk map: a map that displays factors that increase the risk of disease

satellite imagery: electromagnetic data collected about the surface of the earth from satellites

scale: see geographic scale

secondary data: data used by researchers that have not been collected by the researchers themselves (e.g., census data, government reports)

second-line drugs: drugs that are used when the easiest and cheapest therapies no longer work

sense of place: the unique feeling or spirit of a place, associated with its social significance

serosurvey: a survey that tests for antibodies in the blood of a population to see how widespread a disease is in that population

single-payer healthcare system: delivering healthcare using funding from a single source; this source is typically a government-run organization

social capital: social connections and networks that can serve to benefit both individuals and societies

social cohesion: the social bonds that hold a community together

social determinants of health: social, political, and economic conditions in which people live that influence the health status of individuals

social environment: aspects of an individual's surroundings that relate to interactions with other humans, including political, economic, and cultural characteristics

socialized medicine: a national healthcare system that is funded through taxation

socially constructed: the idea that a phenomenon is created by cultural and political understandings of how things should be

space: concerned with defining where things are; notions of location, distance, and area help us to describe space

spatial autocorrelation: the notion that points that are closer together are more likely to have similar characteristics than points that are farther apart

spatial epidemiology: analytical and statistical methods for the study of spatial patterns in disease incidence and mortality

spatial resolution: the smallest unit of space that is recordable in a GIS

spatio-temporal modeling: the observation and analysis of spatial and temporal changes in the earth's environment

spillover event: the spread of a pathogen from its natural host species, causing disease in another species

standardized data: data that have been put on the same scale to better enable comparisons

step migrations: migrating from one place to a destination with stops at other places on the way (e.g., a nurse from Mozambique might initially migrate to South Africa before migrating again to the UK)

structural factors: cultural, economic, and political structures of society, such as government legislation, social hierarchies, and poverty

structural violence: the way in which social structures or institutions lead certain individuals to be disadvantaged and thereby unable to meet their basic needs or otherwise put in danger

structure: overarching frameworks within society (e.g., political and economic) that constrain an individual's decision-making abilities; often used in contrast to "agency"

sub-clinical: a disease that is not identifiable by recognizable symptoms or clinical tests

susceptible: vulnerable to a particular disease

symbiosis: a mutually advantageous relationship between two organisms

synthetic: human made

thematic map: a map that depicts the spatial distribution of data on a particular theme or topic (e.g., infant mortality rates, location of hospitals)

therapeutic landscapes: places that are believed to encourage healing

thrifty gene hypothesis: the idea that past periods of scarcity have influenced our genetic makeup by selecting for genes that efficiently use available resources

thunderstorm asthma: a sudden increase in asthma-related symptoms following a thunderstorm; thought to be associated with the inhalation of micro-fragments of pollen leading to an allergic reaction in susceptible individuals

transactional sex: a sexual encounter in which a person offers sex in return for material or other support or influence

transmissibility: the ease with which a disease can spread

universal healthcare or universal coverage: healthcare systems designed to provide access to healthcare for all members of society

urban penalty: the collective experience of city living that leads to worse health than in rural areas, including factors such as overcrowding, decaying infrastructure, and pollution

vaccine hesitancy: the phenomenon of parents being reluctant to vaccinate their children because of concerns over the safety of vaccines and/or the principle of vaccination

vector: an organism that transmits a pathogen between hosts, such as a mosquito or fly

vertical transmission: transmission of a disease from mother to offspring during gestation, birth, or breastfeeding

viremia: the existence of viruses in the blood

virgin soil epidemic: an epidemic that spreads quickly through a population with no previous exposure to that disease; term coined by Alfred Crosby (1976)

virulence: the potential of a pathogen to cause disease; a virulent pathogen may cause worse symptoms and/or symptoms in more people

vulnerability: the idea that underlying structural characteristics such as poverty, lack of political power, or exclusion from social networks put some individuals at greater risk for health-compromising events than others

weathering: accelerated aging associated with the stress of social, economic, and political exclusion

welfare state: a system in which government tries to protect the well-being of its citizens by means of policies such as pensions, subsidies, state provision of healthcare, etc.

well-being: an individual's life satisfaction; a variety of factors may contribute to well-being, including not only good physical and mental health but also spirituality, happiness, and self-worth

zoonosis: a disease of animals

INDEX